Light Metals 2006

Volume 3: CARBON TECHNOLOGY

LIGHT METALS 2006 VOLUME 3: CARBON TECHNOLOGY

TMS Member Price: $64 TMS Student Member Price: $54 List Price: $94

Related Titles

- *Light Metals 2005*, edited by H. Kvande
- *Light Metals 2004*, edited by A. Tabereaux
- *Light Metals 2003*, edited by P. Crepeau

HOW TO ORDER PUBLICATIONS

For a complete listing of TMS publication offerings, contact TMS for a free copy of the Publications@TMS catalog or visit the on-line TMS Document Center (*http://doc.tms.org*). Through the TMS Document Center, customers will find:

- An easy and convenient way to purchase publications
- Access to TMS member resources, such as the on-line version of *JOM* and *TMS Letters*
- In-depth publications information, including complete descriptions, tables of contents, and sample pages
- Award-winning landmark papers and re-issued out-of-print titles
- A vast selection of resources from which to compile customized publications that meet the customer's unique needs

MEMBER DISCOUNTS

Members of TMS (The Minerals, Metals & Materials Society) receive a 30% discount off list prices for all TMS publications. In addition, TMS members also receive free monthly subscriptions to the journals *JOM* (both in print and on-line formats) and *TMS Letters* (published exclusively in on-line format), discounts on meeting registrations, and more. To begin immediate savings on TMS publications and to receive additional member benefits, become a TMS member today. Fill out a membership application when placing a publication order at the on-line TMS Document Center (*http://doc.tms.org*) or contact TMS for more information:

- Phone: 1 (800) 759-4867 (within the U.S.) or (724) 776-9000 (elsewhere)
- Fax: (724) 776-3770
- E-mail: membership@tms.org or publications@tms.org
- Web: *www.tms.org*

Light Metals 2006

Volume 3: CARBON TECHNOLOGY

Proceedings of the technical sessions presented
by the TMS Aluminum Committee
at the 135th TMS Annual Meeting,
San Antonio, Texas, USA
March 12-16, 2006

Edited by
Travis J. Galloway

A Publication of **TMS (The Minerals, Metals & Materials Society)**
184 Thorn Hill Road
Warrendale, Pennsylvania 15086-7528
(724) 776-9000

Visit the TMS web site at
http://www.tms.org

ISBN Number 978-0-87339-617-2

If you are interested in purchasing a copy of this book, or if you would like to receive the latest TMS publications catalog, please telephone (800) 759-4867 (U.S. only) or (724) 776-9000, EXT. 270.

LIGHT METALS 2006 VOLUME 3
TABLE OF CONTENTS

Carbon Technology

Anode Raw Materials

Greenmill/Rodding

Anode Baking

Cathode Properties/Refractory Materials

Cathode Preheating/Wettable Cathodes

LIGHT METALS 2006 VOLUME 1
TABLE OF CONTENTS

Alumina and Bauxite

Solids/Liquid Separation

Bauxite and Bauxite Characterization

Bayer Digestion Technology

Joint Session of Alumina and Bauxite & Aluminum Reduction Technology

Plant Design, Operation and Maintenance

Precipitation Fundamentals

LIGHT METALS 2006 VOLUME 2
TABLE OF CONTENTS

Aluminum Reduction Technology

Environmental Elements

Cell Development and Operations - Part I

Cell Development and Operations - Part II

Pot Control and Modeling

Inert Anodes - Part I

Cell Development Part III and Emerging Technologies

Fundamentals, Emerging Technologies and Inert Anodes - Part II

LIGHT METALS 2006 VOLUME 4
TABLE OF CONTENTS

Cast Shop Technology

Cast House Operations

Furnace Operation and Refractory Materials

Melt Treatment, Quality and Product Properties

Shape Casting and Foundry Alloys

Casting, Solidification and Cast Defects

Cast Processes and Chain Analysis

Recycling

Aluminum Recycling

PREFACE

The past year has presented significant challenges for the aluminum industry with dramatic increases in energy and raw materials prices compressing profit margins despite the positive impact of higher metal prices. This has occurred against the backdrop of continued brownfield expansions and greenfield feasibility studies and plans for both aluminum smelters and alumina refineries driven by world-wide demand led by China. In this environment, the strategic importance of innovation from R&D and implementation of technological improvements throughout the integrated production process from bauxite mining through refining, smelting and downstream applications can not be overemphasized. TMS continues to serve as the key forum for disseminating scientific and technological information to maintain the competiveness and sustainability of the aluminum industry.

For **Light Metals 2006**, the papers cover the full range from theoretical aspects to practical applications of technology for integrated aluminum manufacturing in keeping with the TMS Annual Meeting's theme of "Linking Science and Technology for Global Solutions". There is a special Joint Session of Alumina and Bauxite & Aluminum Reduction Technology with a focus on aspects of alumina quality of common interest to customer and supplier. There was also for the first time a Cast Shop Operations session for which authors made presentations but did not write formal papers. In total, 161 papers are included in this year's **Light Metals** proceedings, which compares very favorably with the number of previous years' contributions. Of this number, 80% percent were contributed by authors outside the United States, which continues the trend in globalization of aluminum technology.

In continuation of the tradition begun last year with the plenary session on the role of technology in the global primary aluminum industry, a plenary session was presented this year for Aluminum Fabrication featuring eight corporate leaders from around the world.

For the first time, the Light Metals proceedings are being published exclusively in CD format for attendees. Print volumes of selected symposia are also available. The on-line Conference Management System was expanded this year to a CMS-Plus version that greatly facilitated the preparation and editing process. New features included on-line editing and communication between authors and session chairs. While the CMS-Plus system continually evolves to meet the needs of our constituency, this facility greatly enhances the preparation of abstracts and manuscripts as well as the editing and session organization process.

Appreciation is expressed to the authors of the numerous papers and presentations and to their organizations, which provided support. Special acknowledgment and thanks are also due to the Symposium Organizers: Dr. Jean Doucet (Alumina and Bauxite), Stephen Lindsay (Reduction Technology), Morten Sorlie (Carbon Technology), and Dr. Rene Kieft (Cast Shop Technology and Operations) and to the Session Chairs for their tireless work in organizing and editing the papers. The TMS staff led by Alexander Scott, Executive Director, provided their traditional excellent support; particular acknowledgment is due to Stephen Kendall, Publications Manager, for his skills and dedication in producing this volume and to Christina Raabe, Manager- Technical Programming and Continuing Education, for her excellent facilitation in organizing the program. The continuing support of the Aluminum Committee in its oversight and strategic direction is also acknowledged and appreciated.

Travis J. Galloway
Vice-Chairman Aluminum Committee
Editor **Light Metals 2006**

EDITOR'S BIOGRAPHY

TRAVIS J. GALLOWAY
LIGHT METALS 2006 EDITOR

Travis J. Galloway is Technical & Engineering Director- Bauxite Mining/Alumina Refining for Century Aluminum. He began his career in 1964 with Reynolds Metals Company at the Hurricane Creek alumina refinery in Arkansas where he became Chief Process Engineer. He later served in various other technical management positions with Reynolds including Manager-Alumina Division Technology in Corpus Christi, Texas, Technical Manager- Worsley alumina refinery in Australia, and Technical Director- Raw Materials Division at the corporate headquarters in Richmond, Virginia. In 2001 he joined Century Aluminum at its Hawesville, Kentucky aluminum smelter. He obtained his B.S. from the University of Arkansas and M.S. from Rice University, both in chemical engineering. Additional graduate education includes an MBA from the University of Arkansas at Little Rock and a Master of Engineering Management from Old Dominion University. He is a Registered Professional Engineer and is a senior member of the American Institute of Chemical Engineers. He has served as a TMS Subject Chair and Session Chair for Alumina and Bauxite and became a member of the TMS Aluminum Committee in 2004.

PROGRAM ORGANIZERS

Alumina & Bauxite

Jean Doucet *obtained his PhD in Chemistry from the University of Montréal in 1974. He then began his career with Alcan in Jonquiére, Canada as a Research Chemist and then held various positions with technical development groups working in Alcan's Jonquiére Alumina Refinery, their specialty chemical group and at an Aluminium Fluoride plant. Jean then became responsible for the Research and Development program for the Bauxite and Alumina Division; in 1995, Jean transferred to Montréal to become the Technology Licensing and Intellctual Property Manager for the same division. From 2002 to 2005, Jean operated out of Brisbane, Australia and in January 2006 returned to Montréal as Director Knowledge Managment and Intellctual Property.*

Dag Olsen *holds a Bachelors degree in Petrochemistry from the Telemark Technical College (1976) and a Masters degree in Chemical Engineering from the Norwegian University of Science and Technology (1982). He joined Norsk Hydro in 1984 as a process engineer at the petrochemical complex at Rafnes, Telemark. In 1990 he joined the newly established Alumina & Bauxite group at Hydro's Corporate Research Centre in Porsgrunn where he worked on various projects and studies in addition to R&D work on Alumina Quality. Dag is currently working in the Alumina & Bauxite department of Hydro Aluminium Primary Metals where his main responsibilities are technology follow-up, R&D cooperation and business development projects.*

Aluminum Reduction Technology

Stephen J. Lindsay *has served in numerous technical and managerial capacities at Alcoa's locations in Massena, NY, Alcoa, TN and Knoxville, TN over the past 26 years. He currently has technical support responsibilities that include; technical support for European operations, plus alumina and metal purity responsibilities that span Alcoa's smelting operations worldwide. He is currently based in Knoxville, TN.*

His articles on alumina and metal purity have been published in ***Light Metals 2005****, the proceedings of the* ***8th Australasian Smelting Technology Conference*** *in 2004 and the 7th* ***International Alumina Quality Workshop*** *in 2005. Steve has also contributed to the TMS Industrial Electrolysis Course and Short Courses as well as acting as a guest speaker for the UNSW-UA Graduate Level programs in 2003 and 2004.*

Carbon Technology

Morten Sorlie *graduated MSc in Extractive Metallurgy, Norwegian University of Science and Technology, Trondheim, Norway in 1974 and received a PhD in Inorganic Chemistry, same place, 1978. Worked as Post Doctoral Fellow at Oak Ridge National Laboratory, TN, 1978-1980. Spent the following year at Institute of Inorganic Chemistry, Norwegian University of Science and Technology, Trondheim, Norway, starting up the cathode materials research there. Employed by Elkem since 1982. Present position is Corporate Specialist in Elkem Aluminium ANS, stationed at the Elkem research facilities in Kristiansand, Norway, and Adjunct Professor at Institute of Materials Technology, Norwegian University of Science and Technology, Trondheim, Norway. Responsibilities in Elkem Aluminium include cathodes and cathode materials, anodes and anode raw materials. Morten Sorlie has authored/co-authored more than 100 papers in international journals and conference proceedings, including several in TMS Light Metals. He is also a co-author of the textbook* **Cathodes in Aluminium Electrolysis**.

Todd W. Dixon *received his B.S. in Chemical Engineering from the Colorado School of Mines in 1990 and is licensed Professional Engineer in Louisiana. He has worked with Conoco (now ConocoPhillips) since graduation. In his career at Conoco, he has been involved with the design, operation, and troubleshooting of delayed cokers and rotary kilns processing petroleum coke. Currently, he is a Process Team Leader at the ConocoPhillips refinery in Borger, Texas. Previously, he was the Operations Manager at the Venco - Lake Charles calcining plant. He has also served as a TMS Session Chair.*

Cast Shop Technology

Rene Kieft *received his Ph.D in Mechanical Engineering from the Eindhoven University of Technology, The Netherlands in 2000 where he worked heat and fluid flow. He joined Corus Research & Technology to work on modelling of aluminium casting, focussing first on mnetal distribution systems and mechanical deformation during DC casting. After that he was involved in the European projects on building a virtual aluminium production chain (VIR[*] project). Besides casting he also worked on purification of aluminium. Within the European Molten Aluminium Purification project he worked together with different partners on different purification concepts. In 2003 he became knowledge group leader of the Corus Research group on Molten aluminium Processing.*

Gerd-Ulrich Gruen *received his diploma in geophysics from Technical University of Clausthal (Germany) in 1982. After some project work regarding flow in porous media related to deep drilling research he then joined VAW aluminium AG in 1990, where he was responsible for various research activities in the DC casting area mainly focusing on process modeling. This is documented in several papers he presented at previous TMS conferences. One recent activity was the co-ordination of the European wide model development project VIR[CAST], which brought together major European Aluminum producers and leading institutes and universities in the field of microstructure research. He is now with Hydro Aluminium, R&D and actually Program Manager Sheet Ingot in the Competence Center Casting, Alloys & Recycling. Since 2001 Gerd-Ulrich Gruen is a member of TMS.*

Recycling

Gregory Krumdick *obtained his M.S. degree in Bioelectrical engineering from the University of Illinois at Chicago, focusing on complex control systems. In 1990, he joined Argonne National Laboratory and has worked as an Engineer in designing and managing the construction of numerous pilot plant systems for the Process Evaluation section. Starting with industrial control systems, his field of interest moved to metallurgical processes where Greg has working on numerous aluminum projects including electrodialysis of aluminum saltcake waste brines, molten aluminum oxidation and the development of inert anodes for the production of aluminum.*

Cynthia K. Belt *is Energy Engineer for Aleris International working with the plants from the former IMCO Recycling, Commonwealth Aluminum, ALSCO Metals, Alumitech, and Ormet in identifying Best Practices in energy, throughput and recovery and implementing them throughout the corporation. Cindy is trained as a Black Belt in Six Sigma. She has worked for Aleris and the former Barmet and Commonwealth Aluminum since 1996. Prior to this, Cindy worked in engineering project management at Ekco Housewares, Crown Cork & Seal, and The Timken Company. She has over 24 years of engineering experience.*

Cindy earned her Bachelor of Science, Mechanical Engineering degree from Ohio Northern University with additional graduate work in materials at Case Western Reserve University and Akron University. She has written two published papers on energy. Cindy is a member of the Recycling Committee for TMS.

AULUMINUM COMMITTEE
2005-2006

Chairperson
Halvor Kvande
Hydro Aluminium AS
Oslo, Norway

Vice-Chairperson
Travis J. Galloway
Century Aluminum Co.
Hawesville, KY, USA

Past Chairperson
Alton T. Tabereaux
Alcoa Inc.
Muscle Shoals, AL, USA

LMD Chairperson
Ray D. Peterson
Aleris International
Rockwood, TN, USA

JOM Advisor
Pierre P. Homsi
Pechiney Group
Voreppe, France

TMS Staff Liaisons
Warren Hunt and Gail Miller
TMS
Warrendale, PA, USA

MEMBERS THROUGH 2006

Lars Arnberg
Norwegian Univ. of Sci. & Tech.
Trondheim, Norway

Andre R. Bolduc
Alcan Inc.
Monreal, QC, Canada

Jay N. Bruggeman
Alcoa Primary Metals
North Charleston, SC, USA

John J.J. Chen
University of Auckland
Auckland, New Zealand

David D. DeYoung
General Motors Corp.
Pontiac, MI, USA

Les C. Edwards
CII Carbon LLC
New Orleans, LA, USA

Seymour G. Epstein
Aluminum Association
Silver Spring, MD, USA

Ann Marie Fellom
Light Metal Age
S. San Francisco, CA, USA

Wayne R. Hale
SUAL Holding Management Co.
Davis, CA, USA

Peter V. Polyakov
Non-Ferrous Metals & Gold Academy
Krasnoyarsk, Russia

Wolfgang A. Schneider
Hydro Aluminium AS
Bonn, Germany

Martin Segatz
Hydro Aluminium GmbH
Neuss, Germany

Geoffrey K. Sigworth
Alcoa Inc.
Rockdale, TN, USA

Fiona J. Stevens McFadden
University of Auckland
Maungaraki, New Zealand

Murat Tiryakioglu
Robert Morris University
Moon Township, PA, USA

Barry J. Welch
Welbank Consulting
Orakei Aukland, New Zealand

MEMBERS THROUGH 2007

Johannes Aalbu
Hydro Aluminium AS
Ovre Ardal, Norway

Yousuf Ali Mohammed Alfarsi
Dubal Aluminium Co. Ltd.
Dubai, United Arab Emirates

Milind V. Chaubal
Sherwin Alumina Co.
Corpus Christi, TX, USA

Paul N. Crepeau
General Motors Corp.
Pontiac, MI, USA

David B. Kirkpatrick
Kaiser Aluminum & Chem. Corp.
Gramercy, LA, USA

Amir A. Mirchi
Alcan Inc.
Montreal, QC, Canada

Gary B. Parker
Wise Alloys LLC
Muscle Shoals, AL, USA

Tor Bjarne Pedersen
Elkem Aluminium ANS
Farsund, Norway

Ramana G. Reddy
University of Alabama
Tuscaloosa, AL, USA

Michael Hal Skillingberg
Aluminum Association
Arlington, VA, USA

Mark Taylor
University of Aukland
Auckland, New Zealand

MEMBERS THROUGH 2008

Thomas R. Alcorn
Noranda Aluminum Inc.
Florence, AL, USA

Chris M. Bickert
Pechiney Group
Neuilly Sue Seine, France

Corleen Chesonis
Aloca Inc.
Alcoa Center, PA, USA

James W. Evans
University of California
Berkeley, CA, USA

Markus W. Meier
R&D Carbon Ltd.
Sierre, Switzerland

David V. Neff
Metaullics Systems Co. LP
Solon, OH, USA

Ray D. Peterson
Aleris International
Rockwood, TN, USA

Alton T. Tabereaux
Alcoa Inc.
Muscle Shoals, AL, USA

MEMBERS THROUGH 2009

Todd W. Dixon
ConocoPhillips Inc.
Bouger, TX, USA

Gerd Ulrich Gruen
Hydro Aluminium AS
Bonn, Germany

Dag Olsen
Hydro Aluminium Primary Metals
Porsgrunn, Norway

PROGRAM ORGANIZERS
Morten Sorlie
Elkem Aluminium ANS
Kristiansand, Norway

Todd W. Dixon
ConocoPhillips
Sulphur, LA, USA

CARBON TECHNOLOGY

Anode Raw Materials

SESSION CHAIR
James Metson
University of Auckland
Auckland, New Zeeland

Light Metals 2006 *Edited by Travis J. Galloway* ***TMS (The Minerals, Metals & Materials Society), 2006***

VISCOSITY MODIFICATION AND CONTROL OF PITCH

Melvin D. Kiser, Michael B. Sumner, Brian K. Wilt, D. Chris Boyer
Marathon Petroleum Company LLC
Catlettsburg, KY 41129

Abstract

The addition of low concentrations of certain oxygenated compounds, particularly esters, to bitumen has been shown to have a profound effect on viscosity. The bitumen in question includes but is not limited to petroleum pitch, coal tar pitch and asphalt cements. A cost effective source of esters has been found to be biodiesel. This paper will provide general information on biodiesel and detail its effect on the viscosity of bitumen as a function of concentration. Other topics of discussion are effects on other hydrocarbons.

Introduction

Petroleum pitch competes with coal tar pitch in many applications where the pitch is used as a carbon source and/or a binder. As such, critical properties are: (a) flow properties as measured by softening point and/or viscosity, and (b) carbon yield as measured by ASTM D 2416, Coking Value by Modified Conradson Carbon [1]. Cutback oils have been used to modify the flow properties of petroleum products such as pitch and asphalt. Historically, these have been petroleum based, non-oxygenated hydrocarbons such as diesel fuel or various types of fuel oils.

A question arose concerning the effect of oxygenated compounds such as ethylene glycol on pitch viscosity. This work indicated ethylene glycol has a limited ability to reduce viscosity of petroleum pitch. Similar observations were observed for certain ester compounds. Recently, the use of methyl esters of fatty acids derived from either soybean or animal fats has received attention as a supplement to conventional diesel fuel supplies in the United States. Based on the chemistry of the fatty acid esters, use of biodiesel was investigated as a viscosity modifier for bitumen type materials [2].

Bitumen includes a broad class of materials including but not limited to petroleum pitch, coal tar pitch and asphalt cements such as those derived from vacuum residuum or solvent de-asphalting operations. The characteristics of these types of hydrocarbon include low volatility, high flash point and high viscosity. Most of these types of hydrocarbons exhibit Newtonian flow behavior. Pitches (petroleum or coal tar based) are essentially mixtures of multi-ring aromatic compounds. The petroleum residua are classified as non-aromatic bitumen.

Biodiesel is an alternative to petroleum-based diesel fuel made from renewable resources such as vegetable oils or animal fats. Chemically, it comprises a mix of mono-alkyl esters of long chain fatty acids. A lipid transesterification production process is used to convert the base oil to the desired esters and remove free fatty acids. After this processing, unlike straight vegetable oil, biodiesel has combustion properties very similar to those of petroleum diesel, and can replace it in most current uses [3]. However, it is at present most often used as an additive to petroleum diesel, improving the otherwise low lubricity of pure ultra low sulfur petrodiesel fuel. It is one of the possible candidates to supplement fossil fuels. It is a renewable fuel that can replace petrodiesel in current engines and can be transported and sold using today's infrastructure. Biodiesel use and production is increasing rapidly, especially in Europe, the United States, and Asia, though in all markets it still makes up a small percentage of fuel sold. A growing number of fuel stations are making biodiesel available to consumers, and a growing number of large transport fleets use some proportion of biodiesel in their fuel.

Unlike petrodiesel, biodiesel is biodegradable and non-toxic, and it reduces emissions when burned as a fuel. The most common form uses methanol to produce methyl esters, though ethanol can be used to produce an ethyl ester biodiesel. A byproduct of the transesterification process is the production of glycerol.

Typical properties of biodiesel are detailed in Table I.

Table I
Properties of Biodiesel

Property	Value
Flash Pont, Pensky Marten, °C	120
Sulfur, wt%	0.006
Cetane	52
Total Glycerin, wt%	0.08
Sulfated Ash, wt%	< 0.01
Water & Sediment, vol%	< 0.01
Specific Gravity @ 25°C	0.89
Melting Point, °C	-1
Boiling Point, °C	> 259

One characteristic of the ester functional group is a strong infrared absorption at approximately 1710 cm^{-1}. This characteristic has been used to determine ester concentration in bitumen for the purpose of detection and possibly semi-quantitative determination.

Experimental

A commercial biodiesel was characterized by conventional testing protocols (ref: Table I). Laboratory blends of biodiesel or #6 fuel oil in bitumen were made at concentrations from 0.5 to 23 percent. Baseline studies were performed using a non-oxygenated, aromatic petroleum hydrocarbon having similar flow properties. Flow properties of these blends were measured using standard analytical techniques.

The infrared spectra of the blends were measured using a Mattson Galaxy 6020 FT-IR with Spectra-Tech 0002-171 Diffuse Reflectance attachment, using the diffuse reflectance technique. The blends were first ground in an agate mortar, and mixed with dry potassium bromide (KBr) to make 2 wt percent mixtures for spectral measurements. Data reduction was performed using Galactic Industries Grams/32 AI v. 6.00 software. In order to relate the peak intensities in the carbonyl region (ca. 1700-1775 cm^{-1}) to additive levels, all of the diffuse reflectance spectra were subjected to Kubelka-Munk transformation[4] and then offset-corrected at 2000 cm^{-1}. Spectra for the pitch and pitch-biodiesel mixtures were then normalized to the aromatic C-C stretch at 1603 cm^{-1}. This region was chosen for normalization because it is present in the spectrum for pitch but not in the biodiesel spectrum. Derivative spectra were calculated using the point-difference method.

Discussion

Effect on Softening Point

Mettler softening point (ASTM D 3104) is a measure of viscosity. Producers and users of both coal tar and petroleum based pitch typically use softening point to monitor pitch production, shipments and receipts[1]. The effects of varying levels of biodiesel and #6 fuel oil additive was studied as a function of additive concentration and are plotted in Figure 1.

Figure 1

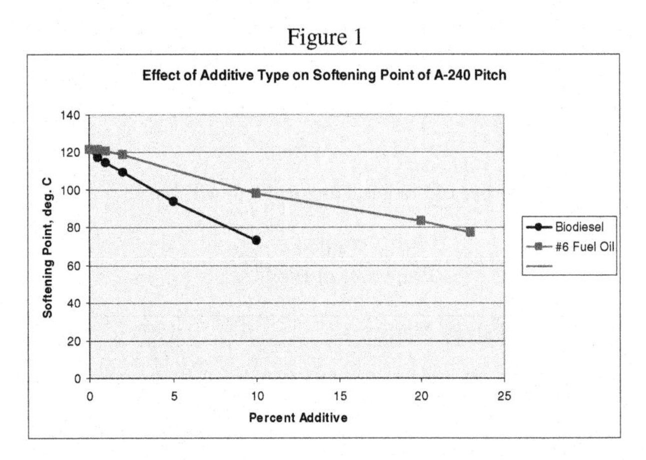

The effect of biodiesel addition to other types of bitumen was determined to be similar. As shown in Figure 2, the rate of change in softening point as a function of biodiesel concentration appear to be similar for two types of petroleum pitch, a coal tar pitch and a classic petroleum residuum.

Figure 2

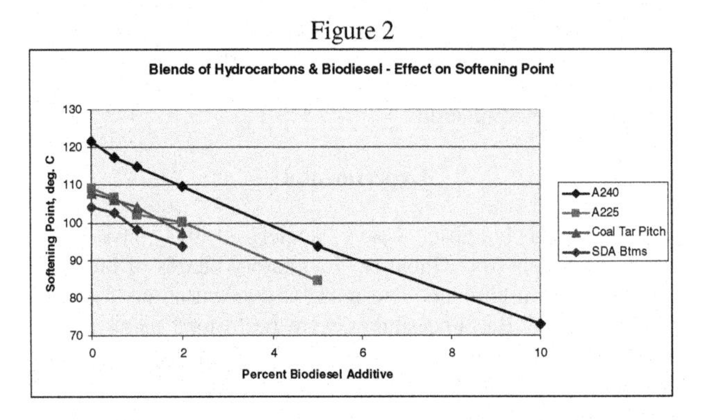

Effect on Coking Value

Most applications involving pitch use the material as a source of carbon. This is particularly true for applications where pitch is used for producing graphite electrodes, anodes for aluminum production and carbon / carbon composites. For most applications, the coking value of pitch is a linear relationship with the softening point. As shown in Figure 3, use of biodiesel enables one to change the relationship of softening point to coking value.

Figure 3

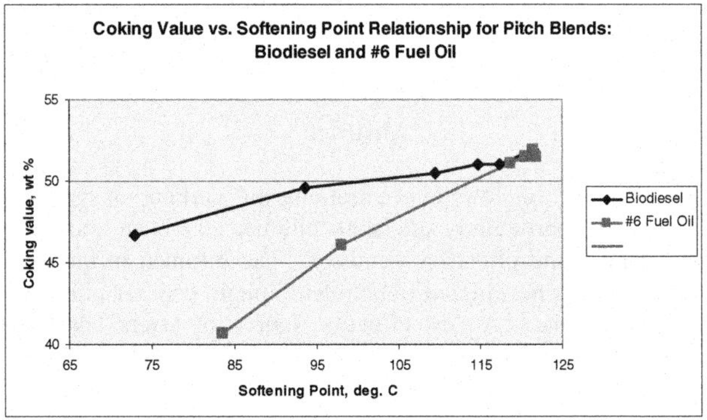

Effect on Viscosity

The effect of biodiesel on the viscosity of the bitumen can be dramatic. Figure 4 represents the baseline case and contains viscosity curves for blends of A-240 pitch and no. 6 fuel oil. Figure 5 contains similar viscosity curves for A-240 pitch and biodiesel.

Figure 4

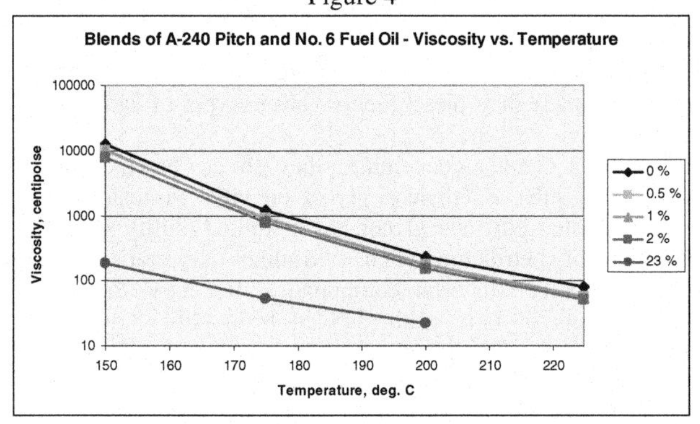

Figure 5

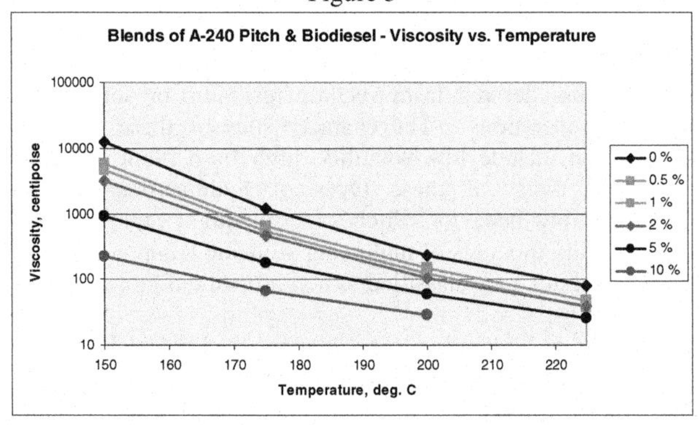

Figures 4 and 5 show that the viscosity is significantly more sensitive to biodiesel than to similar levels of no. 6 fuel oil. Addition of biodiesel to bitumen at a level of 10 percent has about the same effect as the addition of no. 6 fuel oil at a level of 23 percent.

Figures 6 through 9 contain respective viscosity curves for blends of biodiesel and the following bitumens: a second petroleum-based pitch (A-225), a coal tar pitch, a petroleum residuum produced by solvent de-asphalting, and one petroleum residuum (asphalt) produced by vacuum distillation.

Figure 6

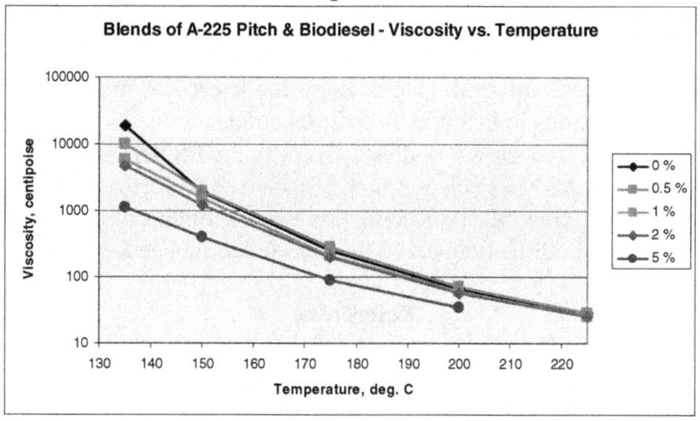

Blends of A-225 Pitch & Biodiesel - Viscosity vs. Temperature

Figure 7

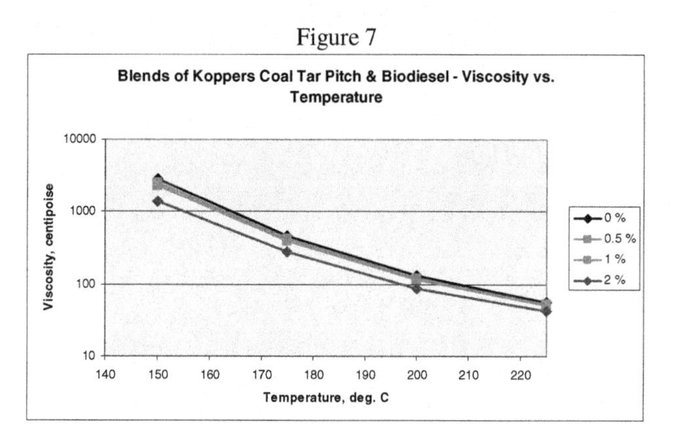

Figure 8

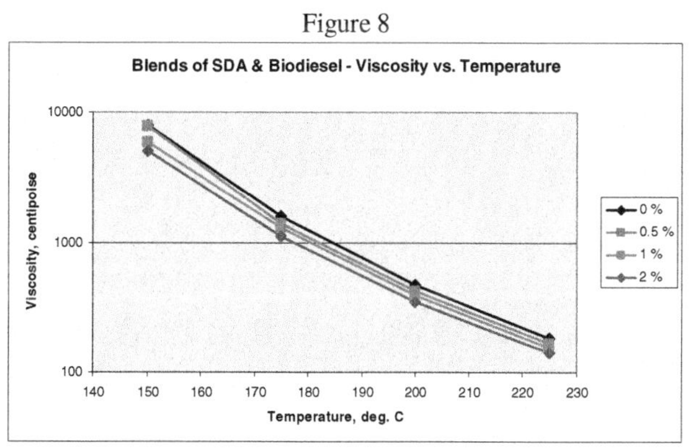

Figure 9

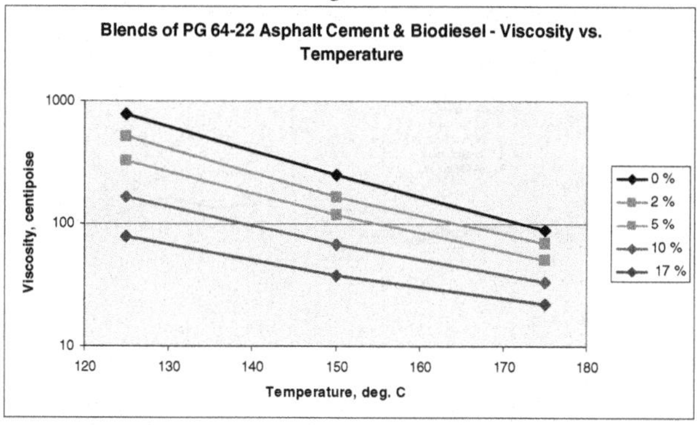

Blends of PG 64-22 Asphalt Cement & Biodiesel - Viscosity vs. Temperature

Detection of Biodiesel by Infrared Spectroscopy

The ester functional group exhibits strong absorbance in the infrared spectra between 1700 and 1750 cm^{-1}. While the differences between the IR spectra of the bitumen / biodiesel blends are small, these differences are detectable using standard FTIR data reduction techniques, including derivative treatment of the spectral curves. Figure 10 illustrates the diffuse reflectance spectra after Kubelka-Munk transformation of the raw reflectance spectra and normalization of the spectra to the pitch absorption at 1603 cm^{-1}. Figures 11 and 12 respectively illustrate the results after first- and second-derivative treatment. Using these techniques, concentrations of biodiesel are detectable at 0.5 wt percent, and correlate well with the level of biodiesel addition. Each of the following figures includes spectra for A-225 pitch with and without various levels of added biodiesel, and a biodiesel spectrum for comparison.

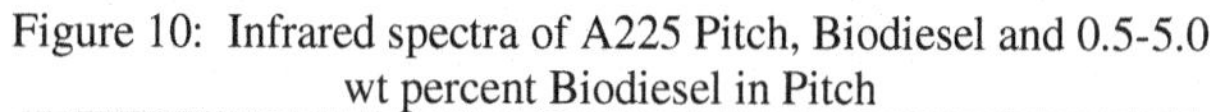
Figure 10: Infrared spectra of A225 Pitch, Biodiesel and 0.5-5.0 wt percent Biodiesel in Pitch

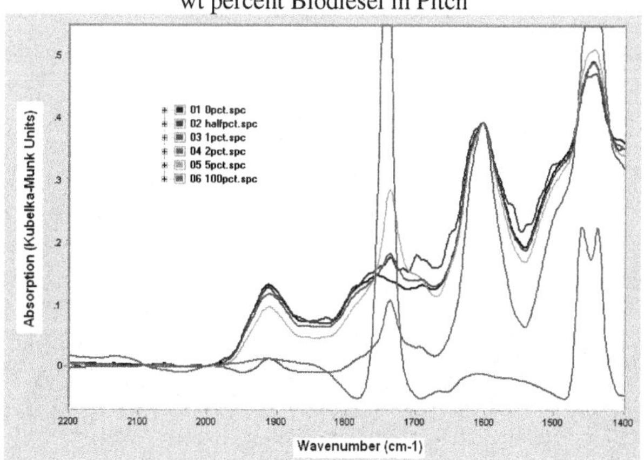

Figure 11: First Derivative Infrared spectra of A225 Pitch, Biodiesel and 0.5 – 5.0 wt percent Biodiesel in Pitch

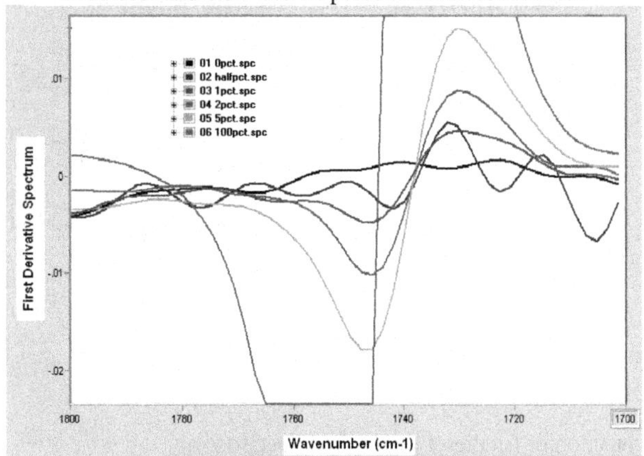

Figure 12: Second Derivative Infrared spectra of A225 Pitch, Biodiesel and 0.5 – 5.0 wt percent Biodiesel in Pitch

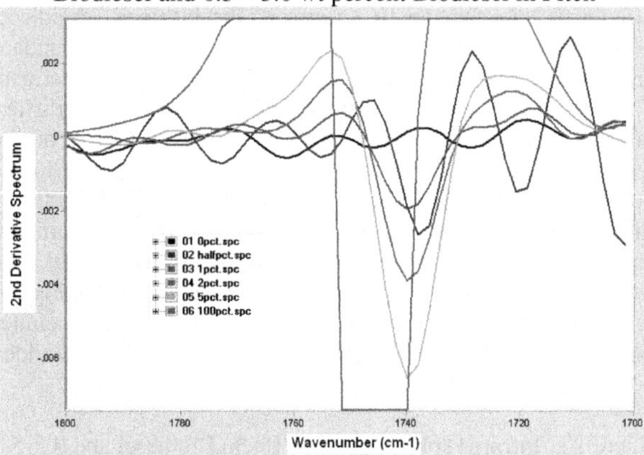

Conclusions

- Oxygenates have been shown to have a significant effect on the viscosity of various types of bitumen.
- Oxygenates with ester functional groups have been shown to have the most significant effect.
- The effect on viscosity is observed for several classes of hydrocarbons, regardless of general chemical composition.
- The levels required to reduce softening points by 10°C are such that minimal effect on coking value is observed.
- One can use infrared spectroscopy to detect the presence of ester functions indicative of biodiesel content.
- Data presented shows biodiesel viscosity modifications of A225 and A240 Petroleum pitch. Since data collection, a new pitch, Marathon M-50, is being produced. Similar biodiesel viscosity modification would be expected in this new product.

References

(1) ASTM Annual Book of Standards, Vol. 5.01, 5.02, 5.03 and 5.04.
(2) US Patent 6,827,841 B2
(3) Schuchardt, U., Sercheli, R., Vargas, R. M., *J. Braz. Chem. Soc.* **1998**, 9, 199.
(4) Smith, B.C., *Fundamentals of Fourier Transform Infrared Spectroscopy*, New York: CRC Press, **1996**, 114-117.

LABORATORY ANODE COMPARISON OF CHINESE MODIFIED PITCH AND VACUUM DISTILLED PITCH

Robert H. Wombles, John Thomas Baron
Koppers Inc., Harmarville Technical Center, 1005 William Pitt Way, Pittsburgh, PA 15238

Keywords: Pitch, Chinese Pitch, Modified Pitch, Mesophase Pitch, Coal Tar, Anode

Abstract

The process currently used to produce the large percentage of 100°C to 110°C softening point anode binder pitch in China today is a heat treating process which produces a product called modified pitch. Modified pitch has all the characteristics of a pitch produced by heat treatment including 3% to 10% mesophase content. The presence of mesophase in binder pitch introduces problems with binder pitch and coke mixing and the resultant properties of the baked anode. This paper will present the results of a laboratory anode study comparing the properties of laboratory anodes produced with Chinese modified pitch and Chinese vacuum distilled pitch. The results indicate that the use of vacuum distilled pitch results in significant improvements in anode properties.

Introduction

The first coal chemical recovery ovens were installed in the United States in 1893. By 1915, by-product ovens accounted for 97 percent of the metallurgical coke produced in the United States. Since the building of by-product ovens, coal tar pitch has been the binder of choice for the aluminum, commercial carbon, and graphite industries.[1]

Current projections are that aluminum consumption will grow 3 to 4% per year for the foreseeable future. That means that the demand for coal tar pitch as an anode binder will increase at an equivalent pace. This increase in demand for coal tar binder pitch will bring forward some interesting problems in obtaining a reliable supply of high quality pitch. These problems are intensified by the events in the metallurgical coke industry in North America and Europe discussed in the following paragraph.

In the late 1980's and early 1990's there were a rash of closings of United States and European coke ovens due to economic and environmental pressures. These coke oven closings have left coal tar pitch suppliers and users searching for strategies to cope with the shrinking supply of coal tar. The accelerated demand for coal tar binder pitch will simply make the search for acceptable quality binder pitch more difficult.

The good news is that there are sufficient supplies of coal tar in the world to meet this increasing binder pitch demand. Figure 1 gives the amount of coal tar produced in the world on a yearly basis versus the amount of tar necessary to produce the binder pitch needed. As the figure illustrates, there is a considerable excess of coal tar in the world. Most of this excess coal tar is being produced in China.

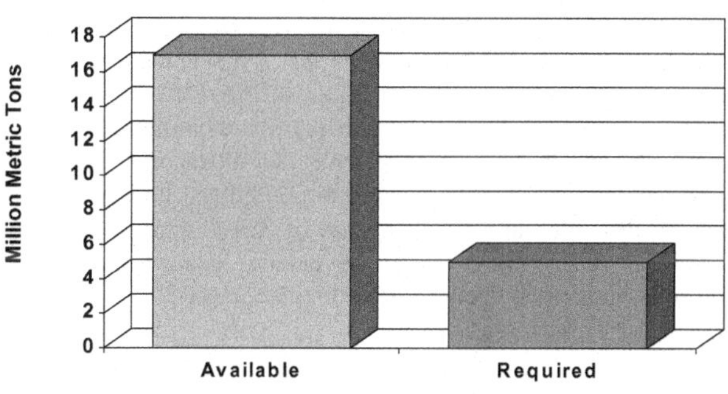

Figure 1 – World Coal Tar Supply

Two types of coal tar pitch are currently being produced in China. These pitches are 1) a pitch with a softening point of 85-90°C used mostly by Soderberg smelters and 2) a modified coal tar pitch with a softening point of 105-120°C used by prebaked smelters. The modified pitch is produced by a thermal treatment process which is discussed in the next section. Typical modified pitch production exhibits significant variations in pitch physical properties. The purpose of this paper is to demonstrate the potential effects on anode quality by using a modified pitch rather than a straight vacuum distilled pitch by presenting the results of a small scale laboratory anode study.[2]

Coal Tar Pitch Manufacturing

Coal tar is a by-product of the coking of coal to produce metallurgical coke. Coal is heated to a temperature of approximately 1100°C in a coke oven to produce coke (the primary product) and by-products such as, coke oven gas, coal tar light oil, and coal tar. A material balance for the coking process is given in Figure 2. As the figure indicates, typical yields are 70% solid products and 30% liquid products. The yield of coal tar, the feedstock for producing coal tar pitch, from a ton of coal is 30–45 liters (8–12 gallons).

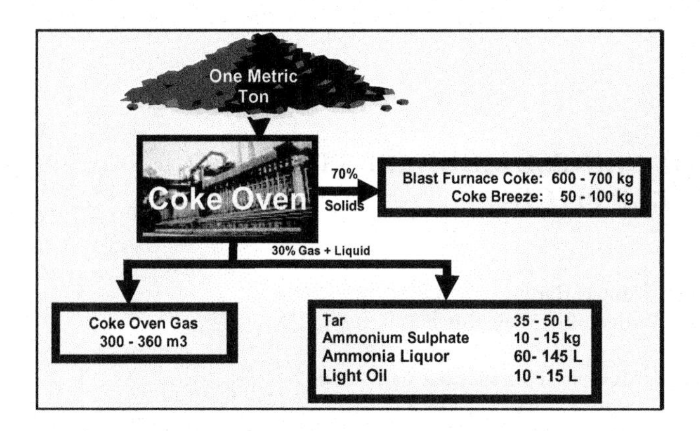

Figure 2. Coking Process Material Balance

Coal tar pitch has many uses, but the majority of the pitch produced is used as a binder for petroleum coke to produce anodes and graphite electrodes. Coal tar pitch is produced from coal tar by a distillation process. In North America and Europe the majority of the anode binder pitch is produced by vacuum flash distillation. In this process the tar is first atmospherically distilled to produce a product called soft pitch which has a softening point of 80-90°C. The soft pitch is then distilled under vacuum to produce a pitch with a softening point of approximately 110°C. The process conditions necessary for producing this pitch are a temperature of 325°C and a residence time of 5 minutes. Figure 3 gives the material balance for coal tar pitch production using vacuum flash distillation.

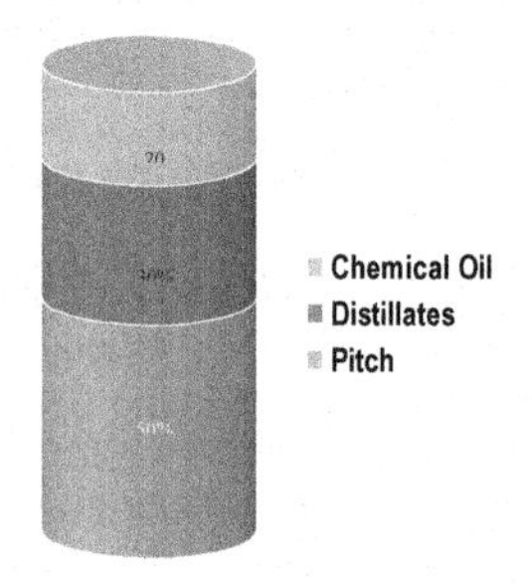

Figure 3 – Material Balance For Coal Tar Distillation

The big advantage of vacuum flash distillation is that it allows production of 110°C softening point pitch without subjecting the pitch to excessive thermal exposure. Excessive thermal exposure can result in the production of mesophase in the resultant binder pitch which has the potential to result in anode physical property degradation. The following observations have been reported for pitches which contain mesophase:

1) Mesophase spheres remain solid at temperatures up to 300°C.
2) The presence of mesophase causes inefficient mixing of pitch and aggregate yielding low green densities.
3) Mesophase spheres break during mixing and coat the aggregate producing an adverse effect on strength and reactivity of baked anodes.
4) Green pastes made with mesophase containing pitches have higher viscosities.
5) Pitch containing mesophase does not wet coke well.[3]

The majority of the 110°C softening point pitch produced in China is produced by a process called pitch modification. In this process the tar is first atmospherically distilled to produce a product called soft pitch which has a softening point of 80-90°C. The soft pitch is then subjected to a thermal treatment which raises its softening point to the desired level. For a pitch with a softening point of 110°C the necessary treatment conditions are a temperature of 405°C for a time of seven hours.

Vacuum flash distillation and pitch modification produce pitches with very different properties. Table 1 gives the physical properties of pitches produced using vacuum flash distillation and pitch modification from similar Chinese coal tars. As the table indicates, pitch modification produces a pitch with high quinoline and toluene insolubles content. These results are typical of pitches which have been thermally treated. The table also indicates that the modified pitch contains a significant quantity of mesophase.

Table 1. Properties of Two Types of Pitch

Property	Modified Pitch (Heat Treated - Heat Soaked)	Vacuum Flashed Pitch
Softening Point, °C	113.2	109.5
Toluene Insolubles, wt.%	31.7	22.1
Quinoline Insolubles, wt.%	9.3	5.6
Conradson Coking Value, wt.%	55.8	51.9
Mesophase, vol.%	6.2	0.0

In order to investigate the differences in the properties of anodes produced with vacuum flash distilled and modified pitches a laboratory anode study was conducted. The remainder of this paper is dedicated to presenting the results of that study.

Laboratory Scale Anode Study

A laboratory anode study was performed comparing the properties of laboratory anodes produced using vacuum flash distilled and modified pitches. The modified pitch used for this study was produced by blending 50% soft pitch and 50% modified pitch. This pitch blend is currently being used by a smelter in China. The smelter in China reported that they were using the pitch blend because the modified pitch alone gave poor mixing. The properties of the pitches used in the study are given in Table 2.

Each of the pitches was submitted for petrographic analyses by ASTM D4616. The results are given in Table 3. As the results show, the modified pitch contained a significant amount of mesophase.

Table 2 – Physical Properties of Pitches in Laboratory Anode Study

Property	Method ASTM	Method ISO	Soft Pitch	Modified Pitch	Blended Pitch	Straight Run Vacuum Distilled Pitch
Softening Point, °C	D3104	5940	91.3	111.7	101.1	109.9
Quinoline Insolubles, wt.%	D2318	6491	10.8	8.6	9.3	5.2
Toluene Insolubles, wt.%	D4072	6376	25.4	29.5	26.6	22.0
Beta Resins, wt.%	TI-QI	TI-QI	14.6	20.9	17.3	16.8
Coking Value, wt.%	D2416	6998	51.8	55.6	53.7	52.7
Density, g/cc	D71	6999	1.30	1.28	1.31	1.30
Mesophase, vol.%	D4616		2.6	2.9	3.7	0.2
Viscosity, cps	D5018	8003				
150°C			542	5760	1490	2860
160°C			308	2330	692	1170

Table 3-Petrographic Analysis of Pitches in Laboratory Anode Study

Pitch Type	Vacuum Distilled Pitch, vol.%	Blended Pitch, vol.%	Soft Pitch, vol.%	Modified Pitch, vol.%
	2005-0610	2005-0743	2005-0730	2005-0731
Continuous Binder Phase	92.4	83.8	79.9	87.4
Aggregate Binder Phase	1.0	2.2	3.3	2.4
Total Binder Phase	93.4	86.0	83.2	89.8
Normal QI	2.6	2.5	1.8	1.1
Coarse QI	0.7	0.3	0.2	0.4
QI in Aggregate	3.1	7.5	12.2	5.8
Total QI	6.4	10.3	14.2	7.3
Mesophase				
<2 microns	0.2	1.9	1.2	1.8
2-4 microns	0	0.8	1.0	0.9
4-10 microns	0	0.8	0.3	0.2
+10 microns	0	0.2	0.1	0
Total Mesophase	0.2	3.7	2.6	2.9
Grand Total	100.0	100.0	100.0	100.0

Table 4 – Anode Production Protocol

Mixing and Molding Conditions	
Materials	Not Preheated
Mixing Time	25-30 Minutes
Mixing Temperature	150°C +/-3°C
Molding Temperature	145-150°C
Vibrating Time	70-80 Seconds
Baking Protocol:	
0-600°C	10°C/hr
600-1170°C	25°C/hr
1170°C	20 hr hold
Center Retort Temperature~	1100-1120°C

Sized coke fractions were received from a Chinese smelter and were weighed and mixed using the following standard anode coke formation:

Coarse	16.0 %
Intermediate	23.0 %
Fine	21.0 %
Super Fine Powder	40.0 %

A series of cylindrical laboratory size anodes (10.2 cm in diameter and 15.2 cm in height) was prepared using each of the two pitches. The details of the anode mixing, molding, and baking protocol are given in Table 4.

After baking, the laboratory anodes were measured for baked apparent density, coking value in situ and shrinkage. The baked anodes were then cored and tested for typical anode properties. Plots of the results of the testing are given in Figures 4-15.

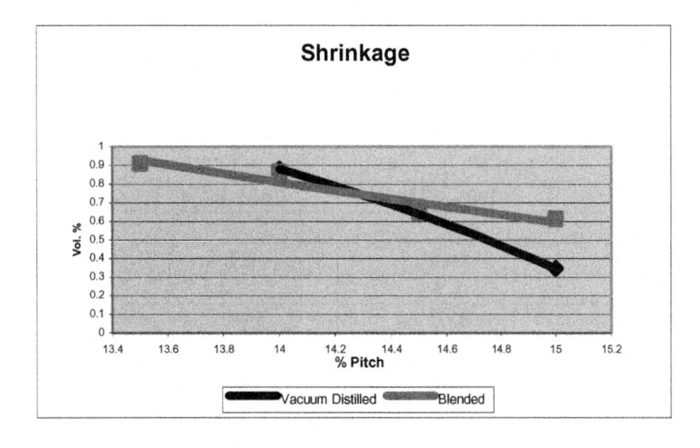

Figure 4. Block Shrinkage

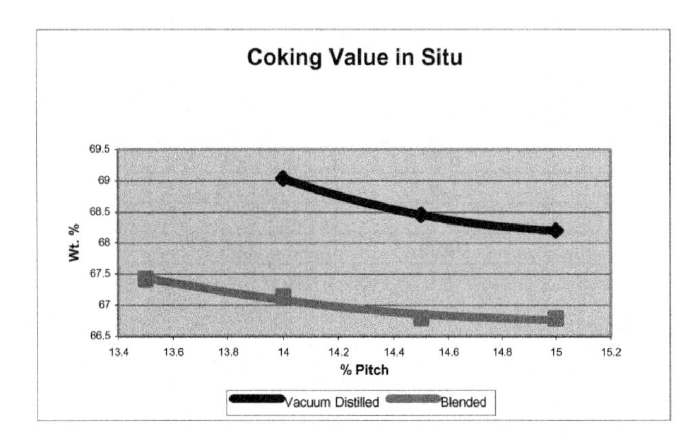

Figure 5. Block Coking Value In-Situ

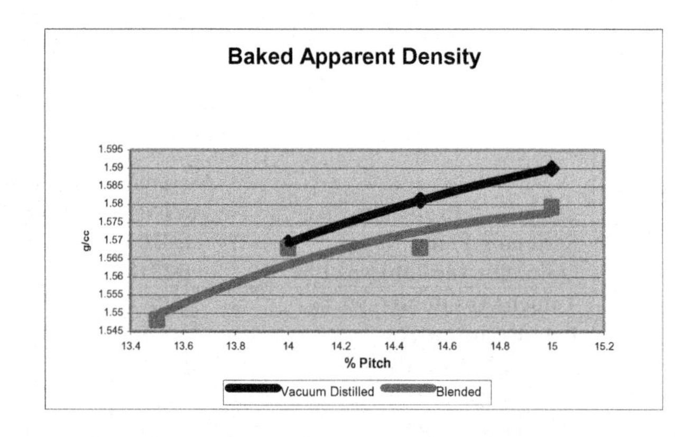

Figure 6. Block Baked Apparent Density

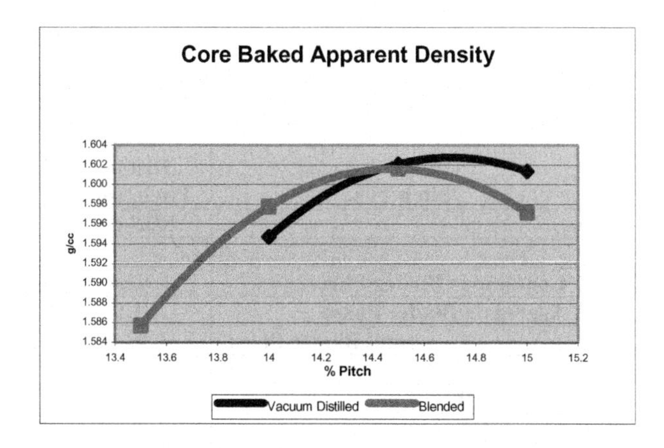

Figure 7. Core Baked Apparent Density

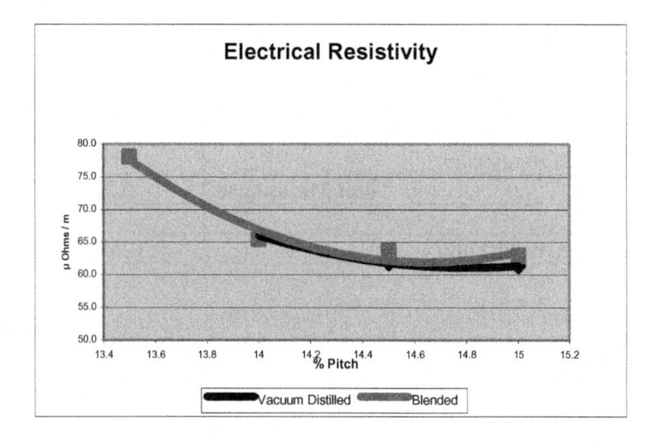

Figure 8. Electrical Resistivity

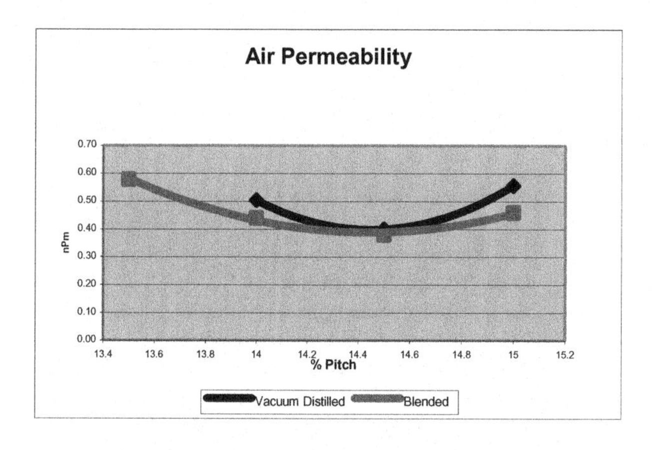

Figure 9. Air Permeability

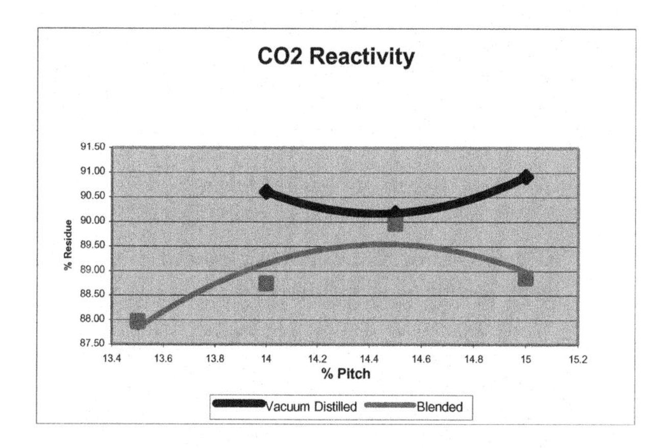

Figure 10. CO_2 Reactivity

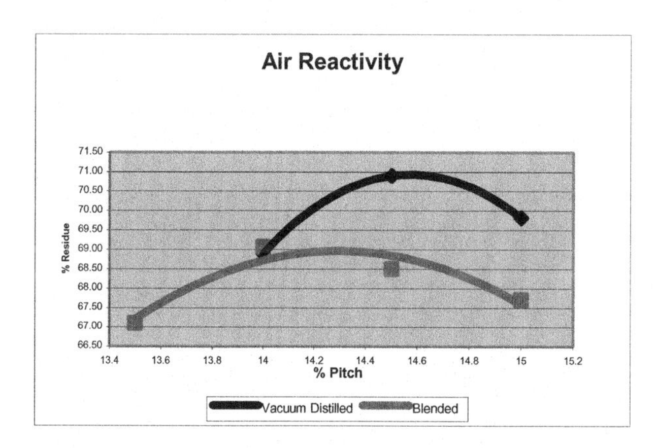

Figure 11. Air Reactivity Weight Loss

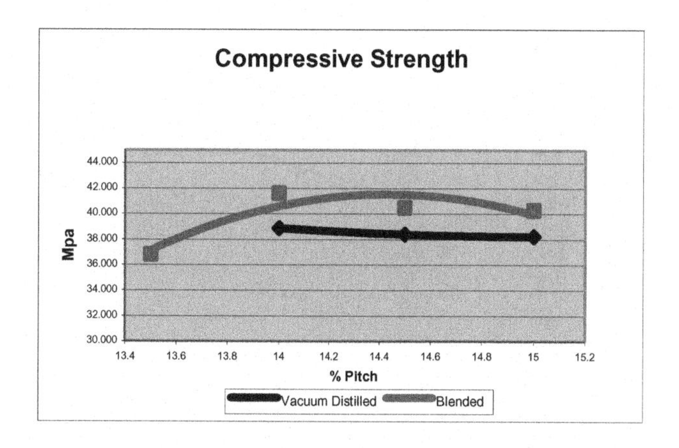

Figure 12. Compressive Strength

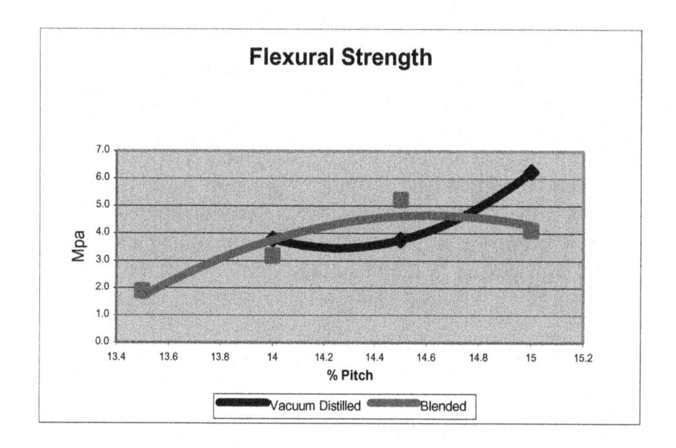

Figure 13. Flexural Strength

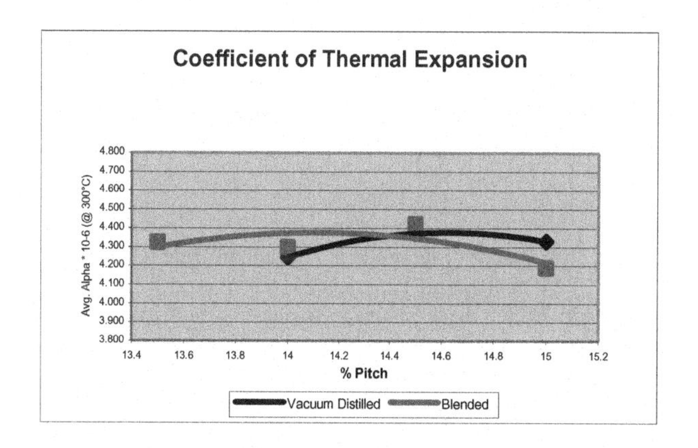

Figure 14. Coefficient of Thermal Expansion

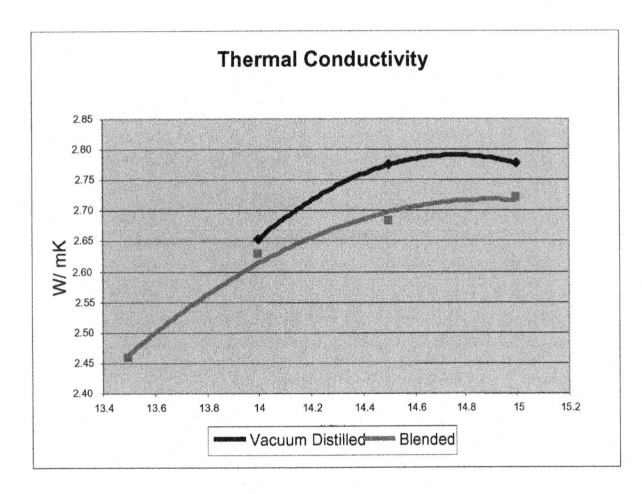

Figure 15. Thermal Conductivity

Discussion

The predominant method of pitch manufacturing in China results in excessive thermal exposure to the pitch. This results in the formation of mesophase spheres. Characterization of both the soft and modified Chinese pitches given in Table 2 indicates that both pitches contain mesophase. Several problems which can result from the use of binder pitch containing mesophase have been discussed in the literature. These potential problems are:

1) Mesophase spheres remain solid at temperatures up to 300°C.

2) The presence of mesophase causes inefficient mixing of pitch and aggregate yielding low green densities.

3) Mesophase spheres break during mixing and coat the aggregate producing an adverse effect on strength and reactivity of baked anodes.

4) Green pastes made with mesophase containing pitches have higher viscosities.

5) Pitch containing mesophase does not wet coke well.[3]

Results of the laboratory anode study demonstrate some of the problems which can be experienced when producing anodes with a binder pitch that contains mesophase. The first indication of a problem is the in-situ coking value. Even though the blended pitch has a higher coking value than the vacuum distilled pitch, it gives a considerably lower in-situ coking value. This is probably caused by the mesophase spheres resulting in poor pitch penetration in the coke porosity. This causes the pitch to accumulate on the surface of the coke where it is more easily volatilized during baking. The reduced pitch penetration into the coke porosity results in degradation of several other important anode properties. As the figures indicate, there is significant degradation in anode baked apparent density, CO_2 reactivity, and air reactivity. Degradation in these properties can cause significant increases in carbon consumption resulting in increased cost for aluminum production.

Conclusions

Results of the laboratory anode study reported lead to the following conclusions:

- The softening point of the vacuum distilled pitch was approximately 10°C higher than blended pitch.
- The presence of mesophase in the blended pitch indicates thermal treatment.
- The vacuum distilled pitch resulted in an anode with improved anode baked apparent density.
- The in-situ coking value of the vacuum distilled pitch is higher despite lower pitch coking value.
- Improved anode baked apparent density and in situ coking value for vacuum distilled pitch anodes indicates improved pitch coke penetration.
- Vacuum distilled pitch produced anodes with improved CO_2 and air reactivities.
- Improved CO_2 and air reactivities indicate reduced carbon consumption in the pot.

References

1. Wombles, Robert H., Kiser, Melvin D., "Developing Coal Tar/Petroleum Pitches," TMS Light Metals 2000.

2. Gengqin, Liu, "Chinese Raw Materials for Anode Manufacturing," 1st Edition, 2004.

3. Wombles, Robert H., Sadler, Barry, "The Effect Of Binder Pitch Quinoline Insolubles Content On Aluminum Anode Physical Properties," 8th Australasian Aluminum Smelting Technology Conference, September 2004.

COMPOSITION AND INTERMOLECULAR REACTIVITY OF BINDER PITCHES AND THE INFLUENCE ON STRUCTURE OF CARBONIZED PITCH COKES

S. Madshus[1], T. Foosnæs[1], M. Hyland[2], J. Krane[3] and H. A. Øye[1]
[1]Department of Materials Science and Engineering, NTNU, Sem Sælands veg 12, 7491 Trondheim, Norway
[2]The University of Auckland, Department of Chemical & Materials Engineering, Private Bag 92019, Auckland, New Zealand
[3]Department of Chemistry, NTNU, 7491 Trondheim, Norway
Keywords: Pitch, Carbonization, Intermolecular Reactivity, NMR, Thermogravimetric Analysis, Optical Texture Analysis

Abstract

The structure of the pitch coke, which acts as a binder between the petroleum coke grains, has an important influence on anode properties like strength, thermal and electrical conductivity and resistance towards oxidation by air and CO_2. However, due to the complexity of pitch composition and pyrolysis chemistry, the link between the raw material and the structure of the carbonized pitch coke is often not clear. The present work aims to take a closer look at this link. One of the main factors that govern the growth and coalescence of mesophase, which in turn to a large extent determines the structure of the pitch coke, is the intermolecular reactivity of pitch constituents. The role of hydrogen transfer in stabilizing thermally induced free radicals and thus reducing the intermolecular reactivity is emphasized. A range of coal-tar and petroleum pitches has been investigated by ^1H NMR spectroscopy in order to identify and quantify structures that are considered to either increase or decrease the intermolecular reactivity. Hydrogen transfer properties were evaluated by the reaction between pitch and anthracene (hydrogen donor ability) or tetralin (hydrogen acceptor ability) at 400 °C. The ratio between the hydrogen donor (HDa) and acceptor (HAa) ability, HDa/HAa, was used as a measure of the intermolecular reactivity of pitches. A correlation was found between the HDa/HAa parameter and the release of volatiles during carbonization. The more thermally reactive pitches exhibiting a low HDa/HAa ratio (high intermolecular reactivity) had a high weight loss at modest temperatures whereas for the pitches of lower intermolecular reactivity the major release of volatiles was postponed till the later and critical stages of carbonization (300 – 500 °C). The HDa/HAa ratio of petroleum pitches was found to correlate with the size of optical texture of the resulting green cokes (550 °C). A lower intermolecular reactivity (high HDa/HAa ratio) resulted in pitch cokes of larger more well-developed optical texture to be formed. However, for pitches of coal-tar origin the presence and amount of particulate matter (QI) was found to be the most influencial factor on the size of optical domains resulting in a fine mosaic texture.

Introduction

Coal-tar pitch is the preferred choice of binder material in anode manufacture today. However, the availability of high quality coal-tar is in decline and at least partial replacement by alternative binder sources will become increasingly important in the future. Due to environmental regulations, petroleum pitches are interesting as they generally have lower PAH emissions than coal-tar pitches during baking. The aluminium industry must be prepared to meet the challenges involved in adapting binder pitches from new sources or coal-tar pitches which may be of inferior quality to the pitches available on the market today. An increased understanding of the processes involved in the transformation of a pitch into a coke and the link between raw material composition and properties and the final artifact is thus highly relevant.

The structure of the binder pitch has an important impact on anode properties and in particular on the resistance towards oxidation by air and CO_2. Due to the less severe thermal treatment of the pitch coke binder phase compared to the petroleum coke filler phase, oxidation may take place in the pitch coke preferentially. Selective airburn or oxidation by CO_2 may result in dusting and weakening of the anode causing a higher excess carbon consumption and may also lead to operational difficulties.

Pitch is an extremely complex mixture of numerous, essentially aromatic, hydroaromatic, alkylated and heterocyclic compounds [1]. Traditionally, the suitability of a binder pitch for use in anodes has been defined from parameters like softening point, insolubility in toluene (TI) and quinoline (QI), coke yield, H/C atomic ratio, ash content and density. Although these parameters, which are mostly empirical in nature, give an indication of the pitch quality more information on the carbonization behavior of pitches is certainly valuable.

In 1968 Brooks and Taylor [2] first recognized and advanced the concept and relevance of mesophase as an essential intermediate stage in the formation of anisotropic cokes from the liquid stage. Coal-tar and petroleum pitches pass through a fluid stage during carbonization. In the early stages of carbonization, free radicals are formed due to thermal rupture of C-C and C-H bonds in reactive components. Polymerization occurs mainly via a free radical mechanism leading to molecular size enlargement (aromatic growth) and the formation of oligomeric systems (mesogens) [3]. If the intermolecular reactivity of the pitch constituents is too high, extensive cross-linking and a rapid transformation of pitch molecules through polymerization will occur at a relatively low temperature. In this case, either mesophase will not be formed or the growth and coalescence of mesophase will take place under low fluidity/high viscosity conditions leading to a premature solidification of the pyrolysis system. An isotropic coke or a pitch coke of small optical domains will then be formed. On the other hand, if the pitch has a low intermolecular reactivity, aromatic growth is constrained and the mesogens will have sufficient mobility to stack parallel to each other and establish a liquid crystal system (mesophase). The growth and coalescence of mesophase take place at a higher temperature where the viscosity of the pyrolysis system is at a low level. Eventually, the system will solidify and an anisotropic coke of large well-developed optical domains is formed.

In particular, the presence of alkyl side chains, reactive functional groups such as hydroxyl and carboxyl and possibly also other heteroatoms than oxygen are considered to lead to an increased intermolecular reactivity [4]. If free radicals formed by thermal rupture of bonds in reactive pitch species can be stabilized by hydrogen transfer from within the system, extensive cross-linking at a too early stage is prevented [5]. Then initiation, growth and coalescence of mesophase are facilitated and consequently a coke of large well-developed optical domains is formed. Hydroaromatic rings and naphthenic rings in hydroaromatic species are considered to be principal hydrogen donor groups [6]. Oxygen acceptor sites are believed to deplete the supply of donatable hydrogen and leave radicals free to recombine. The intermolecular reactivity of a pitch is thus dependent on both the amount of reactive species and the ability of the pitch to stabilize free radicals by hydrogen transfer.

The present work aims to describe and explain the link between pitch composition and carbonization behavior based on NMR spectroscopy, "traditional" pitch parameters, hydrogen transfer properties and thermogravimetric analysis. The structure of the pitch cokes is analyzed by optical microscopy.

Experimental

Materials

The series of pitches used in this study includes five coal-tar pitches labeled CTP1-5 and four petroleum pitches labeled PP1-4 Pitch properties are given in Table 1. In addition, a QI-free coal-tar pitch labeled CTPN supplied by GrafTech International has been studied. The pitch labeled CTP1N, which is also QI-free, has been prepared by GrafTech International by filtration of CTP1.

Table 1. Pitch properties.

Pitch	Softening Point (°C)[1]	Coking Value (wt%)[2]	QI (wt%)[3]	O (wt%)*	H/C†
CTP1	110.7	61.0	12.9	0.63	0.54
CTP2	119.7	60.1	9.6	0.25	0.55
CTP3	187.2	78.1	25.0	0.44	0.49
CTP4	118.4	59.7	6.6	0.86	0.56
CTP5	116.5	60.4	4.7	1.24	0.60
CTPN	120.0	60.4	0.6	0.23	0.57
CTP1N	114.2	NA	0.6	0.22	0.58
PP1	123.1	54.4	0.1	NA	0.67
PP2	112.3	48.1	0.3	NA	0.78
PP3	120.3	50.4	0.2	NA	0.77
PP4	109.2	46.9	0.0	1.67	0.79

1. Mettler softening point, ASTM D 3104, 2. ISO 6998, 3. Quinoline Insolubles (QI), ISO 6791, * Elemental analysis. Oxygen content by difference. The sum of C, H, N and S exceeded 100 wt% for PP1, PP2 and PP3, † Atomic ratio.

The elemental analysis (Table 1) was performed for this study by Instituto National del Carbón (INCAR) in Oviedo, Spain. The carbon, hydrogen and nitrogen contents were determined with a LECO-CHN-2000 unit while the sulphur content was determined with a LECO S-144-DR unit. Carbon, hydrogen, nitrogen and sulphur contents are not shown due to limited space. The oxygen content was estimated from the difference between 100 wt% and the sum of the other elements.

Nuclear Magnetic Resonance (NMR)

Proton NMR spectroscopy was performed on all of the pitches with the exception of CTP1N. The pitches were dissolved in carbon disulphide using an ultrasonic bath for a period of two hours ensuring a high degree of extraction. The extracts were filtered through a membrane syringe filter and then added to a 5 mm NMR tube. Deuterated dioxane (1,4-dioxane-8d) was used as a lock substance. A pitch concentration of 3 wt% was chosen for the ^{1}H NMR experiments [7]. All spectra were recorded on a Bruker Avance DRX 600 spectrometer using the XWINNMR 3.5 software. The operating frequency was 600.18 MHz. Fourier transformation of the FIDs, baseline and phase corrections and integration of spectra were performed using the MestReC version 4.3.6 software. The chemical shift classification in the present work (illustrated in Figure 1) is basically the same as reported by Guillén, Díaz and Blanco [7].

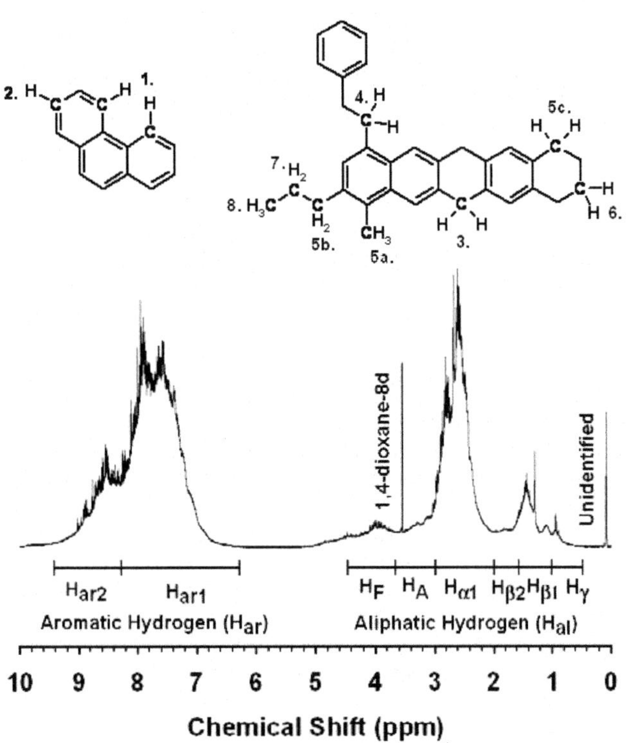

Figure 1. Assignment of chemical shifts in the ^{1}H NMR spectrum of a petroleum pitch (PP1) and illustration of various types of hydrogen. H_{ar2}: 1, H_{ar1}: 2, H_F: 3, H_A: 4, $H_{\alpha 1}$: 5a,b,c, $H_{\beta 2}$: 6, $H_{\beta 1}$: 7 and H_γ: 8. Hydrogen in α-methyl group: 5a, Hydrogen in α-methylene in alkyl side chain: 5b and Hydrogen in α-methylene in hydroaromatic ring (5c).

The H_{ar2} band in Figure 1 represents aromatic hydrogen in sterically hindered positions, highly pericondensed polyaromatic compounds, position next to heteroatoms and hydrogen joined to nitrogen. All other aromatic hydrogens are included in the H_{ar1} band.

For more experimental details on the NMR spectroscopy and chemical shift classification, it is referred to Madshus [8].

Hydrogen donor and acceptor abilities

Anthracene was used as a hydrogen acceptor compound to evaluate the hydrogen donor ability (HDa) [9, 10] while tetralin (1,2,3,4-tetrahydronaphthalene) was used as a hydrogen donor compound to evaluate the hydrogen acceptor ability (HAa) [10, 11]. A mixture of 0.2 g anthracene or tetralin and 0.2 g pitch was added to a pyrex glass reaction tube which was filled with argon gas to avoid reaction with air and sealed. For the hydrogen donor ability test, the reaction tube was heated at a rate of 5 °C/min to 400 °C with no soaking time. In the case of the hydrogen acceptor ability test, the reaction tube was heated rapidly to 400 °C with a soaking time of 8 hours. After cooling to room temperature, the reaction tube was opened and the residue was extracted in CS_2 using an ultrasonic bath. The dissolved reaction product of the pitch and anthracene or tetralin was analyzed by gas chromatography. The hydrogen donor ability (HDa) was calculated from the amount of the hydrogenated products, 9,10-dihydroanthracene (DHA) and 1,2,3,4-tetrahydroanthracene (THA) formed and expressed as milligrams of hydrogen per gram of pitch. The only major product from the reaction between tetralin and pitch was naphthalene. The hydrogen acceptor ability was calculated from the relative amount of naphthalene formed and also expressed as milligram of hydrogen per gram of pitch. For more details on the experimental procedure it is referred to Madshus [8].

Thermogravimetric analysis

The thermogravimetric analysis was performed on a Mettler/Toledo TGA/SDTA 851^e with a sample size of approximately 5 mg. The samples were heated to 1000 °C at a rate of 10 °C/min [12] under a nitrogen gas flow of 50 mL/min. Derivation of the weight loss curves (DTG) was performed using the accompanying STAR^e software. The main parameter derived from the thermogravimetric analysis was the percentage of the total volatile matter (1000 °C) released in the temperature range between 300 and 500 °C (VM300-500).

Optical texture analysis

Each of the eleven pitches (~ 70 g) were heat treated in a pressurized carbonization chamber under an argon gas atmosphere (15 bar) to 550 °C at a slow heating rate. Optical texture analysis was performed on a Leica MeF3A metallurgical inverted reflecting light microscope fitted with polarizing modules and a half-wave retarder plate. The microscope is equipped with a Sony DXC 930P digital video camera and connected to a computer. The main parameter derived from the analysis was the mosaic index which is a measure of the average optical domain size. A coke of small optical domains will yield a high relative value of the mosaic index. A detailed description of the procedure and equipment is given by Rørvik, Aanvik, Sørlie and Øye [13].

Results and discussion

Nuclear Magnetic Resonance (NMR)

Hydroaromatic rings are considered to lead to a low intermolecular reactivity. The presence of alkyl side chains, on the other hand, is expected to lead to a high intermolecular reactivity. According to Greinke [14], it is generally believed that the initiation of polymerization in pitch primarily involves the reaction of the less stable alkyl group compared to the more stable aromatic hydrogens. The length of the alkyl side chains may also be an important factor influencing the intermolecular reactivity. Ida et al. [15] found that the presence of longer alkyl side chains resulted in a rapid polymerization which in turn led to a low fluidity of the reaction system at relatively low temperatures and consequently a fine mosaic structure of the coke.

The distribution of aliphatic hydrogen is shown in Figure 2.

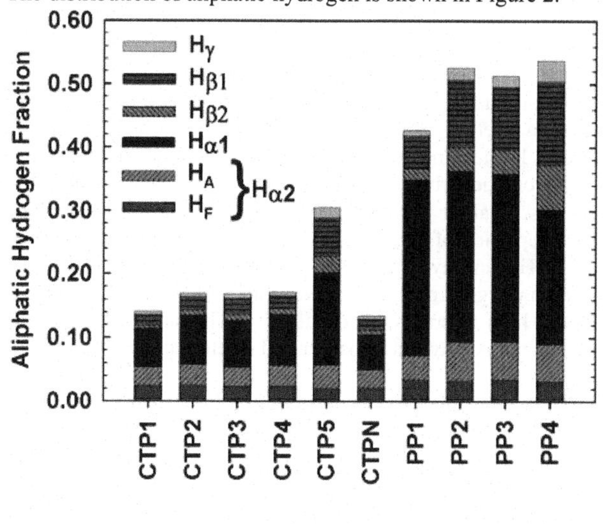

Figure 2. Distribution of the aliphatic hydrogen fraction obtained from integration of the ^{1}H NMR spectra.

The petroleum pitches have a less aromatic character than the coal-tar pitches which is reflected in a higher fraction of aliphatic hydrogen. This is mainly due to a higher amount of methyl and ethyl substituents ($H_{\alpha 1}$ and $H_{\beta 1}$). Longer alkyl side chains (i.e. mainly propyl) are also present to a certain extent and is reflected in the H_γ fraction. However, a proportion of the $H_{\alpha 1}$ band is due to hydrogen located in hydroaromatic rings. This proportion is considered to be equal to the $H_{\beta 2}$ fraction (see Figure 1).

Hydrogen located in the H_A band increases the intermolecular reactivity as thermal rupture of an ethylene bridge would lead to the formation of free radicals. On the other hand, the H_F band may represent a source of donatable hydrogen as the release of the two hydrogens on the dimethylenic groups would result in a more thermally stable aromatic structure being formed. However, hydrogen located in methylene single bridges linking two aromatic structures, Ar-CH_2-Ar', is not potentially donatable as these bridges are expected to undergo cleavage leading to free radical fragments. It is not possible to distinguish hydrogen in dimethylenic bridges from hydrogen located on a methylene single bridge as the methylene hydrogen in both cases will have a similar chemical shift in the H_F band.

From the distribution of aliphatic hydrogen, PP2 and PP3 are expected to have similar intermolecular reactivities. Petroleum pitch 1 (PP1) is more aromatic, has a smaller amount of ethylene bridges (H_A), has a less hydroaromatic structure which is reflected in the low proportion of $H_{\beta 2}$ and is richer in short alkyl side chains (i.e. methyl groups) when compared to the other pitches of petroleum origin. This pitch is considered to have the lowest intermolecular reactivity of the petroleum pitches despite an expected lower supply of donatable hydrogen. The last petroleum pitch, PP4, has a quite high supply of donatable hydrogen, but is also considered to have a considerably longer average alkyl side chain length due to the high proportion of hydrogen located in the $H_{\beta 1}$ and H_{γ} bands. Petroleum pitch 4 (PP4) was found to be particularly rich in oxygen (1.67 wt%) (Table 1). Oxygenated functional groups are expected to deplete the supply of donatable hydrogen. Due to the high oxygen content and long average alkyl side chain length, PP4 is considered to have a high intermolecular reactivity.

With the exception of CTP5, the coal-tar pitches show a fairly similar distribution of aliphatic hydrogen. Most of the hydrogen is located in methyl substituents ($H_{\alpha 1}$) whereas the amount of hydrogen located in methylene (H_F) and ethylene (H_A) bridges linking aromatic structures is almost the same for all of the coal-tar pitches. Coal-tar pitch 5 is distinguished due to a considerably higher proportion of hydrogen located in longer alkyl side chains ($H_{\beta 1}$ and H_{γ}) and hydroaromatic structures ($H_{\beta 2}$). This pitch also has a high oxygen content (1.24 wt%) and is considered to have a relatively high intermolecular reactivity. The structure of CTP5 resembles a mix between a coal-tar and a petroleum pitch.

There was not found a large variation in the distribution of aromatic hydrogen (not shown due to limited space) between the pitches. However, PP4 had a lower proportion of hydrogen located in the H_{ar2} band than the other pitches. This may indicate a higher concentration of relatively low molecular weight compounds in this pitch.

Hydrogen donor and acceptor abilities

With the exception of PP4, the petroleum pitches showed higher hydrogen donor abilities than the pitches of coal-tar origin, reflecting the more hydroaromatic structure and thus larger supply of donatable hydrogen. The petroleum pitches, with the exception of PP1, exhibited higher hydrogen acceptor abilities than the coal-tar pitches. Longer alkyl side chains, which are abundant in PP2 and PP3 and in particular in PP4, are considered to be less thermally stable than shorter substituents (i.e. methyl groups) and lead to the formation of free radical fragments which will abstract hydrogen from tetralin.

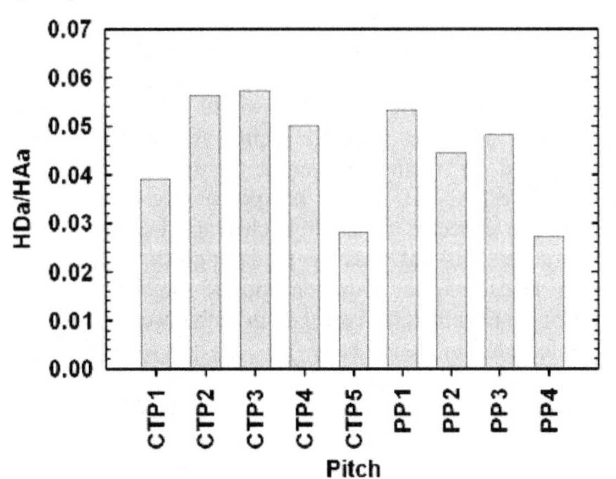

Figure 3. Ratio between hydrogen donor ability and hydrogen acceptor ability, HDa/HAa.

A low intermolecular reactivity is considered to be reflected in a relatively high hydrogen donor ability (HDa) and at the same time a relatively low hydrogen acceptor ability (HAa). The ratio between the hydrogen donor and acceptor ability, HDa/HAa, is used as a measure of the intermolecular reactivity of the pitches studied in the present work and is shown in Figure 3. The HDa mainly reflects the modifying ability of the pitch to stabilize free radicals whereas the HAa is considered to represent the transformation of the pitch itself. A similar HDa/HAa ratio has been used by Yokono and co-workers [16] to reflect the modifying ability of a pitch (hydrogen donor) in the co-carbonization with coal. They found that a higher ratio of hydrogen donor to hydrogen acceptor ability was correlated to improved fluidity conditions of the pyrolysis system and larger size of optical texture of the resulting coke.

Thermogravimetric analysis

The processes occuring during thermal treatment of pitches are reflected in the release of volatiles. A relation between the amounts and timing of volatiles release and the intermolecular reactivity of pitches would therefore be expected.

A correlation was found between the HDa/HAa ratio and the percentage of the total volatile matter (1000 °C) released between 300 and 500 °C, VM300-500 (Figure 4).

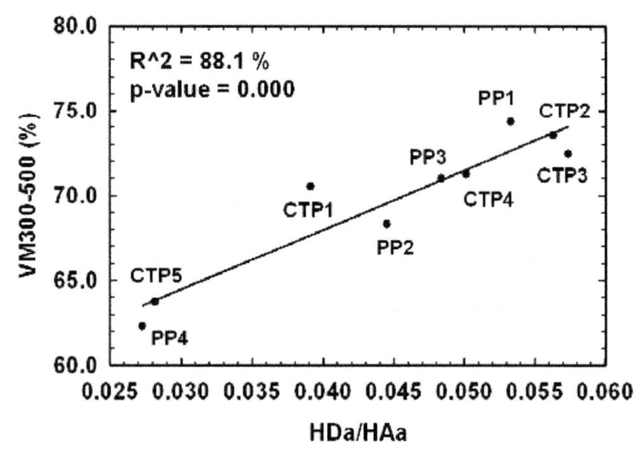

Figure 4. Correlation between HDa/HAa ratio and the percentage of volatile matter released between 300 and 500 °C (VM300-500).

Reactive pitches exhibiting a low HDa/HAa ratio like PP4 and CTP5 have a high activity at relatively low temperatures. Low boiling point pitch molecules and fragmentation species resulting from thermal rupture of weak bonds to longer alkyl side chains and bridges between aromatic structures lead to a relatively high amount of volatiles being released below 300 °C. If on the other hand the pitch has a low intermolecular reactivity, fragmentation species are stabilized by hydrogen transfer and retained in the pyrolysis system. The resulting thermally stable molecules of relatively low molecular weight may then act as solvating vehicles maintaining a low viscosity in the system and may also be important as hydrogen shuttling agents. When the system has reached a critical stage for mesophase growth and coalescence, these smaller thermally stable molecules (non-mesogens) are eventually distilled resulting in a high amount of volatile matter released at relatively high temperatures above 300 °C and an ordered anisotropic coke structure will be formed. For the more reactive pitches, the high loss of volatiles during the early stages of carbonization will result in a more rapid increase in viscosity and the resulting coke will be less ordered with smaller optical domains.

Results from the thermogravimetric analysis indicate that important reactions which in the later stages of carbonization govern the initiation, growth and coalescence of mesophase occur to a significant extent already at modest temperatures of less than 300 °C.

Structure of pitch cokes and impact on anode performance

Sample images of optical textures of green pitch cokes (550 °C) are shown in Figure 5 below.

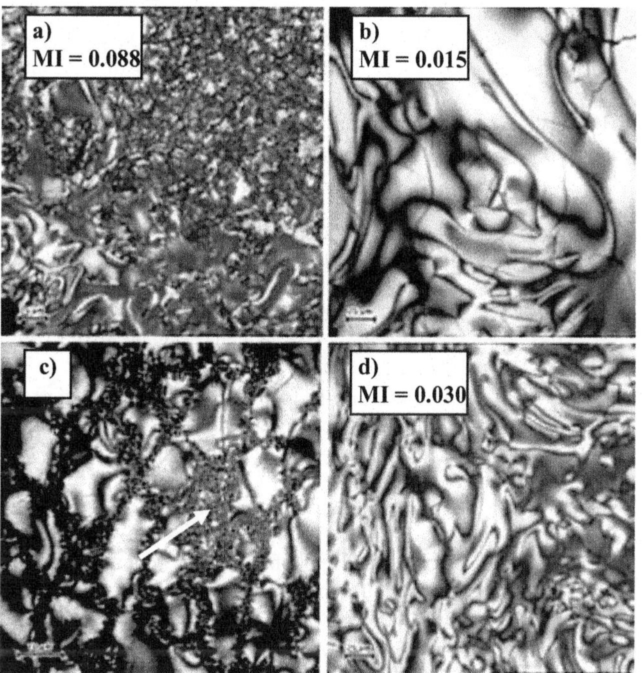

Figure 5. Optical texture of green pitch cokes. The scale bar is 25 µm in images a, b and d and 10 µm in image c. a) CTP1 (12.7 wt% QI), b) CTP1N (0.6 wt% QI), c) CTP1 (12.7 wt% QI) at a higher magnification and d) PP4. MI = Mosaic Index [13].
QI cluster indicated by white arrow (c).

A correlation was found between the mosaic index and the HDa/HAa ratio for the petroleum pitches (Figure 6). The pitches exhibiting a low HDa/HAa ratio, i.e. a high intermolecular reactivity form cokes of smaller optical domains, i.e. a higher mosaic index. However, such a correlation was not found for the pitches of coal-tar origin.

Particulate matter (QI particles) in coal-tar pitches has a major influence on the growth and coalescence of mesophase and thus on the optical texture of the resulting coke as demonstrated by QI removal from CTP1 (see Figure 5 a and b). By carbonization, the QI free CTP1N gives a coke of large optical domains. The QI containing CTP1 on the other hand, gives a mixed structure of predominantly fine mosaic coke with inclusions of larger and more developed domains. The anisotropic domains are separated and surrounded by the QI particles ("black dots" in Figure 5c) which tend to form clusters.

The QI containing coal-tar pitches formed cokes exhibiting a much higher mosaic index than the petroleum pitches, irrespective of the intermolecular reactivity. However, the two QI free coal-tar pitches formed cokes of larger optical domains (lower mosaic index) than the pitches of petroleum origin. Excluding CTP3, a correlation was found between the QI content and the mosaic index for the coal-tar pitches (Figure 7).

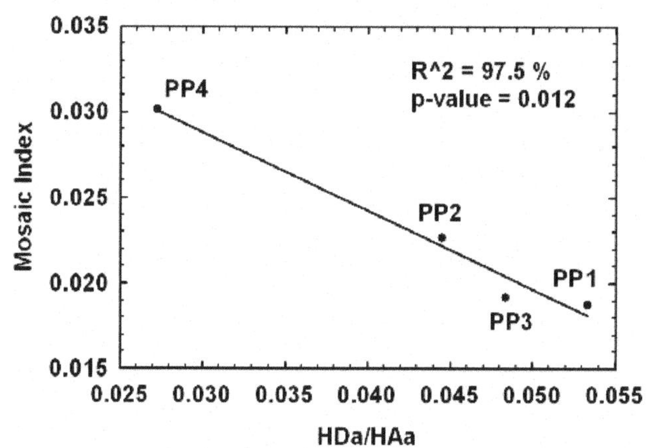

Figure 6. Correlation between mosaic index and HDa/HAa ratio for petroleum pitches.

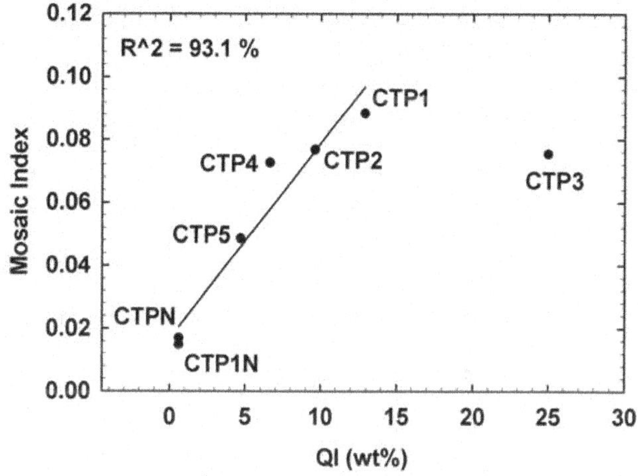

Figure 7. Relationship between QI content and mosaic index for coal-tar pitches.

Coal-tar pitch 3 (CTP3) is distinguished due to its high softening point, high coking value and high QI content (Table 1). This pitch is likely to have received a severe thermal treatment which may have caused a substantial aromatic growth. A large part of the QI fraction of this pitch may therefore be secondary QI, i.e. mesophase. The pitch coke obtained from CTP3 did not seem to have a much higher amount of particulate matter (primary QI) than for instance CTP1 (12.7 wt% QI) from the optical microscopy investigation.

The structure of the pitch coke will certainly have an important influence on the physical and chemical properties of the binder phase and thus on anode performance. In particular, the reactivity towards air and CO_2 is important as a selective oxidation of the pitch coke binder is often observed in anodes. The optical domain size of the pitch coke is probably an important factor in addition to porosity and pore size distribution and the amount and nature of metal impurities (catalysts) influencing the air and CO_2 reactivity. A finer mosaic texture is expected to lead to an increased rate of oxidation. Another factor which may influence the oxidation characteristics of coal-tar pitch cokes is the difference in reactivity between QI particles and the anisotropic domains. The QI particles are generally more disordered (isotropic) and are expected to be more susceptible to oxidation [17]. Metal impurities and oxygenated functional groups are usually concentrated in the QI phase and may be additional factors which lead to selective oxidation. A high QI content may thus ultimately lead to higher excess carbon consumption. A further study of the influence of QI particles and optical texture on the oxidation behavior of pitch cokes is planned.

Conclusion

The following main conclusions were derived from this work:

- Proton NMR spectroscopy provides a relatively rapid method to obtain detailed structural information on the periphery of pitch molecules. Spectra of high resolution were achieved.
- The hydrogen donor (HDa) and acceptor ability (HAa) tests were succesful in differentiating the pitches. The ratio between these two parameters, HDa/HAa, was chosen as a parameter reflecting the intermolecular reactivity of pitches. A high ratio is connected to a low intermolecular reactivity whereas a low ratio is connected to a high intermolecular reactivity.
- Considerations on the intermolecular reactivity of pitches from ^{1}H NMR spectroscopy in conjunction with elemental analysis could generally be related to the HDa/HAa ratio. In particular a high amount of longer alkyl side chains and a high oxygen content as determined by elemental analysis were considered to lead to an increased intermolecular reactivity. Hydroaromatic rings were considered to be the primary source of donatable hydrogen which may lead to a decreased intermolecular reactivity.
- A correlation was found between the relative weight loss between 300 and 500 °C (VM300-500) and the HDa/HAa ratio. Pitches which had a high intermolecular reactivity experienced a relatively high weight loss before 300 °C. For the pitches of lower intermolecular reactivity (high HDa/HAa ratio), the weight loss was postponed to the more critical stages of carbonization thus resulting in a higher VM300-500 parameter.
- For the petroleum pitches, a correlation was found between the size of optical texture in the green cokes obtained and the HDa/HAa ratio. A high HDa/HAa ratio (low intermolecular reactivity) was connected to a green coke structure of larger more well-developed optical domains. However, for the pitches of coal-tar origin, the most influential factor on the coke structure was the particulate matter (QI) which reduced the size of optical domains. A correlation between the HDa/HAa ratio and the size of optical domains was therefore not found for the coal-tar pitches.

Acknowledgements

The authors would like to thank The Norwegian Research Council, The Norwegian Aluminium Industry and The Norwegian Ferroalloy Industry (FFF) for financial support through the PROSMAT and CarboMat programs. We would also like to thank M.Sc. Stein Rørvik (SINTEF) for valuable assistance with the optical texture analysis. Dr. Richard L. Shao at GrafTech International is thanked for supplying the QI free pitches.

References

[1] Zander M. "On the composition of pitches", *Fuel, 66 (1987),1536-1539.*

[2] Brooks J.D. and Taylor G.H. "The Formation of some Graphitizing Carbons", In Chemistry and Physics of Carbon. Walker P.L.J. 4 (1968),243-285. Marcel Dekker, Inc., New York.

[3] Zander M. "Chemistry and Properties of Coal-tar and Petroleum Pitch", In Sciences of Carbon Materials. Marsh H. and Rodríguez-Reinoso F. (2000),205-258. University of Alicante, Alicante.

[4] Marsh H. and Menendez R. "Carbons from pyrolysis of pitches, coals, and their blends", *Fuel Processing Technology, 20 (1988),269-296.*

[5] Marsh H. and Neavel R.C. "Carbonization and liquid-crystal (mesophase) development. 15. A common stage in mechanisms of coal liquefaction and of coal blends for coke making", *Fuel, 59 (1980),511-513.*

[6] Clarke J.W., Rantell T.D. and Snape C.E. "Estimation of the concentration of donatable hydrogen in a coal solvent by N.M.R", *Fuel, 61 (1982),707-712.*

[7] Guillen M.D., Diaz C. and Blanco C.G. "Characterization of coal tar pitches with different softening points by 1H NMR. Role of the different kinds of protons in the thermal process", *Fuel Processing Technology, 58 (1998),1-15.*

[8] Madshus S., "Thermal reactivity and structure of carbonized binder pitches" (Ph.D. Thesis, NTNU, 2005).

[9] Yokono T., Marsh H. and Yokono M. "Hydrogen donor and acceptor abilities of pitch: proton NMR study of hydrogen transfer to anthracene", *Fuel, 60 (1981),607-611.*

[10] Machnikowski J., Kaczmarska H., Leszczynska A., Rutkowski P., Diez M.A., Alvarez R. and Garcia R. "Hydrogen-transfer ability of extrographic fractions of coal-tar pitch", *Fuel Processing Technology, 69 (2001),107-126.*

[11] Pajak J. "Hydrogen transfer from Tetralin to coal macerals. Reactivity of macerals", *Fuel Processing Technology, 21 (1989),245-252.*

[12] Bermejo J., Granda M., Menendez R. and Tascon J.M.D. "Comparative analysis of pitches by extrography and thermal analysis techniques", *Carbon, 32 (1994),1001-1010.*

[13] Rørvik S., Aanvik M., Sørlie M. and Øye H.A. "Characterization of optical texture in cokes by image analysis", *Light Metals (Warrendale, Pennsylvania), (2000),549-554.*

[14] Greinke R.A. "Chemical bond formed in thermally polymerized petroleum pitch", *Carbon, 30 (1992),407-414.*

[15] Ida T., Akada K., Okura T., Miyake M. and Nomura M. "Carbonization of methylene-bridge aromatic oligomers - effect of alkyl substituents", *Carbon, 33 (1995),625-631.*

[16] Yokono T., Obara T., Iyama S. and Sanada Y. "Hydrogen donor and acceptor ability of coal and pitch -factors governing mesophase development from low rank coal during carbonization", *Carbon, 22 (1984),623-624.*

[17] Granda M., Casal E., Bermejo J. and Menéndez R. "The influence of primary QI on the oxidation behaviour of pitch-based C/C composites", *Carbon, 39 (2001),483-492.*

Light Metals 2006 *Edited by Travis J. Galloway* ***TMS (The Minerals, Metals & Materials Society), 2006***

VERTICAL STUD SODERBERG EMISSIONS USING A PETROLEUM PITCH BLEND

Euel Cutshall[1], Linda Maillet[2]
[1]Alcoa, Inc., 900 South Gay Street, Riverview Tower, Knoxville, Tennessee 37902
[2] Alcoa, Inc., 5601 Manor Woods Road, Frederick, Maryland 21703

Keywords: Aluminum, Soderberg, Emissions, Petroleum Pitch

Abstract

Organic emissions and worker exposure levels were determined and compared for individual vertical stub Soderberg cells using 100% coal tar pitch and a coal tar/petroleum pitch blend. Advantages and limitations of using the coal tar/petroleum pitch blend are discussed.

Background

Environmental pressures at all aluminum smelters are increasing as governments place greater emphasis on decreasing the amount of waste products generated, decreasing potentially harmful emissions to air and water, and decreasing exposure of workers to these potentially harmful emissions. This is especially important at vertical stud Soderberg (VSS) smelters which have the potential to generate and emit significant amounts of organic compounds, including polycyclic aromatic hydrocarbons (PAH), during the anode baking and pin pulling processes. These organic emissions originate from the coal tar pitch binder, as it is heated and releases volatile constituents. There are many different technologies available today that can be implemented and that will decrease these emissions. Examples include dry paste, increased casing height, point feeders, and anode top hoods. However, these technologies are all somewhat expensive to implement. Another possibility for decreasing emissions is to use a binder pitch that contains lower amounts of the "regulated" PAHs. A pitch like this is commonly produced by blending the original coal tar pitch (or a coal tar pitch that has been distilled to a higher softening point, removing additional PAH constituents) with petroleum pitch. There has been a significant amount of work reported in the past by the aluminum industry testing petroleum pitch blends [1-7]. In an effort to build upon and complement this work Alcoa tested the use of the Koppers Type B petroleum pitch blend at a vertical stub Soderberg facility. Emissions were documented for individual cells operating with 100% coal tar pitch and the Koppers Type B pitch.

Pitch Used

The petroleum pitch blend selected for testing was the Koppers Type B pitch produced in Denmark. This pitch is formulated using an elevated softening point coal tar component (60%) blended with a lower softening point petroleum pitch component (40%). A comparison of the Type B and the 100% coal tar pitch used is given in Table I.

Table I. Pitch Properties

	Coal Tar Pitch	Type B Pitch
Softening Point, °C	110	116
Coking Value, %	56	54
QI, %	10.4	4.7
TI, %	31.3	23
Sulfur, %	0.4	0.4
Sodium, ppm	184	100
Ash, %	0.10	0.16

Experimental Details

The plant in which the testing was done is a two line VSS operation. Line 1 has 128 cells operating at ~110 kA. Line 2 has 144 cells operating at ~126 kA. Both lines operate with wet paste, the pitch content being 30-31%. Testing of petroleum pitch at the plant began in 1999. After an initial period of testing, half of the Line 2 cells were converted to the Type B pitch. No operational problems were experienced at any time, even during the transition phase from coal tar to Type B. Also, all operating parameters such as cell voltage, current efficiency, etc. are the same for both groups of cells. From an operational point of view the Type B and 100% coal tar pitch cells are indistinguishable.

After half of Line 2 was converted to Type B, two cells (one running 100% coal tar and one running Type B) in the middle of each section were selected for measuring any difference in cell emissions. For each cell two sets of four sampling locations were selected. Each set of four sampling locations was near a corner of the cell. See Figure 1.

One set of four samples was taken about 1 meter above the anode briquettes (as low as possible without interfering with cell operations) while the second set of four samples was taken about 2.6 meter above the anode briquettes (as high as possible without interfering with overhead crane operations). All eight samples were collected at the same time. This procedure was repeated once a week over the next few weeks until there was a total of 48 samples for each cell. The samples were collected on shifts where normal spike pulling would be conducted on the selected cell during the sampling period. No tapping or paste feeding occurred during these sample periods.

Samples were collected by hanging personnel samplers at the designated locations. The samplers consisted of a personnel cassette containing a Teflon® filter and cellulose support pad, followed by an XAD-2 resin tube. The resin tube was connected to a personnel sampling pump via Teflon™ tubing. Sampling was conducted for an 8 hour period at a flow rate of one liter/minute. A photo of the sampling train is shown in Figure 2 and of the cell with the samplers positioned in Figure 3.

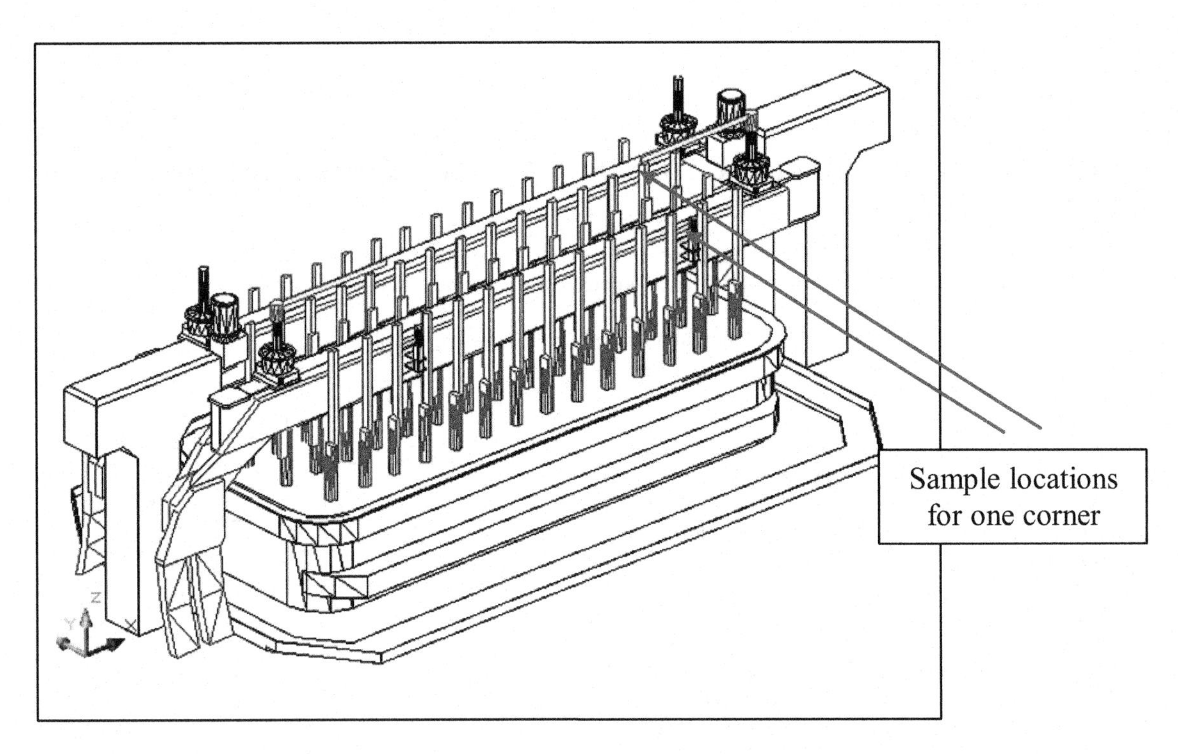

Figure 1. Cell Schematic with Locations of Sampling Points

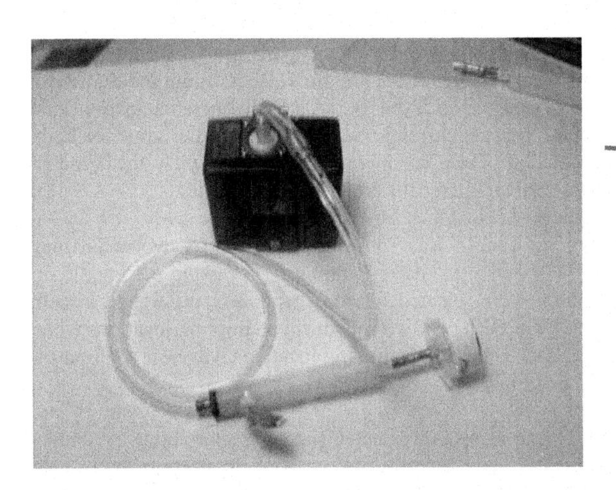

Figure 2. Cassette Sampling Train

Figure 3. Sampling Trains Positioned on Cell

The sampling trains were monitored from time to time to make sure that their temperature did not exceed 40 °C, the maximum allowed to ensure that is no desorption from the XAD resin. The actual temperatures recorded ranged from 22-35 °C.

Upon completion of sampling the cassettes and tubes were protected from light and heat. All sampling media were wrapped with aluminum foil and refrigerated while awaiting shipment to the laboratory. Sampling media were protected from light in the shipping container and shipped overnight to the SINTEF Applied Chemistry Analytical Laboratory in Trondheim, Norway.

Analyses Conducted

The analyses conducted on the samples (filters and XAD tubes) are listed below. A GC/MS was used as a screening tool to positively identify the major components of the pitch samples. A GC/FID was then used to quantify the amounts present.

1. Naphthalene
2. Acenaphthylene
3. Acenaphthene
4. Fluorene
5. Phenanthrene
6. Anthracene
7. Fluoranthene
8. Pyrene
9. Benz[a]fluorene
10. Benz[a]fluorene/Benzdifenylensulfid
11. Benz[a]anthracene
12. Chrysene/4,5-Dimethylpyrene
13. Benz[b]fluoranthene
14. Benz[j&k]fluoranthene
15. Benzo[e]pyrene
16. Benzo[a]pyrene
17. Indeno[1,2,3-c,d]pyrene
18. Dibenzo[a,h]anthracene
19. Benzo[g,h,i]perylene
20. Dibenzo[a,e]pyrene

The sum of the concentrations of 1-20 is the "Total Regulated PAHs" in the included charts since most of the regulated PAHs world-wide are included in this list.

21. 1,6- Dimethylnaphthalene
22. 5-Methylchrysene
23. 3,9-Dimethyl[a]anthracene
24. 7,12-Dimethyl[a]anthracene
25. 1,3-Dimethylphenanthrene
26. 3-Methylcholanthrene

21-26 are PAH compounds to which methyl groups have been added. These compounds are more common in petroleum pitch due to the less severe thermal cracking in its production process.

27. Total Particulate - The total weight of particulate matter collected on the filter portion of the sampling train. The total particulate matter includes all components of airborne dust (alumina, PAHs, etc.).

28. Total Particulate Hydrocarbons - Fraction of the total particulate matter collected on the filter that is soluble in benzene. This is sometimes called the benzene soluble fraction. This includes 1-26 plus any unidentified PAHs.

29. Vapor Phase Hydrocarbons - All of the organic material collected on the XAD-2 tube as measured by total area under the curve from the GC/FID chromatogram. This includes 1-26 plus any unidentified PAHs.

Calculated Values

Particulate PAH - The sum of the particulate phase portion of 1-26. This is a subset of the total concentration collected on the filter, and consists mainly of the PAHs containing four or more benzene rings.

Vapor Phase PAH - The sum of the vapor phase portion of 1-26. This is a subset of the total concentration collected on the XAD-2 tube, and consists mainly of the PAHs containing four or fewer benzene rings.

Total Regulated PAHs (Particulate + Vapor) – The sum of the particulate and vapor phase concentrations of 1-20.

Substituted PAHs (Particulate + Vapor) - The sum of the particulate and vapor phase concentrations of 21-26.

Emission values calculated from the analyses are listed below in Table II.

Table II. Emission Values

All units are mg/m^3 except for BaP	Total Particulate	Total Particulate Hydrocarbons	Total Vapor Phase Hydrocarboms	Particulate PAH	Vapor Phase PAH	Total Regulated PAHs (Particulate + Vapor)	Substituted PAHs (Particulate + Vapor)	BaP, µg/m^3
Coal Tar Pitch Pot	2.97	0.87	0.41	0.23	0.20	0.42	0.011	10.8
Petroleum Pitch Pot	2.77	0.47	0.88	0.085	0.20	0.26	0.026	5.2

The data from Table II are plotted in Figures 4, 5, and 7.

Special analyses were conducted on 11 of the samples for benzene, xylene, and toluene. All three compounds were reported to be below the detection limit for all samples.

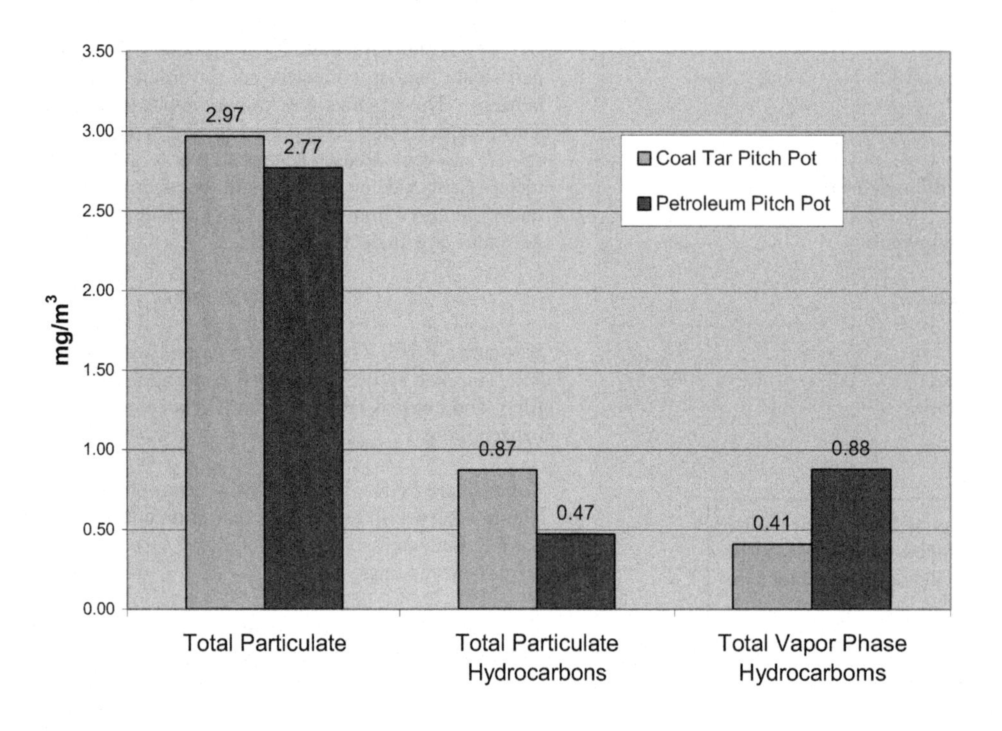

Figure 4. Particulate and Hydrocarbon Emissions

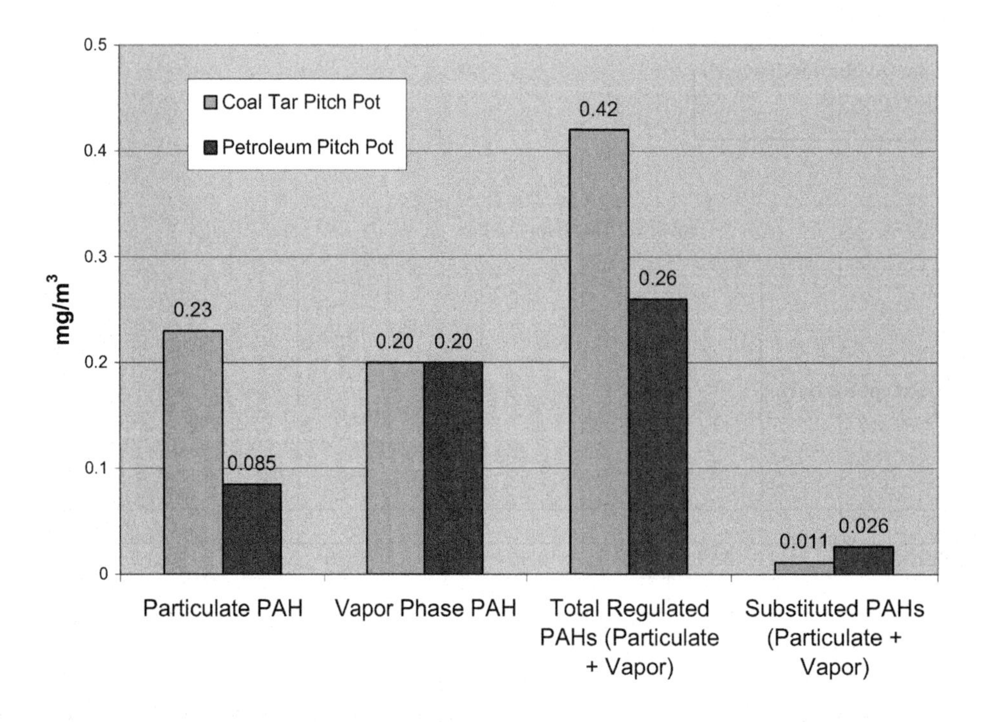

Figure 5. PAH Emissions

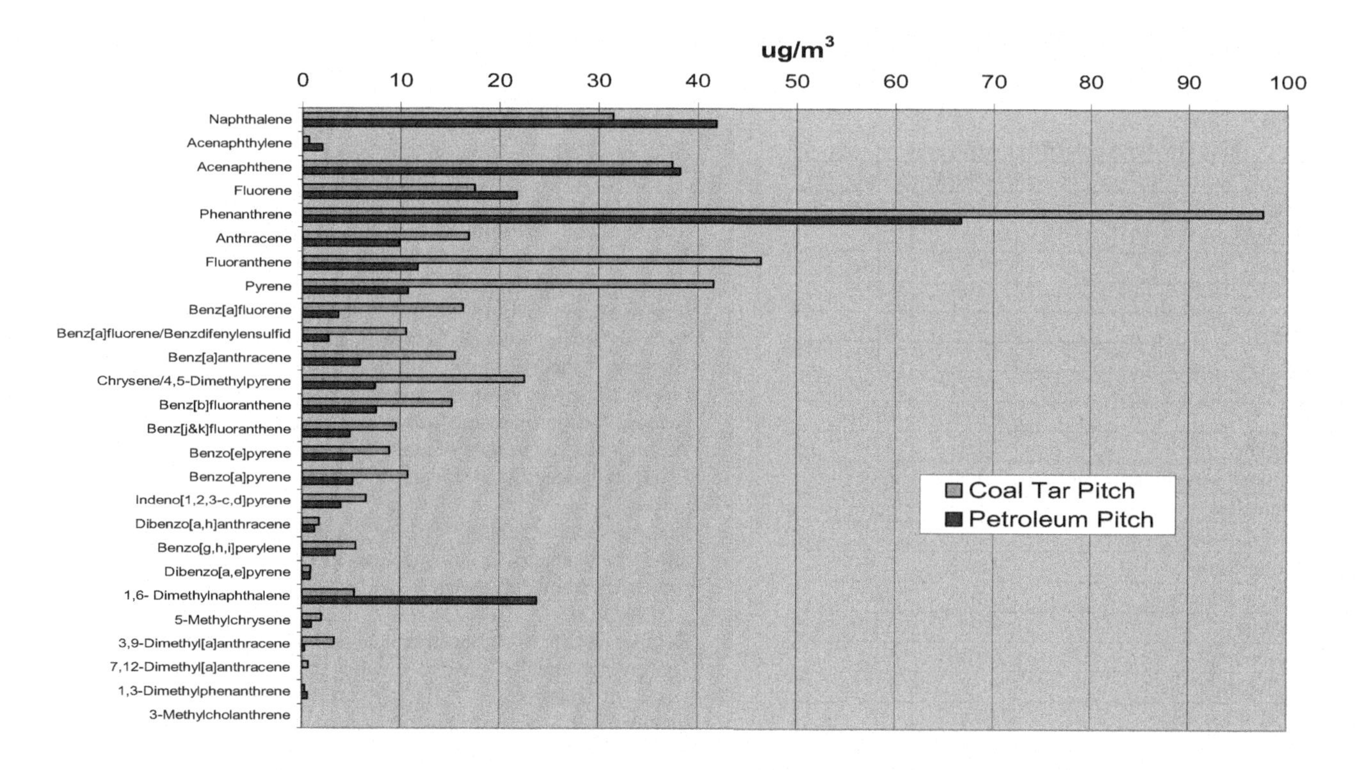

Figure 6. PAH Components

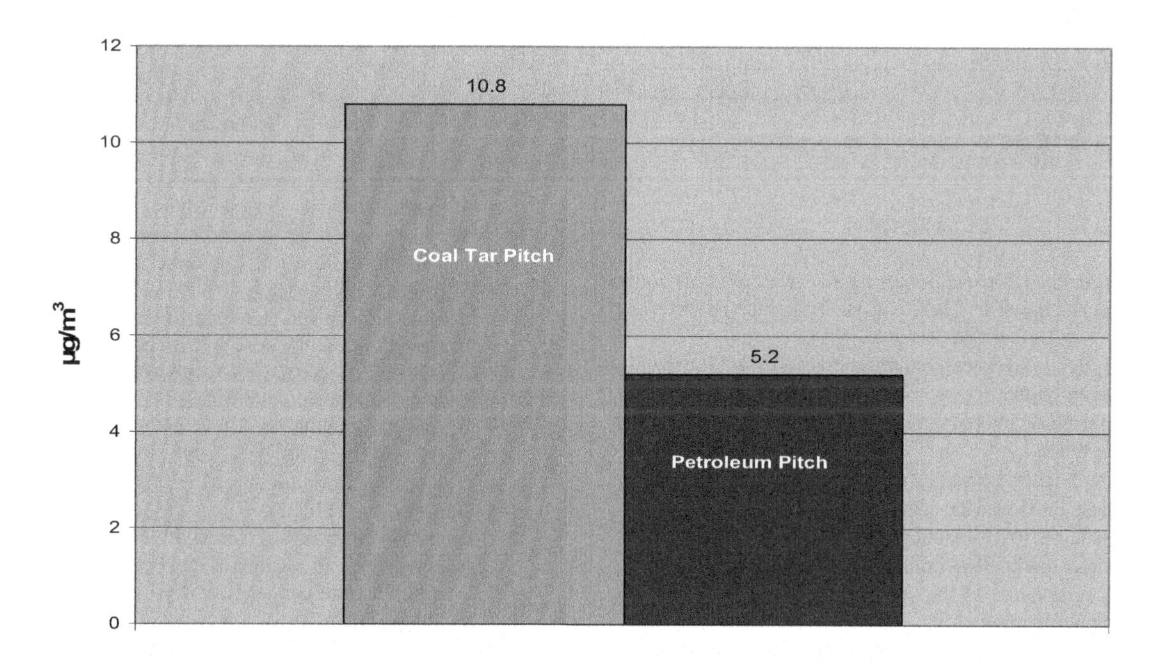

Figure 7. BaP Emissions

Average analyses of the individual PAH compounds in the emissions (sum of particulate + gaseous phases) are listed below in Table III.

Table III. Individual PAH Analyses

	Coal Tar	Petroleum
Naphthalene	31.42	41.86
Acenaphthylene	0.66	2.07
Acenaphthene	37.37	38.16
Fluorene	17.52	21.73
Phenanthrene	97.51	66.69
Anthracene	16.89	9.85
Fluoranthene	46.41	11.76
Pyrene	41.59	10.81
Benz[a]fluorene	16.32	3.74
Benz[a]fluorene/Benzdifenylensulfid	10.60	2.74
Benz[a]anthracene	15.55	5.97
Chrysene/4,5-Dimethylpyrene	22.52	7.49
Benz[b]fluoranthene	15.22	7.58
Benz[j&k]fluoranthene	9.61	4.92
Benzo[e]pyrene	8.95	5.12
Benzo[a]pyrene	10.80	5.21
Indeno[1,2,3-c,d]pyrene	6.58	4.03
Dibenzo[a,h]anthracene	1.78	1.29
Benzo[g,h,i]perylene	5.53	3.43
Dibenzo[a,e]pyrene	0.87	0.83
1,6- Dimethylnaphthalene	5.39	23.77
5-Methylchrysene	1.99	0.98
3,9-Dimethyl[a]anthracene	3.35	0.26
7,12-Dimethyl[a]anthracene	0.66	0.01
1,3-Dimethylphenanthrene	0.28	0.58
3-Methylcholanthrene	0.04	0.00

Data from Table III are plotted in Figures 6 and 7.

Conclusions

- Total particulate matter collected for the two cells was about the same, as expected. There was no difference between the coal tar and petroleum pitches here.
- Total particulate hydrocarbon emissions were 46% less with the petroleum pitch.
- Total vapor phase hydrocarbons increased by 115% using the petroleum pitch.
- Particulate PAH decreased by 63% using the petroleum pitch.
- Vapor phase PAH was the same for both cells.
- Total regulated PAH including both particulate and vapor phases decreased by 38% using the petroleum pitch.
- The BaP component of the emissions was reduced by 52% using the petroleum pitch.
- Substituted PAH including both particulate and vapor phases increased by 52%, as expected since the petroleum pitch is composed of this type of PAH to begin with.
- Looking at the analyses for the individual PAH components the petroleum pitch emissions consist to a greater extent of lower molecular weight, lower boiling point 2-3 ring compounds, both aromatic and substituted. The coal tar pitch PAH components are predominately the higher molecular weight, higher boiling point 4-6 ring compounds.

Summary

The results of this study suggest that the overall amount of total hydrocarbon emissions does not change significantly when using the Petroleum Pitch. Total particulate hydrocarbon emissions are about half as much, but total vapor phase hydrocarbon emissions increase about the same amount. However, the regulated PAHs, those which are usually considered by governmental agencies for control purposes, are reduced by approximately 40% when using this petroleum pitch. Additionally, emissions of BaP, the most commonly monitored PAH, are reduced by just over 50%.

The question remains, "Will the use of petroleum pitch as an anode binder help a plant meet its environmental requirements for hydrocarbon emissions?" The answer is "Maybe."

Regulations for hydrocarbon emissions vary greatly from country to country. In the United States the regulations are based on emissions of POM, polycyclic organic matter, which includes all organic ring structures, aromatic or aliphatic, substituted or unsubstituted. Here the use of petroleum pitch will show no advantage.

However, for other countries such as Canada or those countries in Europe, regulations are based either on a subset of the list of 20 PAHs used in this study or on BaP alone. In these countries the use of petroleum pitch can very well be an advantage.

References

1. M.Skogrand, "A Survey of the PAH Problem in the Aluminium Industry," Light Metals (1991), 497.
2. A.A.Mirtchi, A.L.Proulx and L.Castonguay, "Reduction of the PAH Emissions for Horizontal Stud Soderberg Potrooms," Light Metals (1995), 601-607.
3. M.Eie, M.Sørlie and H.A.Øye, "Evaporation and Vapour Characterization of Low-PAH Binders for Søderberg Cells," Light Metals (1996), 469-475.
4. Acuna, Carola et al., "Petroleum Pitch, a Real Alternative to Coal Tar Pitch as Binder Material for Anode Production," Light Metals (1997), 549-554.
5. Wombles, R.H. and Kiser, M.D., "Developing Coal Tar/Petroleum Pitches," Light Metals (2000), 537-541.
6. Perez, M., Granda, M., Garcia, R., Mionelo, S.R., Menendez, R., and Romero, E., "Petroleum Derivatives as an Alternative to Binder Coal Tar Pitches," Light Metals (2000), 531-536.
7. A.A.Mirtchi, A.L.Proulx, G.Savard, E.Simard, H.Vermette and M.Hamel, "Reduction of PAH Emissions in Alcan Quebec's HS Søderberg Smelters by Evaluation and Conversion to Low PAH Pitch," Light Metals (2002), 571.

CHARACTERIZATION OF GREEN ANODE MATERIALS BY IMAGE ANALYSIS

Stein Rørvik[1], Arne Petter Ratvik[1], Trygve Foosnæs[2]
[1] SINTEF Applied Chemistry, Inorganic Chemistry, Trondheim N-7465 Norway
[2] Norwegian University of Science and Technology, Department of Chemistry, Trondheim N-7491 Norway

Keywords: Image Analysis, Green Anodes, Carbon Technology

Abstract

A method to characterize green (unbaked) carbon composite materials has been developed. The method is based on computer-automatized microscopy and digital image analysis. A reproducible procedure for sample preparation has been developed, as well as a method for distinguishing the pitch from the coke grains and the pores in the microscope images. The method has been used to examine the distribution of the pitch, pores and coke grains in both laboratory made and industrial carbon anode samples. Focus has been on anodes for aluminum production, but the method can also be used for cathodes and other composite carbon materials. This paper explains the principles of the method and presents results from analysis of anodes with variations in binder (pitch) content and coke grain size.

Introduction

It is desirable to study the green, unbaked state of carbon anodes because many of the parameters relevant to anode performance are already established in the green state. There are two principle methods in use for manufacturing carbon anodes for the aluminum industry. One is the prebaked method. The anodes are made as multiple separate blocks covering the electrolysis bath. The anodes are replaced individually when they are consumed too much to allow further use. The other method is called the Søderberg method, and uses large continuous anodes where the carbon paste (mix of pitch and coke) is put on top of a large mantel covering the entire electrolysis bath. The paste then moves slowly downwards while the bottom surface is consumed during the electrolysis. Both these methods require that first heated coal tar pitch is mixed with crushed coke, and then the mixture is mixed to get a homogeneous mass which can be pressed to anode blocks that are heat treated in a baking furnace (prebaked method) or put on top of the continuously carbonizing anode (Søderberg method). The heat treatment, which is done at temperatures up to 1300 °C, causes the pitch to carbonize, creating a solid block of electrically conducting carbon.

There has been very little published work on the characterization of green anodes. Stokka [1] has examined the bulk properties as an indicator of the mixing efficiency. Adams [2] and Adams [3] use optical microscopy and X-Ray microtomography respectively to study the properties of green anodes. The present paper demonstrates a method which is believed to be more suitable for routine measurements than the published methods [2] and [3]. A major difference between the present paper and the work published by Adams [2] is that the latter excludes all particles below 50 µm diameter in the measurements, while the present paper measures all fines down to 1 µm. These small fines particles are important for anode properties.

An useful way to characterize the properties of the green anodes (besides simple bulk techniques such as density and open porosity) is to use quantitative microscopy. A fully automatic method for image analysis of porosity in baked carbon materials was developed by our research group during the Expomat / Prosmat research programs during the 1990s. The method is based on computerized image analysis and optical microscopy, and is capable of analyzing large sample areas (several cm^2). It provides a logarithmic size distribution of pores in the range of 0.5 µm to 10 mm pore radius. In addition to this size distribution, various other parameters can be measured, such as pore surface area, pore connectivity, and pore shape (irregularity, aspect ratio).

The basic technique of the image analysis method was published by our research group by Rørvik and Øye [4], as a method to analyze porosity in baked anodes. Later publications include further developments of the image analysis technique: Rørvik, Øye and Sørlie [5], which shows how to analyze porosity in coke, and Rørvik, Lossius and Øye [6], which shows how to separate pores belonging to the coke and belonging to the binder-phase in baked anodes, using pore shape classification. All the three papers mentioned above deal with baked carbon materials. The present paper shows how the automatic image analysis technique has been further developed to allow analysis of green anodes.

Method Description

Sample Preparation

One of the major challenges in analyzing green materials is the sample preparation. The pitch is extremely soft compared to the coke, and dissolves in the organic solvents used in the commercially available embedding agents and polishing lubricants. This makes embedding and polishing the materials very difficult. ASTM [7] has published a technique for polishing coal but this does not work on the green composite materials because of the use of solvents and the great hardness difference between the coke and pitch. Struers (a commercial supplier of sample preparation equipment and supplies) has published a method [8] which is more suitable for the green anode materials. This method is based on using silicon carbide papers. Using a grinding / polishing agent with grains fixed to a surface is the clue to avoid the soft pitch being torn out from the much harder coke / binder matrix.

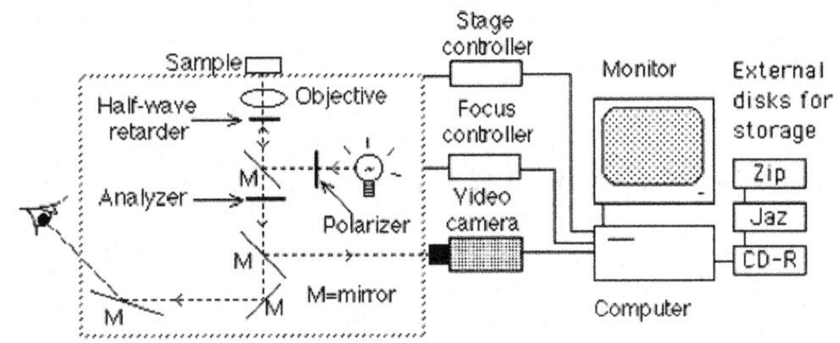

Figure 1: Schematic hardware

Microscopy Hardware

The samples are examined using a standard inverted reflected light metallurgical microscope (Leica MeF3A), equipped with a motorized XY- stage and focus controller. The stage movement and focus is controlled directly by the computer image analysis software. Digital images are acquired using an electronic 3-chip CCD video camera (Sony DCX 950P) with analog RGB output and a frame-grabber card. Newer 1-chip digital cameras provide higher resolution than this camera, but they do not provide the same color accuracy as the 3-chip camera. Color accuracy is very important in the present technique since the pitch is separated from coke by the colors in polarized light. Figure 1 illustrates schematically the current microscope setup.

Image Analysis Software

The image analysis method was implemented using the general image analysis Macintosh™ application NIH image, developed by Wayne Rasband at the National Institute of Health in the US. This software is available in the public domain, and can be downloaded freely from ftp://rsbweb.nih.gov/pub/nih-image/. The source code has been customized by adding support for the microscope hardware and some extra image analysis procedures. The NIH image software has a Pascal-like macro language, which was used to automate the analysis. A commonly used Microsoft Windows-based commercial image analysis software package (Media Cybernetics' ImagePro Plus) has been evaluated as an alternative to NIH image, but it has not shown to be able to provide the same speed and flexibility as the Macintosh / NIH image combination.

The following sections give a more detailed description of the new improvements to the image analysis that allows analysis of the green materials.

Separation of pitch, coke and pores

Besides the sample preparation, the other major challenge when analyzing green materials is how to separate the pitch from the coke grains in the binder phase. Scanning Electron Microscopy does not provide enough contrast to unambiguously separate the pitch from the coke grains (Figure 2).

Using polarized light, it is much easier to distinguish between the pitch and coke. When carbon with a graphitized structure is polished and viewed in polarized light under the microscope, interference colors will appear if the following setup is applied: Two polarizing filters are used. One is placed across the ingoing light path to the sample ("polarizer"), and another one across the outgoing light path ("analyzer"). These two filters are placed in a crossed position (i.e., 90° to each other). This setup will cause no light to be reflected on an optically neutral surface. The textural components in the carbon cause the reflected light to be rotated slightly according to the direction of the components. This causes some light to pass the second filter. The ingoing and outgoing light rays interfere with each other, causing light of different wavelengths to interfere differently. A phase retarder filter made of a thin plate of quartz ("lambda filter") causes the phase of the light to be shifted so that the interference lies in the visible color spectrum. Using this setup, the color of the optically neutral pitch has a dark purple color, while the color of the coke varies from light cyan to yellow depending on the local direction of the textural components. Where the textural components are parallel to the surface, the interference is low and the coke will get a similar purple color as the optically neutral pitch. However, the coke reflects light better than the softer pitch, so the appearance of the parallel coke areas is usually slightly brighter than the pitch.

Figure 2: Scanning Electron Microscope image

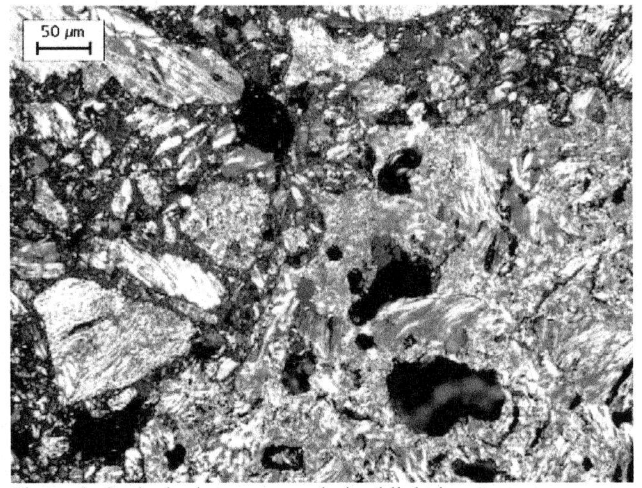

Figure 3: Optical microscope, polarized light image

To separate objects of different colors in an image, the most common way is to apply a threshold to each color channel (red, green and blue) to separate one color from the other. This technique does not work well on these images since the areas of coke with textural components parallel to the surface have the same hue as the pitch. An alternative technique using distances in the three-dimensional color space has shown to give better results.

This method is based on the fact that visible colors can be represented as coordinates in a cubic 3D space, where each axis represents the amount of red, green and blue light in a color. The "distance" between two colors can then be calculated as the Euclidean distance between the two colors' coordinates in this 3D space, as shown in equation (1), where the first color has the coordinate (R_0,G_0,B_0) and the second has (R_1,G_1,B_1):

$$d = \sqrt{(R_1 - R_0)^2 + (G_1 - G_0)^2 + (B_1 - B_0)^2} \quad (1)$$

If (R_0,G_0,B_0) is defined as a reference color (the known interference color of the pitch) and apply equation (1) to each pixel of Figure 3, a new image is created. In this image, the intensity of each point (on a spectrum scale from red to blue) is proportional to the distance between the color in each point and the pitch reference color. This image is thresholded and used as the basis of identification of the pitch.

It is not straightforward to define the pitch reference color. The color varies slightly at each point of the pitch, so it is important to find the color coordinate that best represents the average color of the pitch. If the reference color coordinate is selected wrongly, the analysis under-estimates the amount of pitch in the sample. The selection of the pitch color coordinate is done by selecting an overview image of the entire sample and plotting the number of pixels for each color coordinate in a projected 2D plot (Figure 4). In this figure the clusters of pixels belonging to each of the three "phases" (pitch, coke and pores) are outlined with circles. The values are visualized with spectrum colors, blue is the lowest and red is the highest value. The pitch color is the maximum value, in this case (R,G,B) = (111,43,156).

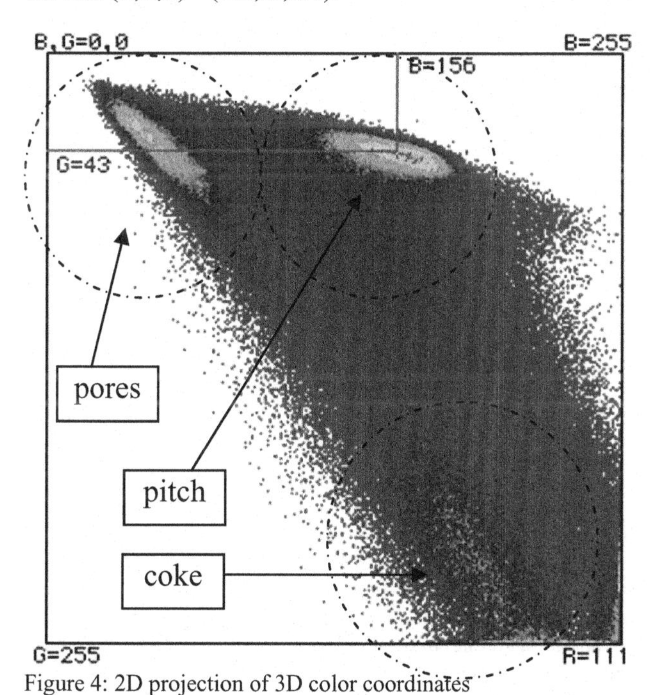

Figure 4: 2D projection of 3D color coordinates

When the proper pitch color is determined, the pitch area can be thresholded and analyzed. The coke is thresholded from the red channel, and the pores are thresholded from the blue channel. This gives us three separate images showing the pitch (Figure 5) coke (Figure 6) and pores (Figure 7).

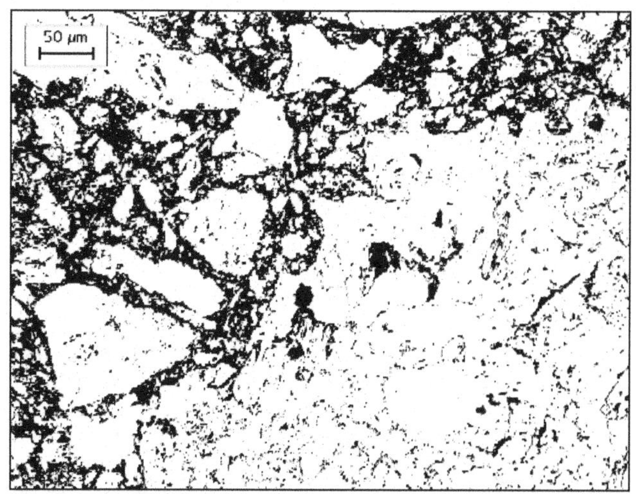

Figure 5: Pitch, thresholded image

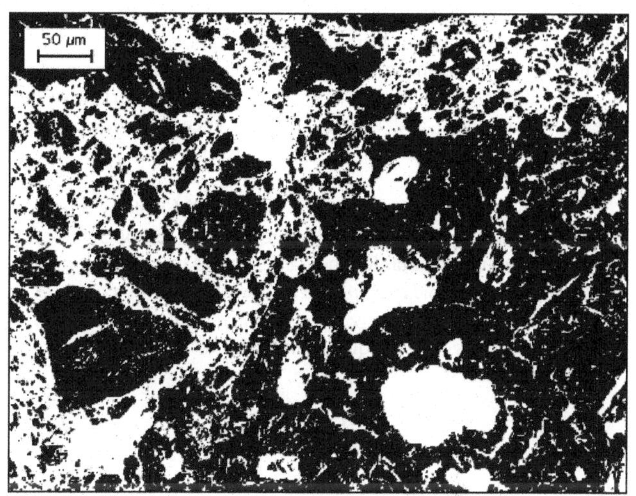

Figure 6: Coke, thresholded image

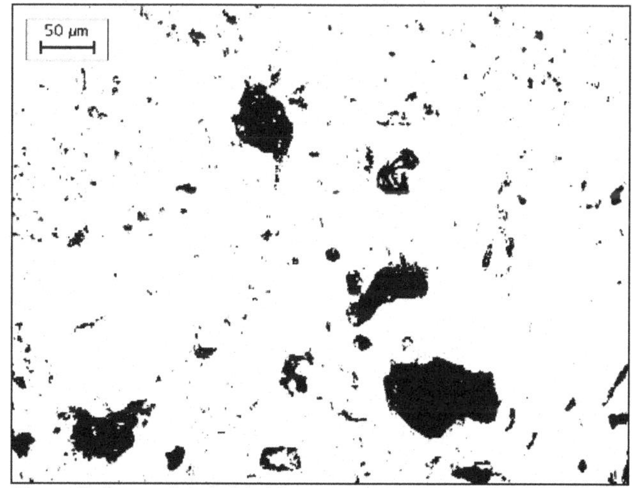

Figure 7: Pores, thresholded image

Object measurements

A common procedure was used to measure the pitch, coke grains and pores. The only difference is the method used to extract the objects to be measured, as described in the previous section. The method is basically the same as the one published by Rørvik and Øye [4]:

A grid of adjacent images, sufficiently large to cover most of the sample is acquired automatically by the computer and stored on disk. The images are then analyzed in batch. When the adjacent frames are analyzed, the objects entirely inside each frame are measured while the objects cut by the image edges are not. These objects are saved and measured after four frames have been analyzed and merged. This process continues recursively, allowing arbitrarily large objects to be measured. There is no upper limit to the measured object size. The procedure is fully automated, and only requires an operator to place the sample on the microscope and start the procedure using the desired parameters. With sufficient storage space, images of a series of samples can be acquired during daytime and be analyzed in batch by the computer during the night. The main challenge is to find the optimal image analysis parameters because the intervals of the thresholds are tight, and depend on how successful the sample preparation is.

The resulting data is merged using Microsoft™ Excel macros and presented using templates. The image analysis outputs the size distribution values as a table where *each value represents the sum of the areas objects with a specified inscribed radius cover, as a percentage of the total analyzed sample area.* The sum of all size distribution values is thus equivalent to the total area coverage of the phase of interest.

Results

Description of the analyzed materials

A series of laboratory made Søderberg samples were analyzed in a previous project to examine the effect the fines (coke dust) content had on the anode properties. These samples were all made using the same pitch level and receipt, but the fines were created using different milling times. This results in different fineness of the fines. Figure 8 shows a plot of the fineness of the fines, represented as fraction particles smaller than 75 μm. On the second axis is a plot of the baked density. Figure 9 shows a plot of the permeability and the compression strength. The permeability is one of the best indicators of optimum anode quality. The permeability is lowest at 60 minutes milling time.

The method has been applied to many different green anode materials, both Søderberg and prebaked; laboratory made and industrial. The reason for presenting results of a series Søderberg samples in this paper is that the provided Søderberg samples had well-defined properties with respect to the raw materials, and all relevant measurement data was provided by our industrial partner.

Figure 10, Figure 11 and Figure 12 shows the size distribution of the pitch "grains", pores and coke grains, respectively. The pitch distribution curves in Figure 10 show that for coarse fines, the thickness of the pitch layers is greater than for fine fines. The coke size distribution curves in Figure 11 shows that there is more of the smaller particles for the samples with longer fines milling time (finer fines). The pore size distribution curves in Figure 12 shows that there are more of the larger pores in the samples with shorter fines milling time (coarser fines). While these results may seem pretty obvious, they prove that the image analysis method is capable of characterizing the true nature of the material.

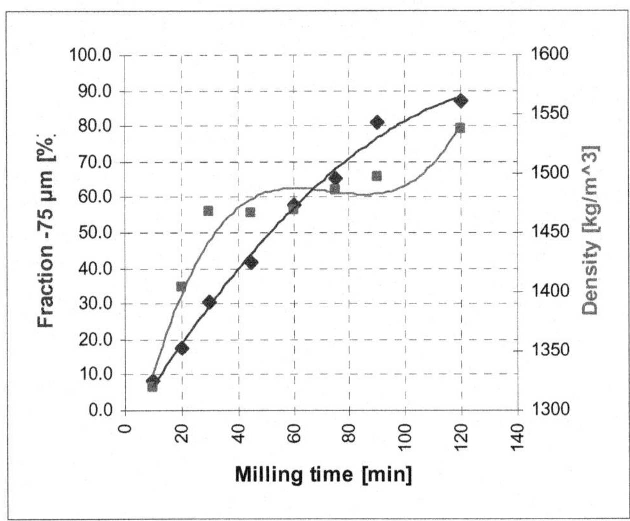

Figure 8: Fines fineness and Baked density vs mill time

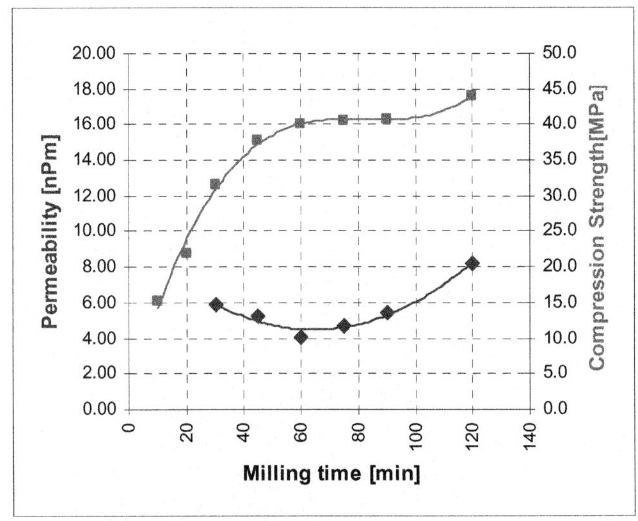

Figure 9: Permeability and Compression strength vs mill time

Correlation to physical properties

The different properties measured by image analysis were compared to the physical data measured for each anode. Figure 13 compares permeability to pitch "connectivity" (as defined by Rørvik and Øye [4]) measured by image analysis. There is no exact correlation but the two properties follow the same trends. The anode with the highest pitch connectivity has the lowest permeability.

Figure 14 compares compression strength to the green porosity measured by image analysis. There is good correlation here. Increased porosity gives lower strength, as expected.

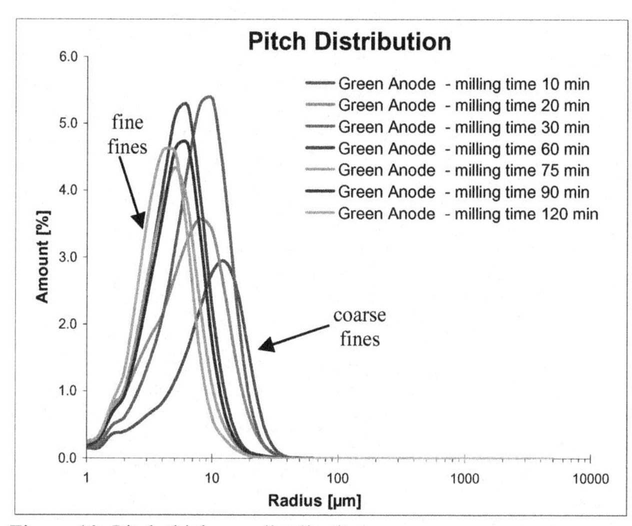

Figure 10: Pitch thickness distribution

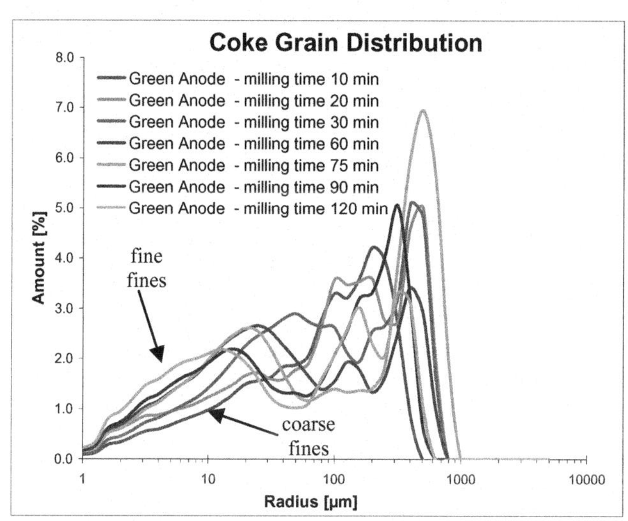

Figure 11: Coke grain size distribution

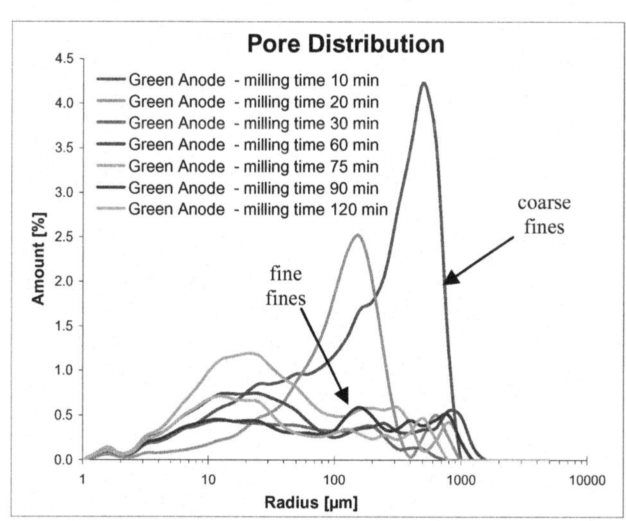

Figure 12: Pore size distribution

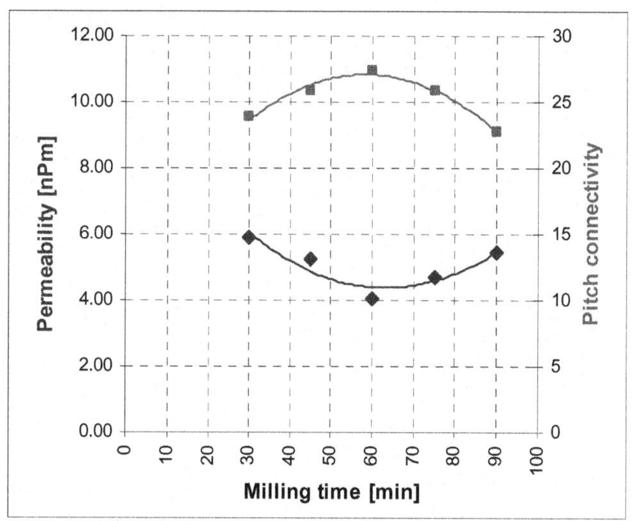

Figure 13: Permeability and Pitch connectivity, measured by image analysis, plotted against milling time

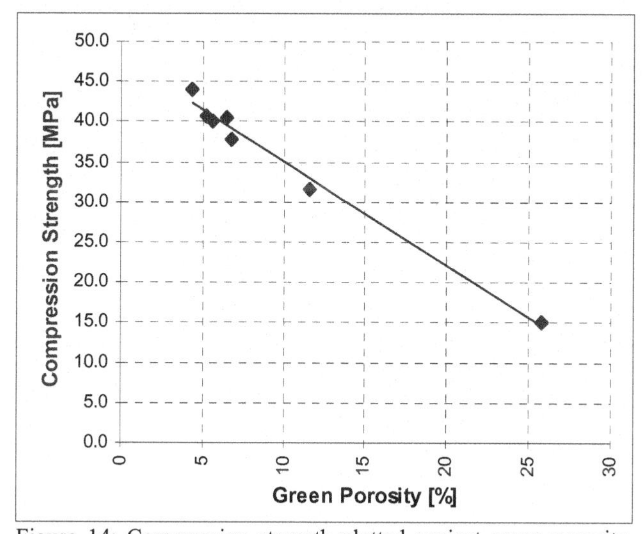

Figure 14: Compression strength plotted against green porosity, measured by image analysis

Figure 15 compares the inter-particular distance between coke grains from image analysis to electrical resistivity, and Figure 16 compares the pitch layer thickness from image analysis to electrical resistivity. The inter-particular distance is defined as the percentage of the binder phase that is shorter than a given distance (in this case 10 µm) to the nearest coke grain. The pitch layer thickness is defined the same way, but excludes the pores. These properties are therefore closely related. Since the porosity is fairly low in the analyzed materials, the values co-vary. Figure 17 shows the correlation between these three properties. A shorter inter-particular distance (higher number of particles close to each other) gives less resistivity. A thicker pitch layer around each grain also gives less resistivity. The distribution of the pitch is important for the electrical properties because it is the pitch coke (pitch carbonized after baking) that carries the electrical current from one coke grain to the next when the grains are not in direct contact.

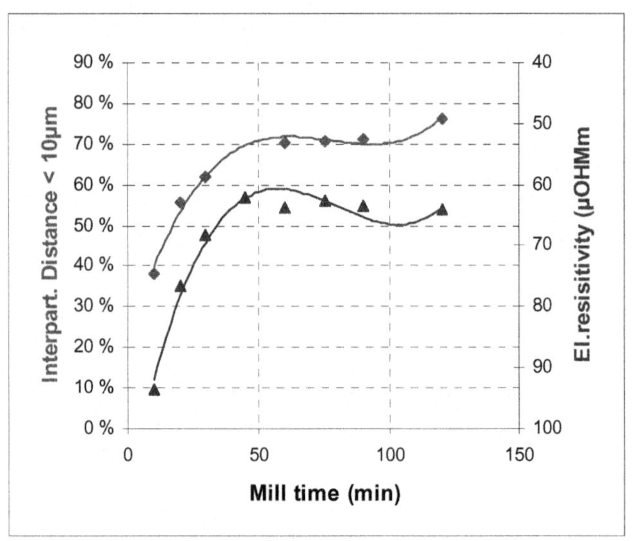

Figure 15: Inter-particular distance from image analysis compared to Electrical Resistivity

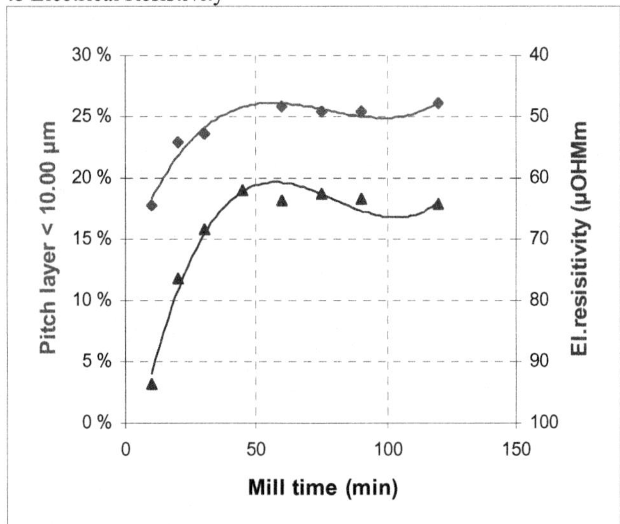

Figure 16: Pitch layer thickness from image analysis compared to Electrical Resistivity

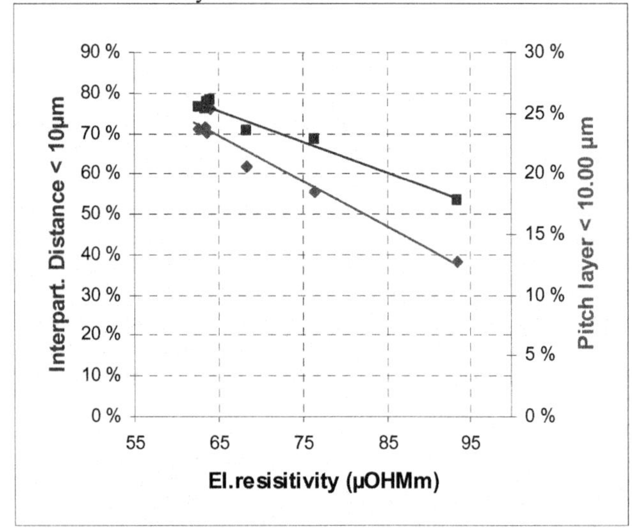

Figure 17: Inter-particular distance and pitch layer thickness from image analysis correlated to Electrical Resistivity

Conclusion

The present paper shows that image analysis is a good method for characterizing green anode materials. The measured properties correspond well with the known properties of the raw materials, and correlates to physical data relevant for anode performance. The three properties that co-vary with the data from the image analysis are permeability, strength and electrical resistivity.

- Permeability is related to the pitch distribution
- Strength correlates with the green porosity
- Electrical Resistivity correlates to the pitch layer thickness and distance between coke particles

Acknowledgements

Financial support from The Research Council of Norway and the Norwegian aluminium industry (via the PROSMAT and CARBOMAT research programs) is gratefully acknowledged. Thanks are also due to Elkem Aluminium ANS for providing anode samples and analysis data.

References

1. Per Stokka: Green Paste Porosity As An Indicator Of Mixing Efficiency. TMS Light Metals 1997

2. Angelique N. Adams: The Use of Image Analysis for the Optimization of Pre-Baked Anode Formulation. TMS Light Metals 2002

3. Angelique N. Adams: The Characterization of Pre-Baked Carbon Anodes using X-Ray Computed Tomography. TMS Light Metals 2002

4. Stein Rørvik, Harald A. Øye: A Method for Characterization of Anode Pore Structure by Image Analysis. The Minerals, Metals and Materials Society (TMS); Light Metals Proceedings 1996, p. 561-568.

5. Stein Rørvik, Harald A. Øye, Morten Sørlie: Characterization Of Porosity In Cokes By Image Analysis. TMS Light Metals 2001

6. Stein Rørvik, Lorentz Petter Lossius and Harald A. Øye: Classification of Pores in Prebake Anodes Using Automated Optical Microscopy. TMS Light Metals 2003

7. ASTM D5671-95 (2005) Standard Practice for Polishing and Etching Coal Samples for Microscopical Analysis by Reflected Light. ASTM International.

8. Microscopy of Pitch, Structure Magazine (Struers Metallographic News) # 15, 1987, p. 14.

Light Metals 2006 *Edited by Travis J. Galloway* ***TMS (The Minerals, Metals & Materials Society), 2006***

NEURAL NETWORK IN COKE CALCINING COMPUTER PROCESS CONTROL SYSTEM

Sinelnikov V.V, Lapaev I.I, Nikandrov K.F.
RUSAL Engineering & Technological Center
Pogranichnikov st., 37, build.1
Krasnoyarsk, Russia

Keywords: kiln, control system, neural network, coke density.

Abstract

Automatic process control system of coke calcination consists of two levels. Control, data transmission and digital control of the kiln temperature conditions are done with the help of regulators on lower level. Upper level of the kiln automatic process control system consists of two principal sublevels: visualization system of calcination; neural models of forecast and control (changes in set points of kiln regulator). Neural network of general regression was chosen as the calcination model. Genetic cone-type algorithm was used for adjustment of network weights. The task of change in regulator set points is narrowed down to the task of optimization theory. Necessity in the implementation of control neural network system consists in fact that regulators alone can not provide fast response to changing process state (for example, coke mixtures consist of a few different chemical parts) because their body of mathematics is destined for control by set points and not their change.

Introduction

Coke calcination is a major anode paste production process because its basic physical-chemical and operational properties largely depend on the quality of calcination.
The automated coke calcination process control system consists of two baxis levels shown in Fig.1.
The lower level based on controllers Simatic S7-633U and Simatic S5-155 monitors, transfers data and discretely control the temperature condition in the calcination kiln by PID regulators.
The upper level of computer process control system (CPCS) consists of two basic sublevels:

- calcination process visualization system;
- neural network forecast and neural control of regulator set point.

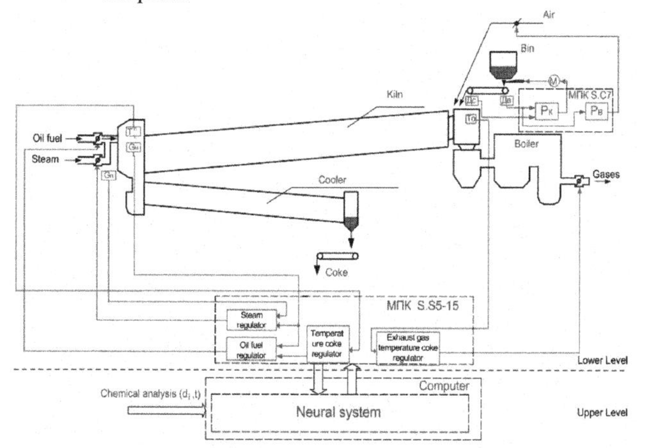

Fig.1. Calcination kiln CPCS level hierarchy

CPCS Lower and Upper Levels

Automatic control systems (ACS) on the whole allow for stabilization of temperature conditions both when the kiln is charged with coke from a single supplier and when the coke supplier is changed, i.e. when temperature conditions are altered. High temperature conditions, kiln lining wear, different suppliers make the model parameters drift and permanently adjust to "new" calcination conditions, i.e. add the adaptation element.
Thus, the problem of constructing a mathematical model has three major lines:

- to choose the neural network type;
- every day to train the network "new" the calcination process information based on process data;
- create control system of regulator set point with use of neural network.
-

The basic idea that forms the mathematical statement of the problem was to reduce the dynamics problem to a static problem by developed automatic stabilization systems and PID regulators of process parameters.

Calcination Identification Problem

Examination of coke calcination process divided the problem of creating density coke model and control system of changing regulator set point into three main sub-problems:

- to collect data on process variables (CPCS data and analyses from the chemical laboratory);
- to choose and develop the process model (neural network);
- to adapt-train the model to process variations (genetic algorithm as a adaptation-training procedure of neural network weights);
- to select and develop calcination kiln control system of changing set point (coordinate descent method).
-

Thus, the structural diagram of the identification problem in its general form is shown in Fig. 2.
From the Fig.2 it is apparent that the fundamental idea to construct the control system of regulator set point is to reduce the problem of inverse operator estimate to the problem of the theory of optimization, i.e. to minimize the quadratic function of error between the model output and required coke density.

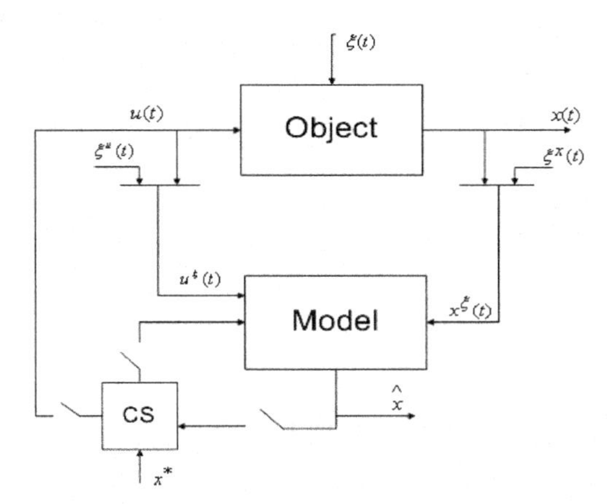

Fig. 2. Calcination identification problem.

Once a day the neural network is trained with new data on the process. For the training procedure we chose a canonic version of genetic algorithm with the following agenda:

- Recombination;
- Mutation;
- Estimation (calculation of function 1);
- Selection;

$$W(\bar{w}) = \sum_{i=1}^{S} \left(d_i - d_s(\bar{z}_i, l_i, \bar{w})\right)^2 \to \min_{\bar{w}}, \quad (1)$$

where d_i are the analyses from the chemical laboratory, d_s is the model output.

The problem of determining the settings for the kiln temperature conditions regulators was formulated as the error minimization problem:

$$W = \left(d^* - d_s\left(\bar{z}, l, z_i^T\right)\right)^2 \to \min_{z_i^T}, \quad (2)$$

d* is the desired coke density at the calcination zone output, z is the input model vector, l is the supplier type, z_i^T are the settings of the kiln temperature profile regulators:

$$\begin{aligned} a_1 \le z_1^{T\pi\kappa} \le b_1; \\ a_2 \le z_2^{To\imath} \le b_2. \end{aligned} \quad (3)$$

where $z_1^{T\pi\kappa}$ is calcination temperature, $z_2^{To\imath}$ is exhaust gas temperature.

Once every 5 minutes the neural network predicts coke density values in the calcination zone; the set points of the kiln temperature regulators are corrected every 8 minutes.

Fig. 3,4,5 show neural model's graphic and chemical analysis of test period 08.12.2004 – 16.02.2005.

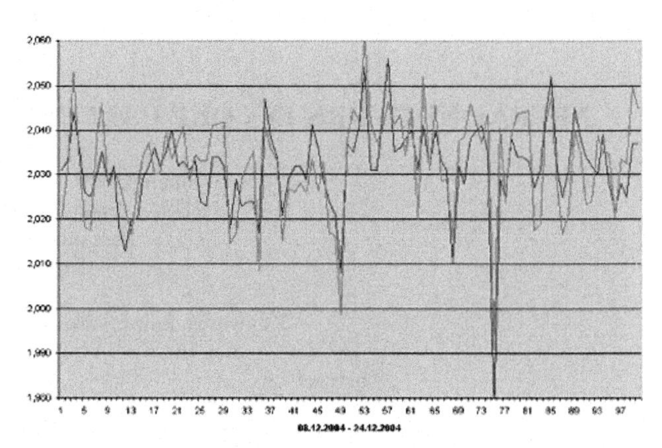

Fig. 3. Forecast of the neural network (red) and analyses of the chemical laboratory (blue).

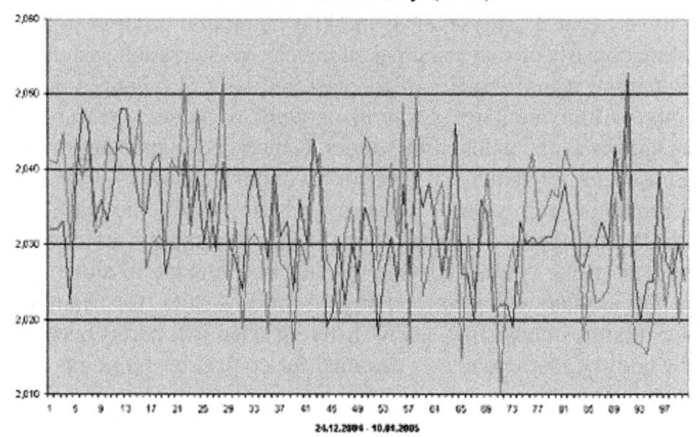

Fig. 4. Forecast of the neural network (red) and analyses of the chemical laboratory (blue).

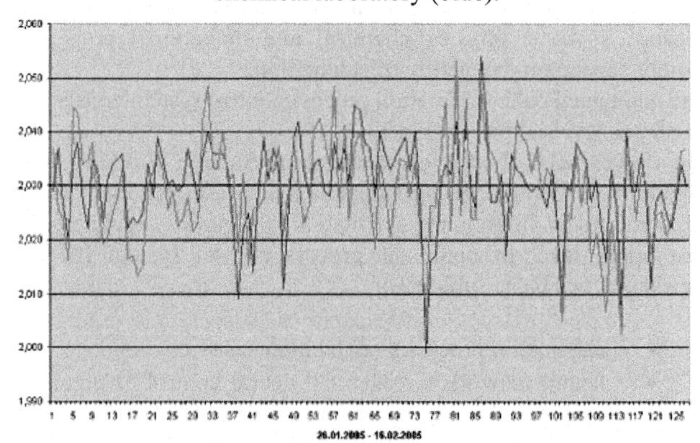

Fig. 5. Forecast of the neural network (red) and analyses of the chemical laboratory (blue).

The form of modeling error is given in formula (4)

$$e = \sum_{i=1}^{N} \left| \frac{d_i^{chem.analysis} - d_i^{neuronet}}{d_{max}^{chem.analysis} - d_{min}^{chem.analysis}} \right| \cdot 100\ \%, \quad (4)$$

where N is the sample size.

Mean modeling error is 15%. This is due to numerous factors; among them is the fact that the training sampling was made on "pure" coke suppliers, while the tests were run on mixed suppliers, this notwithstanding, the mathematical apparatus

exhibited its robustness and adaptability to variable process characteristics.
During algorithm test period the kiln increased its yield by 5% of its maximum load. Increase of kiln load make possible to implement local CPCS control loops (PID law) developed and implemented by "ETC" Ltd. specialists.
The control neural network helped increase high-quality coke yield by 4% of the total volume produced during the test period.

Conclusion

Thus, comprehensive approach to automating the coke calcination kiln and control algorithms developed demonstrated their efficiency in calcining coke mixtures from different suppliers.

References

V.V. Sinelnikov, I.I. Lapaev, K.F. Nikandrov, *Neural network in coke calcination process control system,* Proceedings of the X international conference "Siberian Aluminum - 2004". Krasnoyarsk, 2004, p262-264.
I.I. Lapaev, I.I. Gorbunov, A.I. Murashkin, *Coke calcination process control system in rotary kiln,* Technical economic proceedings of "Russian Aluminum", Krasnoyarsk, 2003, №5, p22-24.

CARBON TECHNOLOGY

Greenmill/Rodding

SESSION CHAIR
Marilou McClung
Century Aluminum of West Virginia
Ravenswood, WV, USA

Light Metals 2006 *Edited by Travis J. Galloway* ***TMS (The Minerals, Metals & Materials Society), 2006***

ANODES PLANT – THE NEXT STEP

THE HIGHT CAPACITY ANODE PLANT

André Molin[1], André Pinoncely[1], Jean-François André[1], Jérôme Morfoise[1];,
Jean-Christophe Rotger[2], Yann El Ghaoui[2], Jérémie Lhuissier[2], Magali Gendre[2]
[1]Solios carbone; 32 Rue Fleury Neuvesel; 69700 Givors, France,
[2]Alcan – LRF; BP114; 73303 Saint Jean de Maurienne, France

Key words: Carbon, Anodes, Coke, Paste, Investment, CAPEX, OPEX, IMC®, Rhodax®,

Abstract

Investors in greenfield primary aluminium smelters nowadays often contemplate the construction and operation of one 300kA (or more) potline as a first phase followed by a second identical potline a few years later for a targeted metal production of 600 000 T/year. Alternatively the 2 potlines can be built in one single phase.
In such situation there is a strong economic incentive, both CAPEX and OPEX, to consider a high capacity green anode plant, 60 t/h or more, rather than two "standard" 35t/h paste plants.
This is now feasible thanks to the two major developments achieved in the last few years : the SCAP Rhodax® process for dry mix preparation and the IMC® process for paste mixing and cooling.
The start-up early 2005 of these 2 processes at 35t/h at respectively ALBA line 5, Bahrain and AOSTAR, China confirmed the high standard of operating and anode quality performance achieved.
The equipment is already available for the design of one single line of 70 t/h capacity achieving the same performance standards.

Introduction

In the beginning of the year 2005 two new 35 tph anode plants went into operation, each of them with an innovative process which answers the producer's expectations:

- the SCAP Rhodax® (Single Crusher Anode Plant) process at the ALBA line 5 paste plant in Bahrain, which relates to the preparation of the dry materials
- and the IMC® process (Intensive Mixing Cascade) at the AOSTAR paste plant in China, which relates to the preparation of the paste.

The purpose of the development of these two technologies was to reduce the operation and capital cost while maintaining a high standard of anodes quality.
This double goal is reached today, however the search for more profitable solutions is still continuing.

The present status

The savings on the capital and operation costs brought by these new processes are resulting from technological optimization without any compromise on anodes quality:

- Reduction of the required number of equipment with the SCAP-Rhodax® process
- Implementation of a more cost effective mixing technology with the IMC® process

Dry material preparation

The comparison between the 2 sketches in the Figures 1 and 2 immediately highlights the obvious savings brought by the new SCAP-Rhodax® process in terms of capital and operating costs.

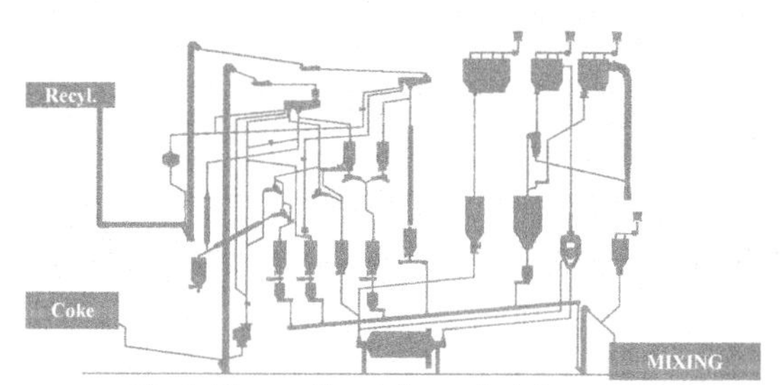

Fig 1. Conventional dry material process.

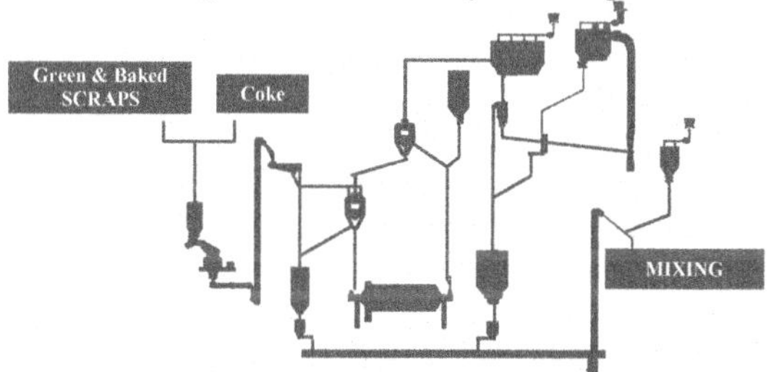

Fig 2. SCAP Rhodax® process.

As exposed in our previous papers [1][2][4][5], the exhaustive analysis of the industrial operation of the first full scale unit operated at the Aluminium Dunkerque Alcan plant in France demonstrated that, apart from these costs savings, "the quality of the SCAP-Rhodax® anodes is at least equivalent or even better than the conventional reference produced with the same raw materials, mixing and forming parameters."

Upon positive validation of the new process, Aluminium Bahrain decided to select the SCAP Rhodax® technology for the AP35 line 5 expansion project, and in spring 2003, awarded Solios for a turn key paste plant of 35 tph rated capacity.

The construction savings that resulted from this first commercial reference have already been summarized in the previous paper [1] but of course there is still room for further improvement in the future, and some have already been identified and will be implemented for the coming Sohar project. Operation and maintenance costs savings are still under investigation as it is too early to get representative figures.

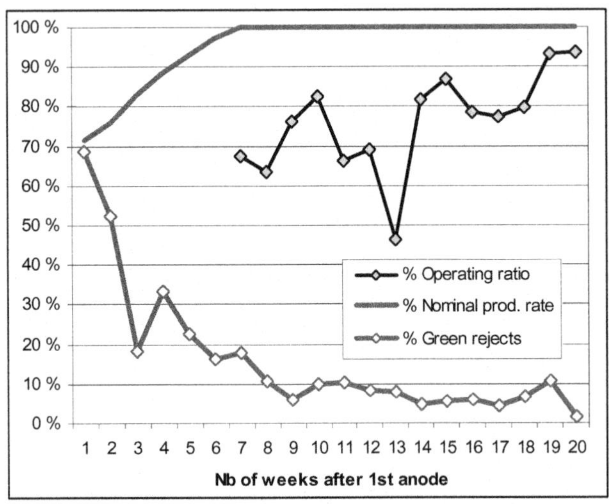

Fig. 3: Alba line 5 paste plant commissioning trends

The first anode was produced on the 15th of January 2005, and 7 weeks later the plant reached its nominal 35 tph paste production rate (Fig. 3). The complete performance tests were satisfactorily achieved only 20 weeks after the first anode date and the corresponding main results are given in the following Table 1.

Table 1: Alba line 5 paste plant performance test main results

Start/Stop dates	May 27th, 2005 / June 1st ,2005	
Total duration	18 shifts	
Paste plant operating ratio	93,8 %	
Paste production rate	35 tph	
Baked scraps content	30 %	
Green scraps content	5 to 10 % (8,4 % average)	
G/S ratio	7,36	
Pitch content	14 to 14,2 % (14,14 % average)	
Good anodes produced	4863	
Green anodes rejection rate	0,69 %	
Green paste rejection rate	0,30 %	
Green anodes height	Standard deviation :	2.5 mm
Green anodes density	Average :	1,634
	Standard deviation :	0,005

The cumulated production of the first 50.000 good anodes was achieved only 25 weeks after the first anode production date, with an overall average green anode density of 1,631 and a standard deviation of 0.005. Such high and consistent level of anodes quality for this prime batch of production illustrates the SCAP-Rhodax® process stability, even before every automated control loops were properly fine tuned.
This participated in the very successful start up of the new pot line within a benchmarking timeframe.

The Table 2 compares the quality of the baked anodes produced with the SCAP-Rhodax® process on the Alba line 5 paste plant, to those produced with a conventional process on line 4 paste plant using the same raw materials and an equivalent mixing and forming process: It appears that the SCAP-Rhodax® anodes show an increased average baked density of +0.015 and a significantly lower air permeability, with at least equivalent other characteristics, thus confirming the results obtained at the Aluminium Dunkerque plant.
The single line on stream operation and the steadiness of the process has also reconfirmed, after the successful experience at Aluminium Dunkerque, that the implementation of a small internal buffer capacity do not deteriorate the operation rate.

Table 2: Compared baked anodes qualities at Alba

	line 4 Conventional	line 5 SCAP-Rhodax
Coke Apparent Density (g/cm3)	1,70	
Baked Apparent Density (g/cm3)	1,570	**1,585**
Elect.Resistivity (microOhm.cm)	5136	5162
Air permeability (nperm)	0,65	**0,38**
CTE (10-6/°K)	5	4,9
Thermal Conductivity (W/m.°K)	3,2	3,2
Flexural Strength (MPa)	13,1	**13,9**
Young's modulus (Gpa)	8,1	**8,7**

Fig. 4 : View of the Rhodax® crusher at Alba line 5

Paste preparation

The comparison between the IMC® process diagrams (fig. 5) of paste preparation and the conventional one (fig. 6) is quite easy: The continuous kneader is replaced by the intensive mixer.
But the economic impact is major: the investment cost for the IMC® is drastically lower than for the conventional continuous kneader. More over a long term experience of operation of both machines (RV23 or RV24 and continuous kneader K600) in 35 t/h paste plants demonstrates a saving of 80 k€/year on the maintenance costs.
However, the saving must be associated with an anode quality which must remain close to the one produced with the conventional process.
An extensive 9-year experience of operations of this mixing technology has been acquired at the formerly Pechiney and today Alcan plant at Alucam [7], at a smaller throughput (12 t/h). This first and successful fully industrial experience of an "Eirich-only" mixing line, combined with the extensive use of the RV24 Eirich intensive mixer as a 2nd cooler on 35 t/h plants, allowed to propose a IMC® mixing line at 35 t/h with 2 RV24. The commercial reference of Qimingxing Aostar in China, went on stream in January 2005. Two others Chinese references are expected to start up by the end of 2005: Hunan Chuagyuan at 35 t/h and Qingai Chalco at 20 t/h.

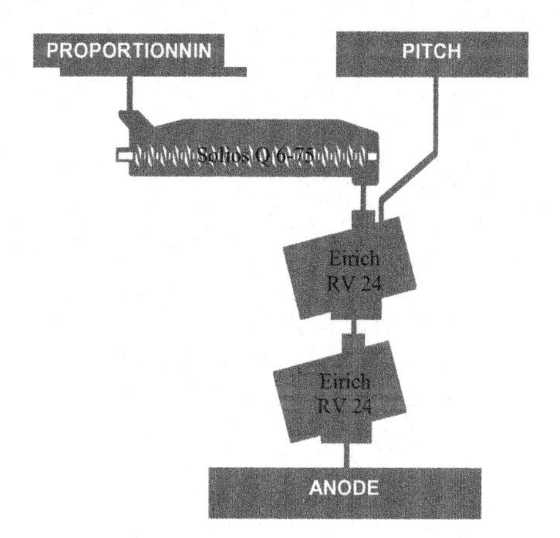

Fig 5. IMC® process.

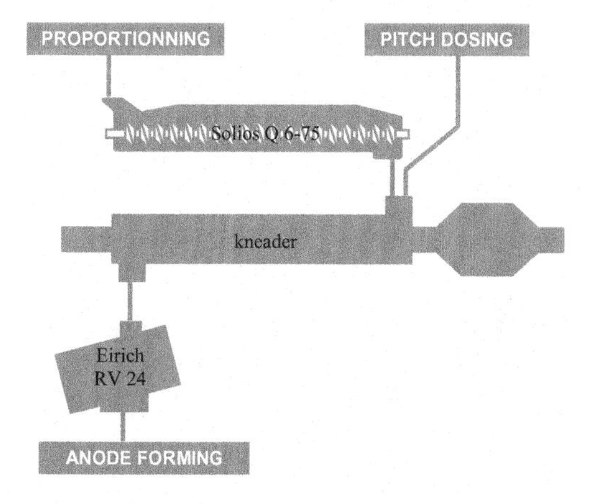

Fig 6. Conventional process.

Solios carbone was awarded a package contract for the supply of the IMC® process section including the coke pre-heating screw, hot mixer, remixer-cooler, pitch fumes treatment system and the associated control system. The coke and butts crushing, the fine grinding and the material proportioning sections was implemented by GAMI.

Pre-heating screw Solios Quatro 6-75

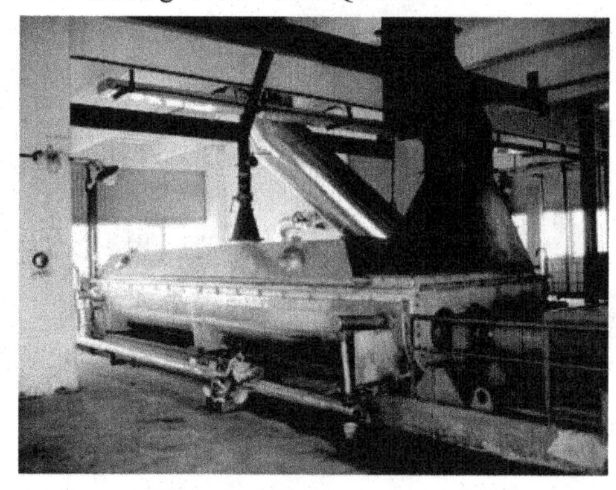

Mixer Eirich Conti RV24

Remixer-Cooler Eirich Conti RV24

Fig 7. The IMC® at Aostar.

The first anode was produced the 24th of January 2005. The start up was carried out in February 2005 at the nominal flow rate of 35 tph.

Fig 8. Aostar (China): The anode plant featuring the IMC®, ready for inauguration.

Immediately since the beginning the anode quality was good with a smooth surface aspect. The green density was found at an average of 1.61 with a good stability.
At the time of the start up the input materials to the IMC® was not stable. The control systems of the various process sections were not interconnected all together and some of them were not totally optimized (it had been recorded variation of ± 5% in the pitch flow rate). At that time it has been also recorded that at the nominal flow rate of 35 tph the fines of the dry mix was coarser than what they should have been.
Thanks to simplicity of the process it was found that the IMC® itself is not sensible at the fluctuations of the input materials: except for the load control and the cooling water flow control no other control loop is required.
The main process operation parameters were the following:

Table 3: Operation parameters

Coke flow rate	30 tph
Coke preheater inlet temp.	20 °C
Water content	< 0.2 %
Pitch flow rate	4.3 tph
Pitch temp	185 °C
Coke mixer inlet temp.	185 °C
Cooler outlet paste temp.	165°C

A period of one week of tests was carried out 5 month later in July 2005. The target was to evaluate the anode quality for confirmation of the results recorded the first period of operation.
At that time the anode quality was in the range of the international standards:

Table 4: Baked anodes main characteristics

Anodes baked density	1.57
Anode resistivity (μΩ.m)	55.15

For these tests, the quality criteria chosen, as a first approach, was the Green Anode Density (GAD). In order to be able to certify the results the anode geometry was accurately measured. And the anode weight was checked with a certified commercial scale.
These tests consisted in producing series of anodes at different mixer load in order to check the correlation between the density and the residence time and the specific energy. Figure 9 shows the relation that has been found between the specific energy and the GAD at various flow rate and filling level of the mixers. The relation between GAD and the retention time follow the same trend. These correlations were found to be similar to those obtained at the Alcan Alucam plant, reinforcing the confidence in the extrapolation from 12 to 35 t/h and over.

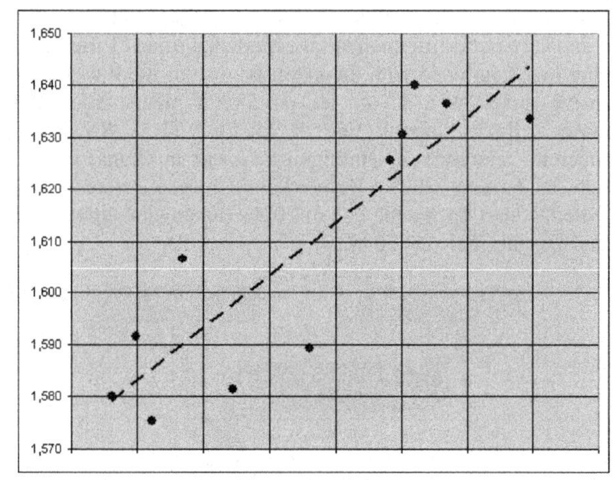

Fig 9. Specific energy impact on GAD

At 35 tph, with a load level of 2400 kg of material in both mixer, the total retention time is more than 8 minutes. The analysis of data recorded during the tests shows that the mixers are not fully loaded and there is still a margin for increasing the mixing energy.

The next step

The next step in the reduction of the costs is the combining of these two technologies in the same installation. This stage is achieved right now with the 35 tph capacity green anode plant of the Sohar project in Oman.

The benefit brought by the combination of these two technologies does not stop there however, as the equipment available allow a great flexibility to design a single paste plant of different capacities, and particularly above 35 t/h.
And in the case more and more considered today by investors where two series of 300.000 tpy of metal are brought into service immediately or a few years later one after the other, a very attractive alternative to two 35 t/h paste plants is to build one single 70 t/h paste plant. This promises savings of around 30% on the investment cost and also drastic reductions of the operation cost.
The anode plant must then be able to produce up to 70 t/h anode with the same level of quality of production and environmental conditions than that obtained by the most up-to-date installations. To make a success of the High Capacity Anode Plant (HiCAP)

development, which elements must be available to be able for processing such flow rate ?

SCAP Rhodax® Sizing

The key equipment for producing up to 70 t/h of paste (62 tph of dry material) are available in the Solios catalogue (Figure 10):

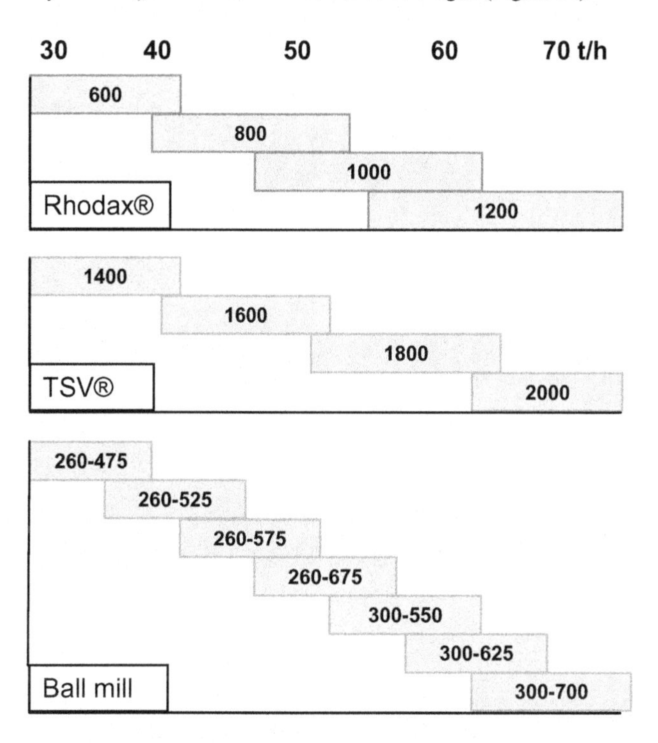

Figure 10. SCAP Rhodax® key equipment range

Other equipment like elevators, conveyors,.. are common technology and available on the market.

IMC® Sizing

Coke pre-heating

For pre-heating 62 tph of dry material a Solios Quatro 8-100 hollow flight screw is available. Refer to the Figure 11.

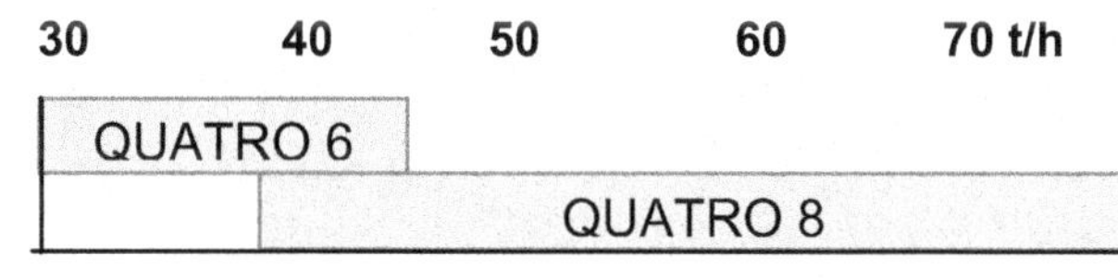

Fig 11. Coke pre-heating equipment range

Mixing and cooling

The mixing and cooling will be done by various combination of RV24 Conti and/or RV32 Conti in series. The combination will be determined according to the total specific energy which is required in order to achieve the specified GAD at the nominal flow rate.

The RV32 Conti mixer is able to contain up to 5200 kg of material. It has already been used for 5 years at the Alcan Alma plant in Canada. The Alma mixing line configuration is not directly comparable to the IMC® as it includes 2 RV32 batch mixers and 1 continuous RV32 mixer/cooler, however it allowed to fully validate the reliability and characterize the efficiency of the RV32 size of mixer.
A nominal flow rate of 70 tph of paste may require to install up to 3 RV32 Conti in series. In that case the maximum mixing residence time will be more than 13 minutes.

Anode forming

The third part of the HiCAP is the anode forming. Assuming the anode size for Alcan AP3X technology (950 kg) the vibrocompactor must be able of producing up to 75 anodes per hours.

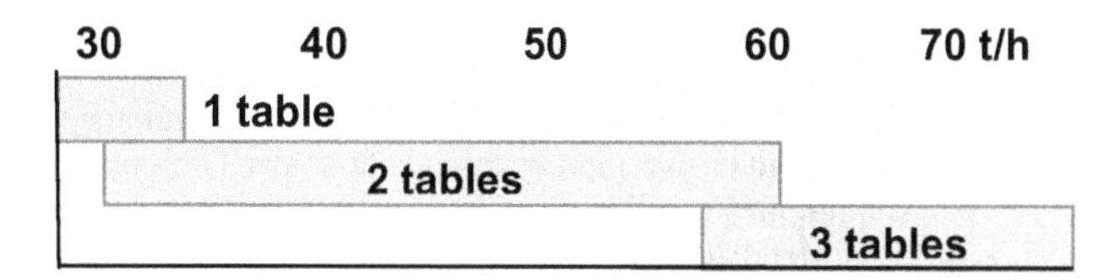

Fig 12. Anode forming equipment range

The maximum capacity will be given by an original arrangement (Figure 6) of three vibrocompacting tables, each capable of more than 30 anodes per hour. Each table is fed by a three position transfer hoper (1) followed by an other one dedicated to each table (2) moving along the perpendicular direction.

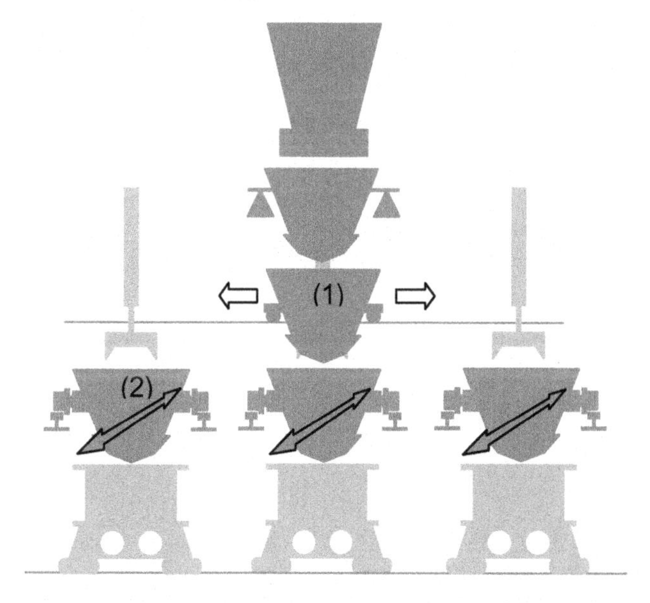

Fig 13. Three tables vibrocompactor feeding .

Conclusion

For their future projects, investors have now in hands a wide range of green anode plant capacity for feasibility optimization. The investment cost can be reduced by more than 30% for a total capacity of 70 t/h. Optimized solutions are also available for lower capacities. The profitability of the investment will be also be enhanced thanks to the drastic reduction of operation costs that is drawn from a single green anode plant solution.

References

1. A.Pinoncely, Jean-François André, Corine Jouault, Yann El Ghaoui, Jean-Christophe Rotger, Christian Dryer
   "The SCAP-Rhodax(R)process for dry mix preparation in anodes plants"
   Light Metal 2005 pp 511 – 515

2. Corinne Jouault, Yann El Ghaoui, Jean-Christophe Rotger, Christian Dreyer & A.Pinoncely
   "Paste plant technology latest developments"
   Australian aluminium smelting technology conference and workshops 2004.

3. B.Hohl, A.Pinoncely, J.Cl. Thomas
   "Anode paste preparation by means of a continuously operated intensive mixing cascade at aostar Qimingxin Aluminium China"
   Light Metal 2004 pp 669 – 674

4. C. Dreyer, C. Jouault, A. Pinoncely & J.F. André
   "The new AP-FCBA paste plant technology"
   Light Metals 2003 pp 547 – 543

5. C. Dreyer, N. Backhouse & A. Pinoncely
   "Featuring the new AP-FCBA paste plant technology"
   Light Metals 2002 pp 577 – 582

6. J.Cl. Thomas
   "New concept for a modern paste plant"
   7th Australasian Aluminium Smelting Technology Conference and Workshops 2001

7. C. Dreyer, C. Ndoumou & J.L. Faudou
   "Reconstruction of the mixing line for anode paste production at Alucam"
   Light Metals 1998 pp 705 – 710

Light Metals 2006 *Edited by Travis J. Galloway* ***TMS (The Minerals, Metals & Materials Society), 2006***

Dynamic process optimization in paste plant

Raja Javed Akhtar (1), Saleh Ahmad Rabba (1), Markus W. Meier (2)

(1) Dubai Aluminium Company Limited,
P.O. Box 3627, Dubai, United Arab Emirates

(2) R&D Carbon Ltd.
P.O. Box 362, 3960 Sierre, Switzerland

Abstract

Dubai Aluminium Company (DUBAL) is one of the largest single site aluminium smelters in the world producing 761,000 MT/year of high quality premium product. It has its own carbon manufacturing plant with a production capacity of 370,000 anodes/year of different sizes and configurations.

As part of a management drive to optimise the current assets, an exercise of Dynamic Process Optimization (DPO) was carried out in the greenmill. R&D Carbon was chosen as the technical partner and worked along with DUBAL operational personnel during the three phases of the DPO. The process parameters were optimised systematically to reduce pitch content, increase paste throughput and increase anode density. A mobile pilot press was utilised to prepare samples for evaluation during the optimisation process.

The optimised parameter settings were implemented on the two anode production lines in the greenmill following a successful 15 cell trial in the potrooms.

This paper describes the DPO methodology, highlights the results achieved in the greenmill together with the subsequent improvement in anode quality and performance.

Introduction

The biggest variable impact to the aluminium production cost are the prebaked anode [1]. Reducing the production cost is an ongoing process in every smelter. Optimization goals may focus on the anode plant to produce cheaper anodes with the same performance in the pots and/or on the potrooms to use anodes of superior quality to produce metal at reduced cost [2].

There are numerous optimization goals in the anode plant that can lead to a reduction of the aluminium production cost provided that the anode performance is not negatively affected:

Lower conversion cost:

- Higher throughput
- Less scrap
- Fewer breakdowns
- Lower energy consumption
- Lower maintenance cost

Lower raw material cost:

- Coke / pitch
- Lower pitch content

Potroom related optimization goals focus on superior anodes to produce metal at lower cost:

- Higher consistency of anode properties
- No anode cracking or slabbing
- No spikes or mushrooms
- Good anode current distribution
- Low anode voltage drop
- High metal purity
- Low consumption figures
- High current efficiency

In order to identify in a systematic manner the area(s) with the greatest potential for optimization, DUBAL contracted R&D Carbon to carry out an audit of the anode plant. A technical assessment was carried out by following the material stream from the raw material intake all the way to the potrooms and the butts recycling station. Every relevant processing step was investigated by sampling the product, analysing its properties and comparing them with the worldwide typical level and benchmark.

The audit conducted by R&D Carbon concluded that DUBAL produces anodes of high quality with limited options for further improvements of their performance in the pots. The high current efficiency of 95.1 % and low anode net consumption of 406 kg/tAl in the CD-20 pots underline this conclusion.

On the other hand the results of the audit showed that there is a potential to reduce the pitch content without negatively affecting baked anode quality, in particular apparent density and air permeability. Prior to the DPO the pitch content in green anode was 15.0 +/- 0.7 % depending on the coke type used. The baked anodes exhibited an apparent density of 1.580 g/cc and an air permeability of 0.6 nPm.

Reduction of the pitch content without adversely affecting the anode properties requires adjustments in the recipe and also to the preheating, mixing and forming processes. A trial and error approach on a full production scale was not feasible due to a high risk of scrap generation and potentially poor anode behaviour in the pots over an extensive trial period. Instead, an approach on a pilot scale was chosen which is described hereafter.

The methodology of the Dynamic Process Optimization (DPO) enabled the determination of the optimum combination of processing parameters with negligible scrap generation in a very short period of time. The implementation of the optimization findings was made within a few weeks. The optimization process was carried out in 3 steps:

1. Plant capability check
2. Dynamic Process Optimization
3. Implementation of DPO recommendations in full scale anode production following positive anode performance confirmation in the potrooms.

DPO Approach, Experimental and Program

DPO Approach

Prior to the actual DPO, a plant capability check was conducted. The purpose of this check was to identify the potential and limitations of the dry aggregate and paste preparation operations and also the forming equipment. This included the following:

- Consistency of the ball mill product during finer fines production
- Capacity and reliability of the fraction scales
- Reliability of the sampling equipment and procedures
- Stability and capability of the Heat Transfer Media (HTM) system to accommodate an increased coke preheating temperature
- Spare capacity of the paste mixer and cooler in terms of intensity and throughput
- Flexibility of the vibro-compactor in terms of temperature, densification and throughput

The principal strategy to reach the target was to utilize the maximum plant potential to increase anode density and then to systematically reduce the pitch level. Accordingly, a new recipe was chosen with a reduced fines content but with the fines being of a finer sizing that has the advantage of less variability in sizing from the ball mill and, as a consequence, less pitch requirement. Finer fines require more intense mixing, hence the mixing intensity needed to be increased. With the implementation of these changes, the forming temperature could be increased, as the paste was less sensitive with the reduced pitch content and better homogeneity. The higher forming temperature allowed a faster densification of the anode block and consequently a resultant higher density.

Experimental

The actual procedure of the DPO [3] involved the production of pilot electrodes with anode paste collected directly after the cooler, (see figure 1). Only one parameter was changed at a time while all others were kept constant. For one set of parameters twelve pilot electrodes of approximately 5 kg each were produced utilising a mobile hydraulic press. Each set was pressed in less than one hour. This allowed the testing of many parameter combinations with the generation of a minimum amount of anode scrap. The short production time reduced the effect of quality variations in the raw materials.

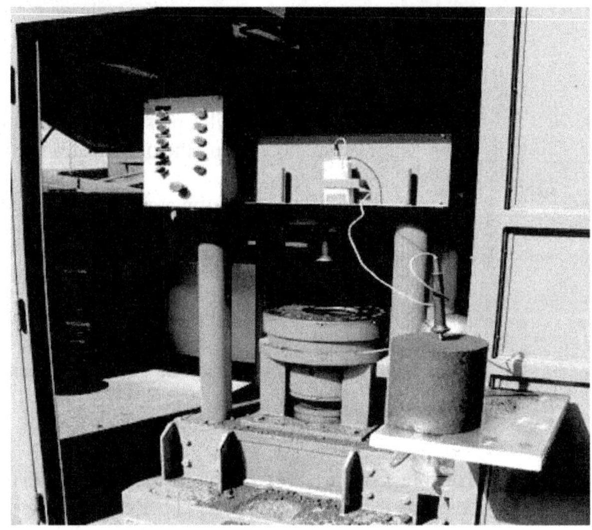

Figure 1: Production of pilot electrodes and monitoring of forming temperature

The pilot electrodes were baked under well controlled conditions in a laboratory furnace to 1150°C with minimal variations. From each pilot electrode, three cores of 50 mm diameter were then drilled. The quality of these cores compared favourably with anodes produced with vibro compaction.

The resulting anode properties were determined according to standardized test methods [4]. The evaluation of the anode properties from each trial enabled the determination of the optimum combinations of process parameters.

DPO Program

Prior to the commencement of the DPO, a consistency test was performed with standard conditions to verify stable operation of the paste plant. This test was essential, as the negative effect of process variations on the anode quality is often underestimated.

The best results were achieved when a combination of parameters were optimised. However, several limitations were respected. The prime limitations were the capacity of the installed equipment and their inherent operational variations. These limitations were identified during the preliminary plant and capability check.

The sequence of optimization trials is summarized in table I. Due to the fast production procedure for the pilot electrodes, the entire optimization program could be carried out in three days.

Table I: DPO program principles

Seq.	Focus	Parameters	Variations
1	Baseline Plant conditions	Fines content Pitch content Mixing conditions Forming temperature Throughput	42% 15.6 - 14.4 % 177°C, 260 kW 145°c 33t/h
2 3 4 5	Finer fines	Fines content Pitch content Mixing power Forming temperature Throughput	31 - 37 % 15.0 - 13.8 % 220 - 280 kW 140 - 150°C 33 - 35 t/h

Whilst the monitoring of the green anode density during the DPO gave an initial indication about potential optimal process parameter combinations, verification was obtained from the analysis and evaluation of the properties of the baked pilot electrodes.

DPO Results

The comparison of properties of baseline anodes before with DPO anodes revealed remarkable improvements, as illustrated in table II. It has to be pointed out that for the entire DPO program no investment in hardware equipment was made, ie. all improvements were achieved by the optimization of the process parameters with the existing equipment.

Table II: Base line vs. optimized settings towards reduced pitch and over-nominal throughput

Parameter / Property	Unit	Base Line Settings	Optimization towards	
			Reduced Pitch level	Over-Nominal Throughput
Fines fineness	Blaine No	2700	4000	4000
Fines content	%	42	31	31
Pitch content	%	15.6	14.4	14.4
Mixing temperature	°C	177	187	187
Mixing power	kW	260	280	280
Forming temperature	°C	140	150	150
Throughput	t/h	33	33	35
Green anode weight	kg	1200	1204	1200
Green anode density	g/cc	1.620	1.625	1.620
Baked anode weight	kg	1145	1158	1155
Baked anode density	g/cc	1.566	1.592	1.586
Air permeability	nPm	0.50	0.49	0.54
Specific electrical resistance	mWm	57	58	59
Flexural strength	MPa	10.5	9.5	8.6
Static elasticity modulus	GPa	4.7	4.0	4.1
Thermal conductivity	W/mK	3.89	3.48	3.38

By using the same raw materials, the pitch content was reduced by 1.2 % abs. Even with the reduced pitch content, the baked anode weight could be increased by 13 kg and the baked anode density by 0.026 g/cc. No significant change was measured for air permeability and specific electrical resistance.

The optimized recipe resulted in a reduction of flexural strength and elasticity modulus. These reductions may be interpreted as deteriorations, but in fact, they were not. Due to the reduction of the elasticity modulus, the anodes become less brittle and thus less susceptible to thermal shock cracking [5]. The level of mechanical strength was still capable of withstanding all arising stresses.

Another positive aspect was the reduced thermal conductivity which has a positive impact on the net anode consumption [6].

Furthermore, the results in table II show that, with the knowledge gained from the pitch optimization trial, the throughput could be increased by 10 % (ie. 3 t/h) over the nominal level without any hardware modifications and without compromising anode quality.

Hence the merit of the DPO was that it gives the option to either generate savings by minimizing the pitch content or maximizing the production capacity without investment on equipment.

Implementation in Routine Operation and Validation in Pots

Following positive laboratory results as illustrated in table 11, the decision was taken to carry out a selected pots trial in the potrooms before implementation of the recommended revised settings for plant wide production. A batch of 2000 anodes each with the original parameters (control) and the revised DPO settings (test) were produced utilising the same raw materials in the Greenmill.

Table 111- Green Anodes Quality Data

Anode group	Production line	Anode No	Green Density (g/cc)		Pitch %		Ball mill product	
Test	1	1987	Avg	SD	Avg	SD	Blaine No.	SD
			1.628	0.005	14.30	0.12	3915	182
Control	1	2000	1.625	0.008	14.89	0.12	2656	19

A reduction of 0.59% binder level was observed without a significant difference in green density.

Both categories of anodes were baked under identical operating conditions in the baking kilns.

Table IV - Baked Anode Quality Data

Criteria	Anode Group	
	Test	Control
Core density g/cc	1.580	1.577
Real density g/cc	2.091	2.099
Air perme npm	0.44	0.58
Elec. Resit. µΩm	52.28	51.79
Flex. Strgh. Mpa	14.89	14.01
Mean Cryst-size	30.18	31.35
CRR %	95.85	98.73
CRL %	4.05	4.22
CRD %	0.10	0.06
ARR %	86.52	85.08
ARL %	11.79	13.30
ARD %	1.69	1.63
Therm. Condition. w/mk	3.38	3.61

Test anode analysis was comparatively better with regards to flexural strength, real density, air permeability and air reactivity loss.

15 pots each for test and control anodes were selected in potline-6 and the trial was carried out for 3 clear cycles.

Table V - Butts parameters

Criteria	Anode group	
	Test	Control
Reduction in area %	6.5	6.9
Avg. Butt thickness (cm)	19.3	19.1
Top exidation (cm)	0.9	0.8
Carbon under stub (cm)	6.95	6.70
End burn (outer – cm)	2.65	2.70
End burn (Inner – cm)	3.10	3.05
Butt weight (kg)	259	241
Consumption per anode (kg)	883	900

18 kgs increase in butt weight was observed following three clear cycles in selected cell operation. No negative impact on pot parameters was recorded.

The confidence gained as a result of the selected pots trial helped in making the decision to implement the revised DPO setting for plant wide production in March/2005.

The actual supply of anodes to the potrooms commenced in May/2005. At the time of writing this article, these anodes have completed four clear cycles of operation without any adverse impact on pot operation.

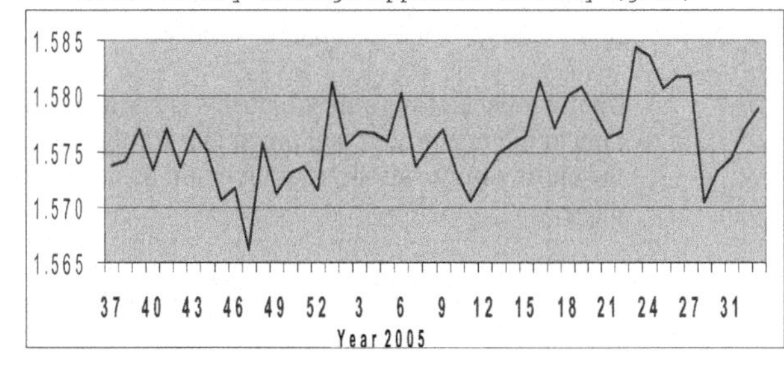

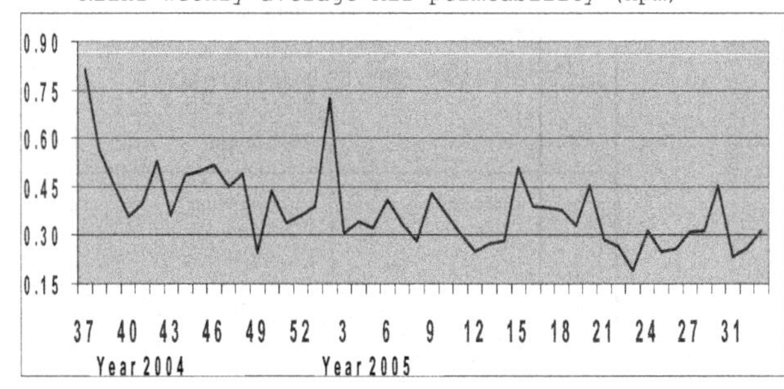

Figure II – Baked Anode Quality Indicators

Conclusions

1. *A reduction of 0.7% in binder was achieved without compromising product quality.*
2. *It was not necessary to undertake any plant modification to achieve the benefits.*
3. *Apparent density and air permeability improved resulting in better anode performance.*
4. *Baking loss reduced from 5% to 4.5% - an increase of 13kgs in baked anode weight.*
5. *The throughput of each green anode marking train could be increased from 33 tph to 35 tph – a benefit of additional 10,000 anodes/year.*
6. *Provided an opportunity to further increase the anode density in the future.*

References

[1] M. Meier, T. Müller, "Influence of Anode Quality on Aluminium Production Cost", 3rd Rodding Conference, Iceland, 2005

[2] R. Perruchoud, "Carbon Plant Optimization", CII Carbon Symposium XX, New Orleans, 2005

[3] U. Bühler, R. Perruchoud, "Dynamic Process Optimization", Light Metals 1995, 707 - 714

[4] R. Perruchoud, M. Meier and W. Fischer, "Survey on Worldwide Prebaked Anode Quality", Light Metals 2004, 573 - 578

[5] M. Meier, "Cracking Behaviour of Anodes", R&D Carbon Ltd, 1996, 233 ff

[6] W. Fischer, F. Keller and R. Perruchoud, "Interdependence between Anode Net Consumption and Pot Design, Pot Operating Parameters and Anode Properties, Light Metals 1991, 681 - 686

Light Metals 2006 *Edited by Travis J. Galloway* ***TMS (The Minerals, Metals & Materials Society), 2006***

IMPROVING CARBON PLANT OPERATIONS THROUGH THE BETTER USE OF DATA.

Keith A. Sinclair[1], Barry A. Sadler[2]
[1]Sinclair Associates, Inc., 12620 Ridgepath Lane, Knoxville, TN 37922 USA.
keith@sinclairassociates.com
[2] Net Carbon Consulting Pty Ltd, P.O. Box 286, Kangaroo Ground, Victoria 3097 Australia.
barry.sadler@bigpond.com.au

Keywords: data traps, specifications, process control, sampling, averages, analysis, improvement

Abstract

In most Carbon Plants an enormous amount of data is collected and used for short and long term control, improvement of plant operations, reporting, and to monitor anode quality for the customer. In the experience of the Authors, however, the way that data are used can be improved considerably - full value is rarely extracted from the information and several "data traps" are common:

- ❑ Control decisions and actions are based only on comparisons to arbitrary specifications
- ❑ Averages are used inappropriately, e.g. with a lack of attention to variability in the data and data is over-aggregated
- ❑ Insufficient attention is given to the impact of sampling methods on the usefulness of data

This paper will demonstrate these data traps, describe their implications, and suggest ways of avoiding these pitfalls using plant examples including information from the sampling and testing of anode cores.

Introduction

A quick review of the weekly or monthly reports generated in a typical Carbon Plant will show that a very large quantity of numbers are collected, utilized in some way, and reported as a part of the running of these facilities. All too often these reports consist of vast tables of numbers (often averages of some kind) with an occasional graph thrown in to emphasize what is considered to be a critical parameter. Only rarely is it evident that full value is being extracted from the data collated in these reports; unfortunately it is much more common to find errors in the way the data is used and interpreted. Of these "data traps", the Authors have encountered the following most frequently:

- Decisions on whether to take control actions (or not) are based solely on comparisons to "arbitrary" specification limits; these limits having little or no technical validity, and are commonly justified with "that is what we have always done". They rarely have a clear connection with the needs of the Customer (i.e. the next stage in the process) but are used as the "magical" trigger point at which action or explanation is suddenly required.
- The over-aggregation of data into averages and use of these averages without appropriate attention to the variation within the data are rampant in Carbon Plants (And the other parts of Aluminium Smelters and many other Process Industries). Averages used in this way obscure variation in the process, thereby hiding legitimate signals that action may be required.
- Inferences regarding process behavior or product quality are made without sufficient attention given to the sampling method and the resulting usefulness of the data. Sampling processes must be designed based on the purpose of collecting the data. It is a common trap to use data specifically sampled for one purpose to make other, unrelated decisions.

In previous publications the Authors have looked at the need for "Statistical Thinking" in Carbon Plant operations [1] and presented the potential opportunities to improve processes by integrating Lean Manufacturing and Six-Sigma methodologies [2]. What follows in this paper is an expansion on the implications of the data traps that have been outlined above, and suggestions on how to avoid the waste associated with these traps using Carbon Plant data as examples.

Control by conformance to arbitrary specifications

This approach to process management is based on the view that as long as a parameter is within set limits (specifications), no action is required. This good value (inside the limits) versus bad value (outside the limits) form of "binary" control model results in action only when the parameter moves outside the limits. There are a number of problems/issues with the binary control model:

I. Specification limits are often set in a very arbitrary way and do not have a valid basis stemming from the impact on the Customer or the process. They are commonly derived from:
   a. History – "always been that way"
   b. Experience – often outdated, and rarely challenged
   c. "Experts" applying general industry standards (also often outdated) without considering the needs of specific Customers or the impact of local conditions.

   With such a model, all results within the limits are all good, are assumed to not cause any loss to the process, and therefore, require no action. (See Figure 1). As will be seen later, this is rarely, if ever, the case in reality. An implication of "within the limits means zero loss" is that a parameter can continually be measured to be just inside a specification limit and this will be considered acceptable with no action triggered. A small drift in the parameter could then move it outside of the limit. Suddenly the result is unacceptable as loss is now deemed to be occurring and action is called for – at the very least an explanation will be required. The reality is that the drift that resulted in the parameter moving beyond the limit could have only a negligibly small impact on the process, and does not warrant special attention.

II. Process variation can be "Common" cause - random variation which is part of the process and does not normally have an assignable cause; hence it should not trigger a response, and "Special" cause – which is not expected as a part of the process and normally will have an identifiable cause that should trigger a response [1]. The binary control model can result in two primary forms of loss to the business. First, it can cause over control or "tweaking" associated with reacting to common cause variation as though it were a special cause. This increases variation in the process and product. Second, it can result in under control, or the lack of action on a special cause change in the process (still within specification limits), leading to problems not being identified and actioned until the damage associated with the process disturbance escalates.

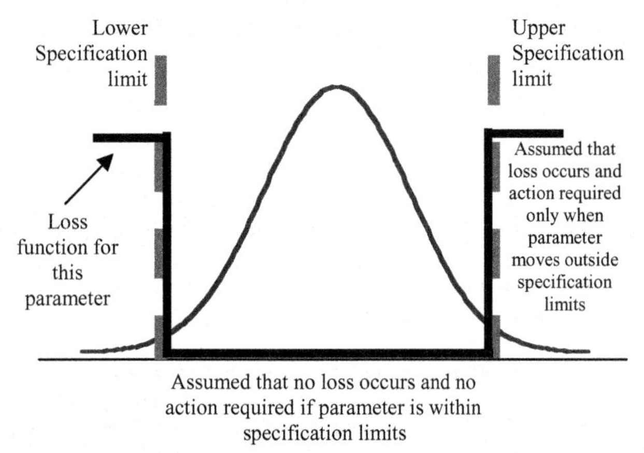

Figure 1. "Binary" control model assumes that as long as a parameter remains within arbitrary specifications no loss is incurred by the process. (See loss function curve) resulting in increased process variation and cost.

The issues associated with binary control can be explored further by using anode core Air Permeability as an example.

"Binary" control of anode core Air Permeability.

Binary control of Air Permeability (AP) assumes that variation in AP has no impact on anode performance in the cell (i.e. on the Customer's process) until it reaches an upper limit, at which point anode performance suddenly deteriorates unacceptably, impacting on the cell performance – i.e. loss occurs (See Figure 2.).

Ideally, the upper specification limit for AP would have been established through rigorous scientific research into the in-cell performance of anodes with different AP values, covering a wide range of specific conditions such as: cell design and operating parameters, anode raw materials, and anode production conditions. The results of this work would then be used to set the specification limits. Unfortunately this is not what actually happens.

In reality, the AP specification limits are set based on standard industry values, and these are normally attributable to historical standards or expert opinion. It is rare for plant data to be used to establish the point at which loss in the cell becomes "economically" unacceptable, i.e. according to binary control, the point at which cell performance is suddenly impacted.

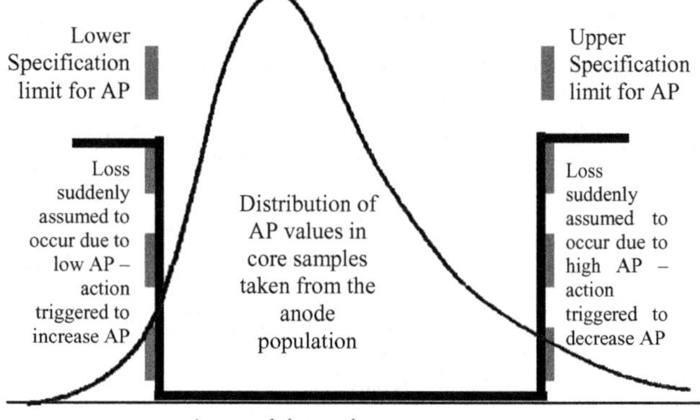

Figure 2. "Binary" control of anode core Air Permeability.

The result of the binary control approach is that AP can be anywhere within the arbitrary specification limits and no action will be taken, even if the measured AP varies wildly or if it tracks very close to one of the specification limits. Should the AP value move outside the specification limits, however, suddenly it is assumed that the AP value is no good, that loss occurs, and that action is required. All of this because the measured AP moved outside an arbitrary limit that has little technical validity or direct correlation with how the anode will perform in the customer's cells. The actions taken to "recover" the situation and get AP back into specification are likely to cause unnecessary variation in other anode properties. There is also the distinct possibility that if no action was taken, the AP values would come back within the specification limit of their own accord - if the move outside the limits was due to common cause variation.

There is an appropriate role for valid specifications; however, the "binary" method of control by comparison to specification should be avoided when making decisions about when action is required on the process. There is a better alternative:

"On target with minimum variation" control of anode core permeability.

There is likely to be an optimum AP value that results in the least loss to anode production and cell operations. AP values less than this optimum may result in loss due to difficulty in baking the anodes. The very low permeability hinders the release of pitch volatiles during baking and can result in internal cracking of anodes if "normal" heat-up rates are exceeded during the anode baking process. This results in loss through poor anode performance in the cells due to internal cracking, and/or through a reduction in baking furnace capacity due to the need to use longer fire cycles to reduce the anode heat-up rate to avoid internal cracking. With significant test work it might be possible to determine the lowest limit for AP based on the longest fire cycle that can be used and still meet production requirements and at what point internal cracking starts to have an unacceptable impact on the cells. The likelihood of this work being done is not high.

Air Permeability values greater than the optimum value cause loss by increasing the rate at which anode carbon is consumed in a non-productive way by gaseous attack in the cell, primarily due to reaction with Carbon Dioxide. A crude estimate of the magnitude of this loss can be made with Fischer's equation for calculating net anode consumption [3]. In this equation, the AP term has a coefficient of 9.3, suggesting that for each 1 nPm increase in average AP, an increase in anode consumption rate of 9.3 kg Carbon/Tonne of Aluminium can be expected. The upper specification limit could be set at the point where the Customer determines that the economic impact on the process from the additional carbon loss due to increasing AP is not acceptable.

A better alternative to "control to specification" is demonstrated in Figure 3, and can be described as "on target with minimum variation". Control decisions and actions are based on a view that economic loss occurs as the process varies from an optimum point regardless of whether it is in or out of specifications. Operational definitions of control points (action limits) are determined from a statistical analysis of process performance.

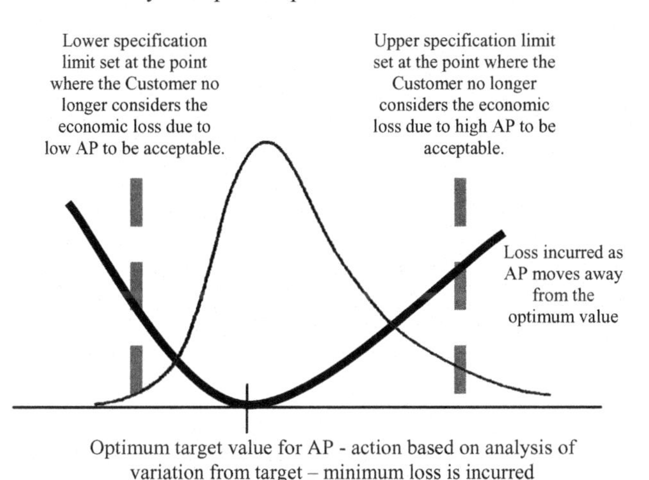

Figure 3. "On target with minimum variation" approach to control of anode core Air Permeability reduces variation and loss (new loss function)

In relation to Figure 3, action based on "comparison of current performance to specification" is not relevant. Appropriate process decisions and actions to achieve on target with minimum variation at the lowest total cost are:

1. Monitor the process through time, using statistically derived action limits based on actual performance behavior.
2. If special causes indicating a process change occur, to either side of target – take action to investigate, understand, and respond to these unique or assignable cause events.
3. If the process is stable (no special cause signals), take no action on specific outcomes, as action will likely result in increased variation.
4. If the process is stable, and DOES NOT meet specification limits, take action to improve the system through process redesign, method changes, engineering actions and so on.
5. If the process is stable and DOES meet the specification limits, determine if further process improvements are economically justified. If so, plan further improvements to reduce variation around an optimum target or to shift the target to a new low cost optimum.

In the "on target with minimum variation" approach, process decisions and actions for stability only require knowledge of the statistically derived limits and a running record of process performance. Specification limits are needed when comparing a stable system to minimum acceptable limits of capability and determining what, if any, further action is required to improve the process.

To properly apply the "on target with minimum variation" approach to process management requires the application of Shewhart's "Principles for Economic Control of Manufacturing Processes" by using control charts [4] to provide the following:

a. Control actions are based on cost effective responses to variation. Actions are triggered when control limits (calculated from process data) are exceeded, indicating that the variation is special cause and investigation is justified to find and remove (or incorporate) the root cause.
b. If the process is not capable of meeting valid specification limits, the focus is on improving the "process design" to reduce variation so that the specifications are achieved. (This contrasts with the "control to specification' approach of reacting to just the points that are outside the specification limits.) Until the process is capable of meeting Customer requirements (i.e. conforming to the specification limits), it may be necessary to reject non-conforming product; this is normally expensive and provides a further financial imperative to improve the process (i.e. reduce variation).
c. The never-ending cycle of analysis, action and improvement inherent in the "on target with minimum variation" approach is directly opposed to the "maintaining the status quo" outcome of controlling to specification. It provides the means to achieve levels of process performance previously unattained, yet necessary in this environment of intense competitive pressure for Carbon Operations.

"On target with minimum variation" - summary

The "on target with minimum variation" approach can be summarized as:

- Aim to achieve a process target value with minimum variation around this target so economic loss to the business is minimized and specification limits are not exceeded.
- Use thorough statistical analysis of the process to determine the control limits that trigger action and the impact of variation on the Customer.
- Take action when control limits calculated from the data are exceeded (or other statistical signals are present).
- If the process does not always result in values that meet the Customer's specification, the focus should be on reducing common cause variation in the process until capability to specifications is achieved.
- The objective is to maintain the process at minimum total cost resulting in maximum business value.

Inappropriate use of averages

A common practice in Carbon Plants is to aggregate data into averages for analysis, decision making – including about when action is required, and for reporting process status. Some of the problems associated with this data trap were covered in a previous publication [1], however the frequency at which the Authors encounter this trap suggests that it warrants further attention.

Reducing variation is a key to improving processes. Averaging data is the best way to hide variation in the data. The issue is immediately obvious – averaging data inhibits process improvement. It is a surprisingly simple fact, but one very often overlooked, that averages tell you nothing about the amount of variation within the data set that makes up the average. A number of issues/pitfalls are associated with inappropriate use of averages:

- What is reported is not what is sent to the Customer. This is due to the practice of using averages to report conformance to specifications or other quality criteria.
- "Phantom" averages – where unusual distributions in the data mean that there is very little, or in some cases no, "average" product in the batch.
- Patterns in data, crucial to understanding variation and enabling process improvement, are hidden by averages.
- Averages alone do not tell you anything about the historical context for the data – providing the historical context of data enables predictions to be made, the essence of effective process management and improvement.

Each of these pitfalls will be covered in more detail using examples drawn from plant operating data.

<u>What is reported is not what was sent to the customer.</u>

It is common for management to report performance to targets, specifications, etc by using shiftly, daily, weekly, or monthly averages of plant data. This practice is a good way to unintentionally or deliberately hide variation. Averaging (and the distortion of results that goes with it) appears to know no bounds. Often daily averages are the average of shiftly averages; weekly averages are the average of daily averages and so on. Taking the average of averages is an even better way of hiding variation.

As an example of "what is reported is not what is sent to the Customer", Figure 4 shows the distribution of paste elongation values (A measure related to paste viscosity at a fixed temperature) reported to the Potrooms customer on a monthly basis in a Soderberg smelter. The comments provided with this data were along the lines of "paste quality has met the required standards." The response from the Potlines was similar to "Thanks for the good paste quality – we can see that paste quality does not appear to explain why we are seeing some dry anode tops".

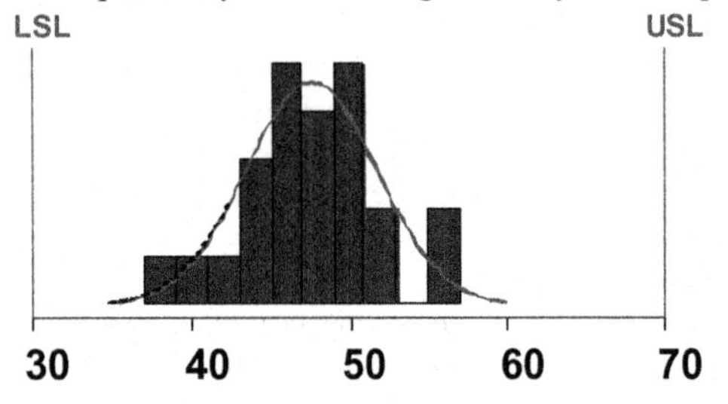

Figure 4. Histogram - paste elongation data as reported to Potrooms.

Analysis of Figure 4 suggests that the Carbon plant comments may appear appropriate – paste elongation (as reported) for the month was well within the Customer's specification limits. The problem is that the data presented to the Customer was a daily average of the elongation values taken every two hours. A very different picture emerges when the individual elongation values from the same monthly data set are plotted on a histogram (See Figure 5). The data plotted in Figure 5 shows that in reality there was quite poor conformance to the Customer's specifications during the month. The affect of averaging reduced the perceived variation to about a quarter of the level seen by the Customer. If this had been reported to Potlines it is not difficult to imagine that the paste quality discussions would have had a different tone and degree of urgency. The reporting of this data would also have suggested that paste quality be pursued as a possible cause of the dry anode tops.

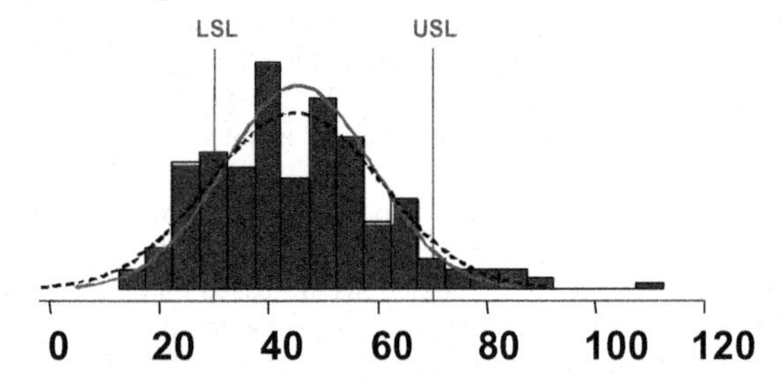

Figure 5. Histogram of individual paste elongation values. These values produced the daily averages plotted in Figure 4.

This problem of averages obscuring the real product quality sent to the Customer is also commonly encountered when examining how anode core property test results are reported to Potrooms.

<u>"Phantom" averages.</u>

When we see an average, a common assumption is that the average represents the most typical value of the parameter being measured, i.e. the average is the most representative value of the items measured. For example, if the Volatile Matter (VM) in green petroleum coke from a particular supplier was measured on each railcar delivered, and the individual railcar VM values were averaged for a week, then it would not be surprising to see this average VM value used to set up the kiln parameters for a calcining run for this green coke. In this example, only the average VM for the weekly green coke batch was reported from the Laboratory to the Kiln Superintendent.

There was one coke source that always caused problems for the Kiln Superintendent – with very unstable kiln operations and high variation in calcined coke Real Density. When investigating the problem, the individual railcar VM analyses were plotted as a histogram (Figure 6.).

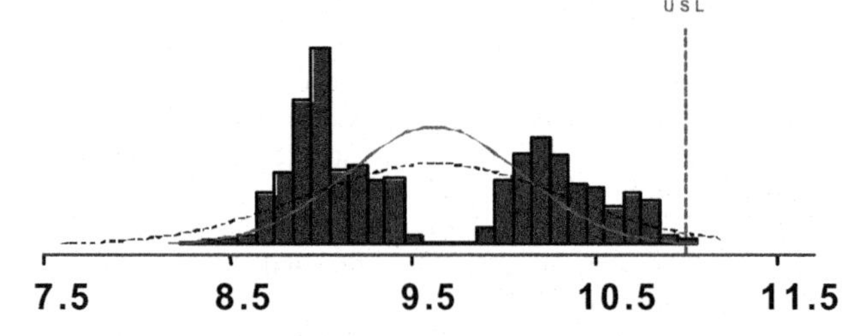

Figure 6. Histogram of individual railcar analyses of green coke VM delivered to a calciner. The average VM reported for this batch of green coke was 9.6%.

The average VM reported to the Kiln Superintendent was 9.6%. This was then used to set up the kiln for the coke batch from this supplier. As can be seen from Figure 6, there was not a single railcar load of green coke from the supplier that had a measured VM of 9.6%. The Superintendent was setting the kiln up for a coke that did not exist. As the slugs of higher and lower VM coke came through to the kiln for calcination, the kiln went out of control due to the variation in VM.

What had happened in this case was that the supplier was using two banks of delayed cokers to produce the green coke. Initially the banks had produced similar green coke qualities, but over time the qualities had drifted resulting in the different VM's evident in Figure 6. As the batch average VM value was all that was reported to the Kiln Superintendent, the divergence in quality was not picked up. The Laboratory had noticed something strange in the results, but had not commented as they had been told once that their job is to just generate the analytical data they are asked to, not to comment on the results. (This is an all too common perception by Laboratory staff and is very wasteful. Perhaps, they should be the first group to raise a warning flag that there might be something unusual in the results they generate.)

Unusual data patterns hidden by averaging.

Unfortunately not all of the data collected and used for decision making and action is reliable. When people are involved in data handling, there can be subtle pressure to not report values outside the limits that trigger action. Averaging the data usually removes any indication that the data is not accurate. The result is that decisions and actions are taken on the averaged (poor) data resulting in increased variation and waste, with opportunities to learn about how to improve the process missed. Plotting the data instead of using averages can, however highlight such data inconsistencies which can then be addressed. Examples of unusual trends in manually reported data are given in Figures 7 & 8.

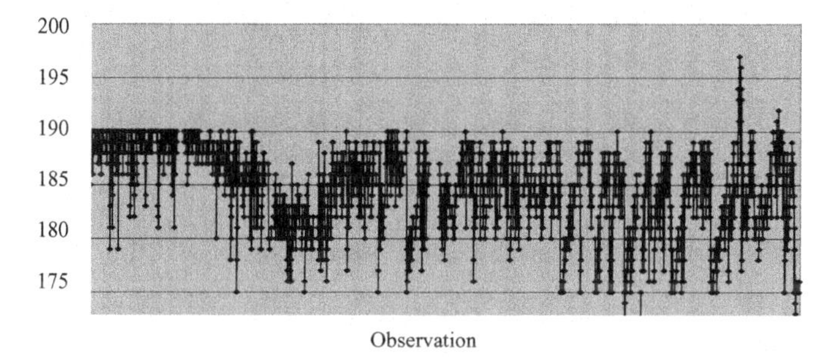

Figure 7. Manually collected mixing temperature (°C) data. Action is triggered if the reported temperature exceeds 190°C.

In both of these examples "unusual" cut-offs in the data can be seen at the point beyond which action is required. It would appear that the data has been manipulated to avoid additional work or other repercussions from exceeding the limits. The nature of these cut-offs in the data and their likely cause is not evident when just averages of the parameters are reported. Using poor data for decision making and control actions is as bad, or even worse, than having no data at all.

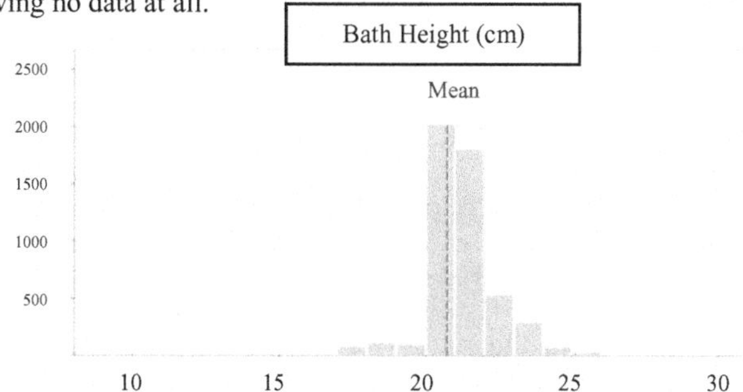

Figure 8. Manually collected bath height (cm) data. Action is required if the bath height is less than 20cm

Using averages without a historical context.

To extract full value from the most recent data, it must be analysed within its historical context. Just looking at the most recent averaged data gives no insight from its historical context. A Green Carbon plant has two Ball Mills for fine coke production. Blaine Index (BI) of the Ball Mill product is used to monitor and control Mill performance. The fine coke quality from the Mills is reported on a weekly basis as shown in Table I:

Table I. Weekly average and standard deviation values for fine coke products from Ball Mills 1 & 2.

Ball Mill 1- Blaine Index		Ball Mill 2 – Blaine Index	
Average	Standard Deviation	Average	Standard Deviation
3095	116	3102	117

The data provided in Table I suggest that the performance of the Mills is very similar, both in terms of the average BI and variation in BI over the week as measured by the standard deviation of the individual values. The conclusions are a lot different, however if the BI data are looked at in their historical context.

An Individual and Moving Range (IMR) control chart of the BI values from Mill 1 is shown in Figure 9. This shows the data to be randomly spread around the average and that under normal conditions we can expect BI values to be within the range 2746 to 3444. This is important as it enables us to predict the future performance of Mill 1 – unless a change is made to the operation of the Mill, we can expect that it will continue to produce fine coke with these properties. It is a relatively stable process with nearly all common cause variation.

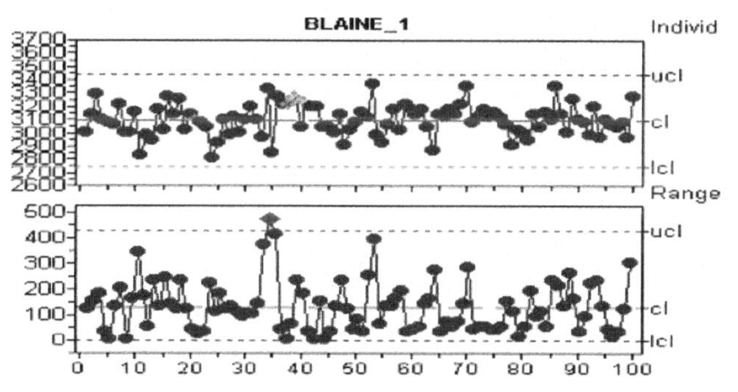

Figure 9. IMR Control chart of Blaine Index values of the fine product from Ball Mill 1 over a week.

If we now look at the performance of Mill 2 within its historical context, we can see different patterns in the Mill product data (See Figure 10.). In this case the BI values are not randomly spread around the average – as a result of instabilities in the Mill operation, there is a distinct pattern to the data with steps in BI. These steps will result in jumps up and down in aggregate pitch demand and variation in green anode quality. As expected with this, the anodes produced with fine coke from Mill 2 would be more variable than those produced from Mill 1 fine coke. This would not be suspected by just looking at the BI averages and standard deviations of the fine products from the two Ball Mills shown in Table II. Predicting the future performance of Mill 2 is more difficult than Mill 1 as the process not stable.

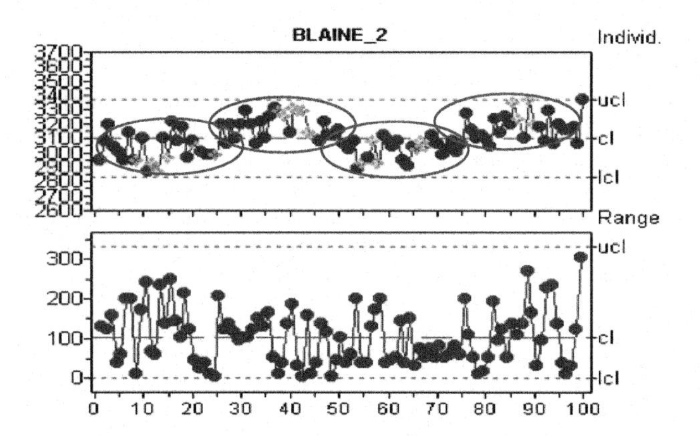

Figure 10. IMR Control chart of BI values of the fine product from Ball Mill 2 over a week. An unstable pattern of steps is evident in the data.

Avoiding the inappropriate use of averages – summary

- Using averages & standard deviations to describe process or product performance is useful however; it requires that the individual values that make up those statistics arise from a single, homogenous population (a stable process), which can only be determined through analyzing data in its historical context.
- Evaluate histograms and control charts plotted from the original data. If the volume of data involved prevents this (e.g. Individual green anode data) use appropriate averages with a measure of variation (standard deviation) and use a histogram to look "inside" the averages.
- Use the appropriate control chart such as average & standard deviation or average & range. Always determine and report conformance to specification limits and other criteria through proper statistical methods for capability analysis and using individual data points, never with averages.

Impact of sampling methods on the usefulness of data.

This data trap is commonly encountered when evaluating anode core property data, and this example will be expanded on. (The whole topic of the validity of anode core sampling strategies is worthy of a more detailed discussion than can be undertaken in this paper. For this reason it will only be covered briefly here, but will be the subject of a future publication.)

Data sampling strategies are determined by the purpose of the data being collected, i.e. what is the question you want to answer? This question is the starting point for designing a sampling strategy – as opposed to simply collecting and analyzing data and hoping it will answer your questions. You should not use data to answer questions for which the sampling strategy was not designed.

In anode core sampling, a question that is rarely answered well is "What is the purpose of sampling and testing cores taken from baked anodes?" There are at least two different answers:

1. To monitor and characterize the quality of anodes sent to Potrooms. In reality, the ability to provide useful data for this purpose is very limited – the turnaround time for core drilling, sample preparation, and testing is too slow compared with the in-cell life of an anode and the frequency of sampling is normally much lower than required to detect an important change in anode properties in a useful timeframe. Anode core properties sampled using typical strategies do not give a good representation of the overall anode quality sent to the Customer. At best these data may help with long-term trend analysis.
2. To monitor on-going process behavior in the Green Carbon plant – to study the impact of "upstream" process changes (such as raw material quality, green anode production conditions) on anode core properties.

To answer question 1, i.e. "What is the quality of the anodes sent to Potrooms?" the sampling strategy must capture all of the variation in green anode properties and all the variation due to the baking process (e.g. section, pit, layer, location in layer, pit condition, "straight" sections, cross-over sections, etc) .

To answer question 2, the sampling plan in the baking furnaces should be restricted to taking core samples from one location within the same pit in selected sections. This helps to "block" out the impact of some of the variation in the baking process.

If the sampling strategy for question 2 is adopted and you try to use the resulting data to give answer 1, the amount of variation in the anode core property results will be understated, masking potential sources of problems.

Alternatively, if the question 1 sampling strategy is adopted, and the results are used to try to give answer 2, the amount of variation in anode core data will be larger, making it difficult to detect changes in anode quality. The effect of upstream process changes are masked by, or confounded with, the variation coming from the baking process.

In either case, using data to answer questions for which the sampling strategy was not designed will result in a poor basis for decision-making and misleading results. This issue is largely ignored when anode core property data are used to try to address the myriad of questions that can be asked about it.

Conclusions

To avoid the data traps outlined in this paper and improve Carbon Plant operations through better use of data:

1. Adopt "on target with minimum variation" for process management instead of the binary "control to specification".
2. Avoid using averages for process analysis, decision making and reporting. Wherever possible use a graphical presentation of the individual data points instead.
3. Only use data to answer the questions that the sampling strategy was designed to answer.

References:

1. K.A. Sinclair, and B.A. Sadler, "Going Beyond SPC – why we need statistical thinking in operations such as Carbon Plants", *Light Metals 2004, TMS (The Minerals, Metals & Materials Society)*, 597-572.

2. K.A. Sinclair, R. Phelps, and B.A. Sadler, "The Integration of Lean and Six Sigma – A powerful improvement strategy for Carbon Plants," *Light Metals 2005, TMS (The Minerals, Metals & Materials Society),* 641-646.

3. R&D Carbon Ltd, *Anodes for the Aluminium Industry*, (R&D Carbon Ltd, Sierre, Switzerland, 1st Ed. 1995), 342.

4. D.J. Wheeler, *Understanding Variation – the key to Managing Chaos.* (SPC Press, Knoxville, Tenn., 1993)

EXPERIENCE REPORT – AOSTAR ALUMINIUM CO LTD, CHINA ANODE PASTE PREPARATION BY MEANS OF A CONTINUOUSLY OPERATED INTENSIVE MIXING CASCADE (IMC)

Berthold Hohl[1], You Lai Wang[2]

[1]Maschinenfabrik Gustav Eirich GmbH & Co KG, Walldürner Str. 50, 74736 Hardheim, Germany
[2]Aostar Aluminium Company Ltd, Meishan, Sichuan Province, China

Keywords: Anode paste preparation, Mixer, Cooler, IMC

Abstract

In 2002, Sichuan Aostar Aluminium started the construction of a 250,000 t/y greenfield smelter including a 190,000 t/y anode plant in Meishan, Sichuan Province, Central China. The smelter uses GAMI 300 kA electrolysis cells. The affiliated anode plant was put into operation in February 2005. It is based on the newly developed Intensive Mixing Cascade (IMC) technology for anode paste preparation combined with vacuum vibrocompaction for anode forming. Design figures and current results of the plant operation are given. Typical features like self-adaptation to low raw material qualities and fluctuations of properties are described and discussed.

Based on the experiences gathered at Aostar, the IMC process does not only significantly reduce both the investment (capex) and operating expenditure (opex) of an anode paste plant but also allows for the production of more than 60 t/h of anode paste in one single line. This perfectly fulfils the future requirements of the primary aluminum smelting industry.

Introduction

The standards to be met by a modern anode plant have drastically changed over the last years. Substantially increased throughput rates along with a considerable cost pressure regarding investment (capex) and operating expenditure (opex) require new technological solutions. As a consequence, EIRICH developed together with Aluminium Pechiney, Solios Carbone and other well-known partners from the primary smelting industry the continuously operated Intensive Mixing Cascade (IMC). Using this technology, anode paste preparation is performed exclusively with intensive mixers; conventional kneaders become unnecessary.

Starting with trials at Alusuisse Chippis in 1993 [1], a pilot plant at Hydro Sunndal [2], and the 12 t/h industrial scale production unit at Alucam Pechiney works in Cameroon [4], the breakthrough was achieved in 2002/2003 with the first order from Aostar Aluminium [5], followed by two other contracts from Chinese customers: Hunan Chuangyuan, and Chalco Qinghai.

Aostar Greenfield Project

Aostar Aluminium is a shareholding company with Sichuan Province Electric Power as the main shareholder. The smelter is expected to be completed with 250,000 t/y of metal capacity and

190,000 t/y of anode capacity. The first potline with a metal capacity of 125,000 t/y was put into operation in 2003. With the completion of the anode plant in February 2005, the total investment up to now amounted to only 1.8 billion RMB finally = approx. 180 Mio EUR (220 Mio USD) – 17 % of the total investment were spent for environmental protection. Before the on-site anode plant was started, anodes purchased on the Chinese market were used.

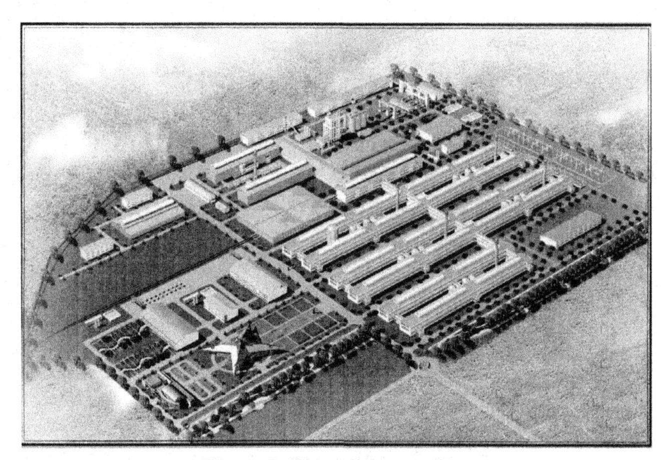

Figure 1. The Aostar smelter

The anode plant with a capacity of 190,000 t/y is not only covering the requirements of own need but also has still 30,000 - 40,000 t/y of capacity for sale on the market. Solios Carbone was awarded the contract for the complete IMC paste plant, including the coke preheating screw, the EIRICH equipment, the pitch fume treatment system and the process control. Aluminium Pechiney, today part of Alcan, acted as a process advisor for the new IMC technology. The two vacuum vibrocompactors have been supplied separately by Outokumpu (KHD). It took only three days to produce the first anode and only six weeks to achieve stable production. Coke and pitch, both of regular quality, are supplied by 2 - 3 suppliers, all based in North-East China.

The anode size is L x W x H = 1550 x 600 x 620 mm. Green anode weight: 950 kg. Two slots are made during vibrocompaction. The anodes, each having 4 studs, are assembled as double anode. In the potline, the anodes are operated at a current density of 0.733 A/cm^2. There are 40 anodes in each cell.

Figure 2. The Aostar anode assembly

Paste Plant Design and Layout

IMC plants are characterized by a tower design compared to long flat platforms in conventional paste plants.

According to the final anode requirement of 160,000 t/y, the paste plant has been designed for 35 t/h of normal and 40 t/h of maximum capacity.

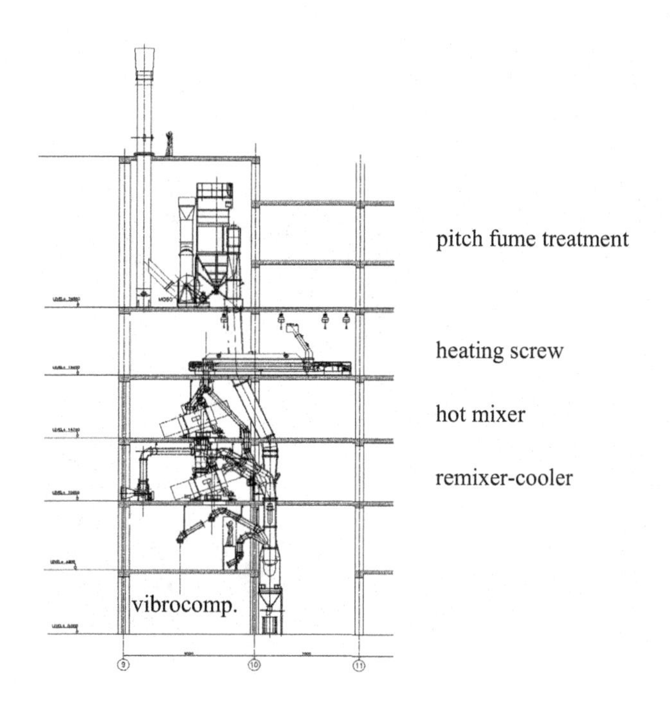

Figure 3. The Aostar paste plant layout

Figure 4. The Aostar paste plant building

Figure 5. Intensive hot mixer EIRICH RV 24

Process and Operation Data

During the first 9 months of regular operation, the main process parameters remained more or less unchanged due to the fact that from the very beginning the anode quality was significantly better than that of the locally purchased "start-up" anodes. The consequence is a significant potential for process optimization, which could be beneficial in future. Main issues of optimization could be to raise the liquid pitch temperature to a minimum of 200 °C and to increase both the retention time and the specific mixing energy of the IMC. The main rotor drive of the hot mixer and the remixer-cooler will then be 200 kW instead of currently 160 kW. The increased mixing intensity should offer opportunities to further improve anode quality and reduce pitch content.

Table I. Process and operation data

Main process parameters		Design	Design	Current operation	Benchmark test	After future optimization
		Operating range	Nominal			
Paste flow rate	**t/h**	**up to 40**	**35**	**35**	**37**	**35**
Dry aggregate flow rate	t/h (%)	up to 35 (...)	30 (86)	30 (86)	31.8 (86)	30 (86)
Dry aggregate inlet temp.	°C	5 to 35	20	20 - 30	32	
Dry aggregate outlet temp.	°C	170 - 180	175	185	186	185
Pitch flow rate	t/h (%)			5 (14)	5.2 (14)	5 (14)
Pitch temp. inlet hot mixer	°C	180 - 190	185	183	186	210
Paste temp. outlet hot mixer	°C	165 - 175	170	183	.....	195
Retention time hot mixer	min	> 4	> 4.5	4	.....	4.5
Paste temp. outlet remixer-cooler	°C	140 - 155	145 +/-2	165 +/- 2	170	......
Retention time remixer-cooler	min	> 4	> 4.5	3.9	....	4.5
Retention time IMC	min	> 8	> 9	7.9	7.6	9
Spec. mixing energy IMC	kWh/t	> 8	> 9	8.29	7.98	10.7
Residual water content	%	< 0.2	< 0.2	< 0.2		< 0.2

From February until July 2005 approx. 40,000 t of green anodes have been produced. The reject rate was 2 % compared to Aostar's target of 3 %. Both mixers are showing a very positive wear pattern. Only one pair of rotor blades per machine has been replaced up to now. Due to the fact that no re-welding of parts takes place inside the machines, the standstill during replacement lasts only 4 to 6 hours! A preventive visual inspection of each mixer is made every four weeks.

Anode Quality

When making an analysis of Aostar's anode quality, it has to be taken into consideration that uncontrolled fluctuations of raw material quality and composition can have a significant influence. Nevertheless, from the very beginning, the IMC-made anodes were showing a better quality than those being purchased on the Chinese market. One of the reasons for that is the long retention time in the IMC, which provides a self-compensating effect. In fact, the new process has a high tolerance versus many kinds of fluctuations. The anode reject rate is only 2 % compared to Aostar's target of 3 %.

Table II. Typical anode properties

		1	2	3
Average green density	kg/dm³	1.61	n/a	n/a
Baked anode density	kg/dm³	1.57	1.56	1.50
Resistivity	μΩm	< 55	59	55

1 = Aostar anode – IMC preparation
2 = Anode purchased on the Chinese market
3 = Chinese industrial standard "Grade 1"

Summary And Conclusions

Initially Aostar selected the new process due to significant savings in the initial investment (capex). Looking on the paste preparation itself, the cost reduction compared to a classical system finally was 19 Mio RMB (approx. 1.9 Mio EUR). Aostar's expectations concerning the anode quality were more than fulfilled from the very beginning. Moreover, the plant still has a significant potential for optimization. Thanks to the highly efficient fume treatment center, Aostar's anode plant figures on a very high rank of environment, health and safety (EHS) not only in China but on an international basis as well. The merge of expertise of three leading companies in the field of anode manufacturing - Solios Carbone, Pechiney and EIRICH - together with Aostar's willingness to implement innovative technologies led to the technical and commercial success of this project.

Two more IMC paste plants are currently under construction in China. As a response to the forthcoming requirements of high-capacity green anode plants in connection with primary aluminum smelters at capacities of more than 600,000 t/y, EIRICH made the IMC equipment available for up to more than 60 t/h in one single line.

References

1. B. Hohl, "New technology for continuous preparation of anode paste," *Light Metals,* 1994, 719-722.

2. F. Aune and P. Stokka, "Production of green anode paste for the aluminium industry" (Paper presented at the Slovak-Norwegian Symposium on Aluminium Smelting Technology, Stará Lesná, Ziar nad Hronom, Slovakia, 21-23 September 1999).

3. J.C.. Thomas, "New concept for a modern paste plant" (Paper presented at the 7th Australasian Aluminium Smelting Technology Conference and Workshops, Melbourne, Australia, 11-16 November 2001).

4. C. Dreyer et. al., "Reconstruction of the mixing line for anode paste production at Alucam," *Light Metals*, 1998, 705-710.

5. B. Hohl, A. Pinoncely, and J.C. Thomas, "Anode paste preparation by means of a continuously operated Intensive Mixing Cascade (IMC) at Aostar Qiminxing Aluminium China," *Light Metals*, 2004.

Light Metals 2006 *Edited by Travis J. Galloway* ***TMS (The Minerals, Metals & Materials Society), 2006***

A STUDY ON QUALITY ASSESSMENT OF ETP COPPER ANODE BARS USED IN ALUMINIUM REDUCTION PROCESS TO IMPROVE THEIR LIFE CYCLE TIME

Y. V. Ramana and Rajnish Kumar
Central Laboratory-R & D, Hindalco Industries Ltd, Renukoot – 231 217, India

Keywords: Anode, Electrolysis of aluminium, Anode rodding, Copper anode bar, Hot shortness, Solid solution, Electrical conductivity.

Abstract

Hindalco consumes about 1600 MTPA of electrolytic tough pitch (ETP) grade copper[1] (Cu+Ag 99.90% min.) for production of 350,000 MTPA of aluminium metal. To reduce the consumption patterns several possible control measures were studied. The present paper deals with the study on quality parameters. Copper bars, in the aluminium reduction process are continuously exposed to different physico-chemical environments like flue gases emanating from pots, load bearing, exposure to high temperature, hot and cold working. To meet the demanding requirement of the complex environment, a comprehensive testing of physical, chemical, mechanical and metallographic properties has been introduced and narrowed down the impurities to trace levels. These measures have resulted in improving the life cycle time of the copper bar to a significant extent. The results have established significant relationships between different physico-chemical and mechanical properties, the patterns of different impurities, especially oxygen and their role in controlling the life cycle time.

Introduction

At Hindalco, aluminium is extracted from alumina (Al_2O_3) by the Hall-Heroult process. The alumina itself is separated from bauxite, a natural ore, by the Bayer process. In the Hall-Heroult process, alumina is dissolved in an electrolyte based on cryolite (Na_3AlF_6) kept at 950-970^{0}C. The process takes place in electrolytic cells (Figure-1), where carbon cathodes (negative electrodes) form the bottom of the pot and anodes (positive electrodes) are held at the top of the cell.

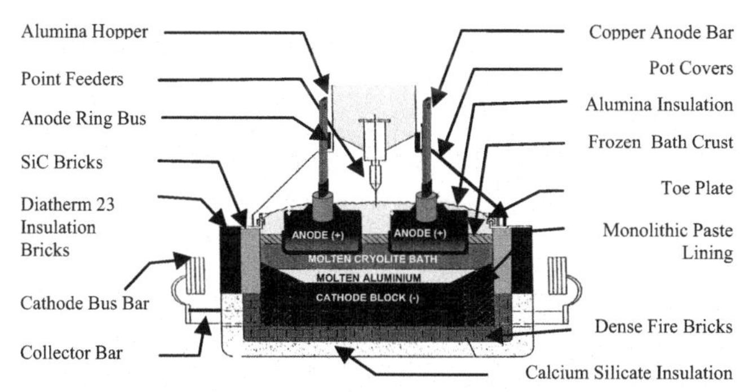

Figure-1: A cross-sectional view of an electrolysis cell used for aluminium reduction.

When direct current is passed, alumina is decomposed into molten aluminium and oxygen. The molten aluminium metal deposited at the cathode is siphoned off periodically, and oxygen, which reacts with carbon anode to form CO_2 consumes the carbon anode.

Hindalco uses pre-baked anodes (of dimension about 40cm w x 67cm b x 56.5cm l) which are formed by blending sized petroleum coke aggregate, crushed spent anodes, and coal tar pitch, molding this into blocks with preformed electrical connection sockets by pressing and firing to about 1100^{0}C in oil fired furnaces. Hindalco anode rodding uses ETP copper bars (of dimension 38mm x 60mm x 1586mm) to connect the anode block (Figure-2) that hold the anode assembly and feed electric current to the cells. The lifetime of one carbon anode is about 600 hours and used anode butts are removed from the cell for replacement. The anode bars are cleaned and reused after replacement of anode butts with fresh carbon anode blocks in subsequent cycle.

Figure-2: Anode rodding by pouring molten cast iron in the carbon anode blocks.

During the operational experience and as indicated in the recent communication[2], it was observed that there is an increase in copper metal consumption after plant expansion and imbibing new technologies resulting in increased cost of production. This led to review of different factors responsible for the increase. Apart from the operational practices, a major thrust area identified was quality of the copper bars since copper bar accounts for about 40 % of the anode making cost.

As the copper metal used in the process earlier was of ETP grade, all the quality norms of the process were driven by electrical conductivity values of the bars sampled as per the established procedures. The other parameters like metallic impurities and physical properties were given second priority except copper content determined through electrolysis. Moreover, different sources were entertained based on the above criteria adding further confusion to the problem. In spite of meeting the quality criteria by suppliers, the failure rate showed continuously increasing trend. Further, frequency of failure in most of the cases observed was at the area just above stiffener where the metal is subjected to different thermo-mechanical stresses, exposure to elevated temperatures and also to hot corrosive gases emanating from reduction process of the pot.

In view of the above, it was envisaged that apart from electrical conductivity for judging the quality, the other metallic impurities particularly low melting elements and physical properties like tensile strength, hardness and ductility should also be given due consideration to compensate the quality criteria. From a detailed analysis, it was noticed that there are different metallic impurities especially of low melting elements like tin, lead and zinc, presence of poisoning elements like bismuth, arsenic, phosphorus etc. in significant concentrations in the metal. The very interesting feature observed is the presence of oxygen both in soluble and insoluble forms in the received copper (Table-1). In the present study, the role of these individual parameters, their interrelationships and their effect in causing failure is discussed.

Methods of Analysis

Tensile testing of the metal is done using Tinius Olsen Machine following the test method of ASTM: B557M-1994. Hardness is measured using Rockwell Hardness Machine following ASTM: E18-02 test method. Electrical conductivity is measured with Sigma Tester EC 2.068 following the test method described in the supplier's manual. Spectrochemical analysis of copper metal samples is done using an Optical Emission Spectrometer of ARL-4460 calibrated using CRMs of either Brammer or MBH following the methods described in the operator's manuals of the system. Metallographic analysis is conducted using Olympus microscope with Leco Image Processing system following ASTM E 112-96 and ASTM E 562 – 01 test methods for grain size and volume fraction determination.

Results and Discussion

During the aggravated and increased failure period, data on different physical and chemical parameters was taken as reference for studying the reasons of failure and the data is presented in Table-1.

Table-1

Chemical analysis of copper anode bars (in ppm)

Element	Sample No.						
	1	2	3	4	5	6	7
Ag	30	78	17	40	43	51	57.8
As	3.2	2.3	0.9	5.3	1.6	3.1	2.0
Bi	2.1	1.4	2	1.9	2.0	0.3	0.3
Cd	0.9	2.1	0.8	1.0	2.3	1.2	2.3
Fe	3.6	3.6	3.6	3.6	7.3	10.8	10.6
Mn	0.5	0.5	0.5	0.5	0.3	0.3	0.3
Ni	74	47	2.6	73	25	5.4	0.9
Pb	92	52	12	170	350	42	30
S	45	28	17	35	25	28	63
Sb	2.8	9	7	9.4	5.7	0.5	5.4
Sn	170	225	14	226	96	86.5	51.5
Zn	300	37	17	340	56	25	387
P	1.4	2.6	0.8	1.2	7.0	0.5	0.5
Se	0.7	0.9	0.4	0.8	0.5	1.0	1.0
Te	1.3	1.5	2.1	0.9	0.7	2.0	3.0
O	425	1290	1155	575	400	2352	2257

The above data on levels of different impurities and very high concentrations of oxygen prompted to redefine all the quality norms to minimize their effect on the life of the bars. As the concentration of some low melting elements like lead was high, it was felt to reduce them to the lowest available to verify and rule out the possible factors that cause the failure. New specifications are made looking into different available sources and their impurity levels as furnished in Table-2.

Table-2

Revised specification of receipt copper anode bars.

Elemental Impurities			
Element	Limit (ppm)	Element	Limit (ppm)
Ag	-	S	< 40
As	< 5	Sb	< 5
Bi	< 10	Sn	< 50
Cd	-	Zn	< 50
Fe	< 20	P	-
Mn	-	Se	-
Ni	-	Te	< 5
Pb	< 25	O	< 600
Copper + Silver		99.9 % min.	
Total impurities excluding silver and oxygen		300 ppm max.	
Physical and Mechanical Properties			
Hardness	RF 54 max in fully annealed condition		
Electrical Conductivity	100.5 %IACS min in fully annealed condition		
Grain Size	150 μm min.		
Freedom from Defects	Free from blisters, slivers, scale, cracks, voids, microstructural segregation and other defects.		

While making new specifications, instead of following only ETP grade copper standard specifications, other things like reducing the levels metallic impurities particularly of low melting elements to near to zero concentrations have been made. Moreover, the metallographic observations like Cu-Cu_2O eutectic, inclusions and grain size have also been incorporated additionally keeping in view of the complex application environment of the copper bar.

After adapting the new specifications and adhering strictly to the revised quality norms as mentioned in Table-2, the rejection patterns are continuously monitored to see their effect on failure and their relation with respective consumption pattern. Figure-3 shows the pattern of the copper bar rejection due to failure plotted for the last two years (2004 and 2005).

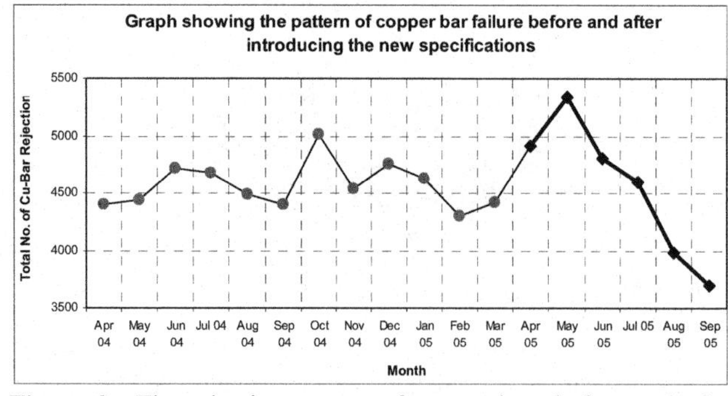

Figure-3:. The rejection pattern of copper bars before and after the new specifications. Bold line from Apr.-05 to Sept.-05 showing decreasing trend in rejection.

The specifications were implemented for the intake copper from April, 2005 and the actual effect started showing only from June, 2005 as only about 10 % of the bars are replaced every month. The peak rejection of 5345 bars reached during May, 05 and then onwards a continuous and progressive downward trend was followed. The trend line also showed the record lowest number of rejection (3500) so far during September, 2005 with scope for further decrease. As there was no change in operational practices during the period of study and only change is in the specifications of the copper used, the decreasing trend is attributed mainly to the change in the quality norms, particularly, the elemental impurities and other deformities in the receipt bars.

The impurities, though present in trace concentrations in copper metal, greatly affect the physical properties like conductivity, strength and ductility. These impurities behave in two different ways depending on their mode of occurrence, i.e., the elements dissolved in solid solution and the elements that are insoluble in solid copper[3].

Data on spectrochemical, physical and metallurgical observations during the period of study was processed to establish the relationships. The data obtained showed interesting trends on the effect of individual parameter on the quality of copper bar. The role and significance of individual parameter studied in controlling the consumption rate is discussed under the above mentioned two classes:

1. Elements that are dissolved in the copper in solid solution:

The elements of this category are distributed almost uniformly throughout the metal by modifying the crystals of pure copper into crystals of solid solution. The most important impurities categorized under this class normally are **silver, arsenic, antimony, cadmium, nickel, iron, phosphorus, aluminium, silicon, tin, and zinc.** These elements usually will have a bad effect in lowering electrical conductivity, but have a beneficial effect on hardness and strength. In the present study Ag, As, Sb, Cd, Ni, Fe, P, Sn and Zn are studied for their effect.

All elements studied under this category showed a positive relationship with tensile strength, and As & Sb (Figures-4 & 5) showed the maximum affinity towards this property. Similarly, except As and Cd all other elements showed increasing tendency towards hardness with increasing levels of their concentration in the copper metal. The established trend of increase could not be observed in the above case may be due to very low concentration (< 4 ppm As and < 8ppm Cd). Ductility has shown negative tendency with all the elements and in case of P and Cd it was insignificant. Electrical conductivity has shown only a negative relation ship with all the elements studied.

2. Elements that are insoluble in solid copper:

These elements separate as the metal solidifies between the crystals of copper. These crystals of pure copper will not greatly reduce the electrical conductivity in comparison to elements in solid solution, but the elements separated along the grain boundary may give scope for weakening of metal strength. The chief impurities studied in this class are **lead, bismuth, selenium, tellurium and sulphur,** the last three being combined as Cu_2Se, Cu_2Te and Cu_2S and also as their oxides. All the parameters studied under this category showed a positive relation with tensile strength and negative relation with ductility (Figures-6 to 9). The impurities of this category such as lead and bismuth are practically insoluble in copper, and form easily fusible eutectics with it[4]. These eutectics precipitate along the grain boundaries and cause trouble when the copper is worked. Further, these two elements seriously impair hot workability by causing a form of embrittlement known as hot shortness. Cold workability is also reduced, particularly by bismuth, which readily segregates at grain boundaries[8].

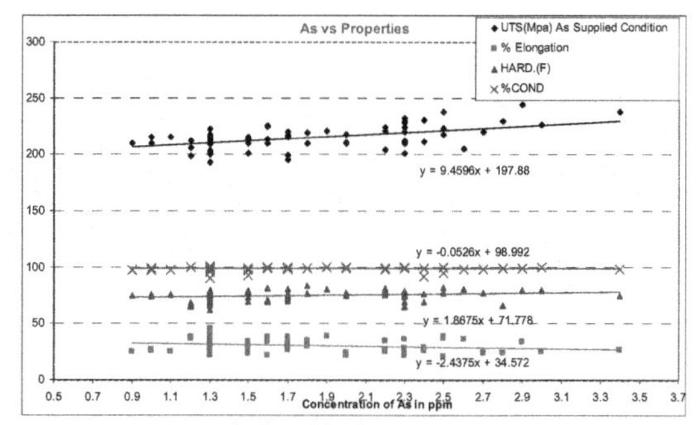

Figure-4: Relationship between As vs Properties.

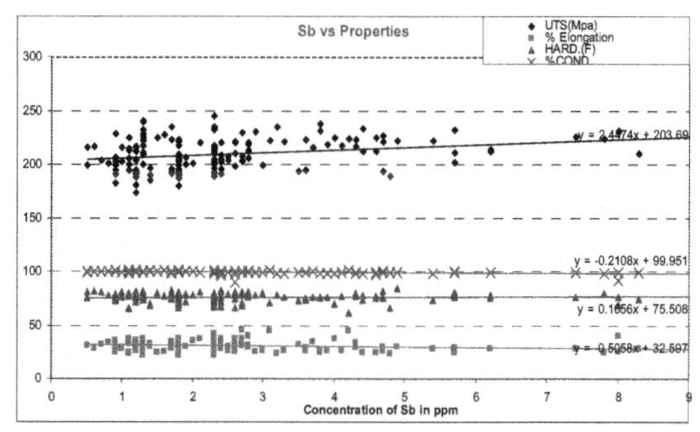

Figure-5: Relationship between Sb vs Properties.

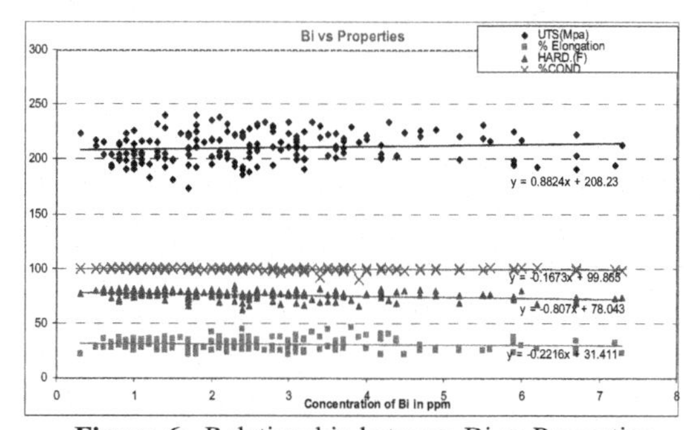

Figure-6: Relationship between Bi vs Properties.

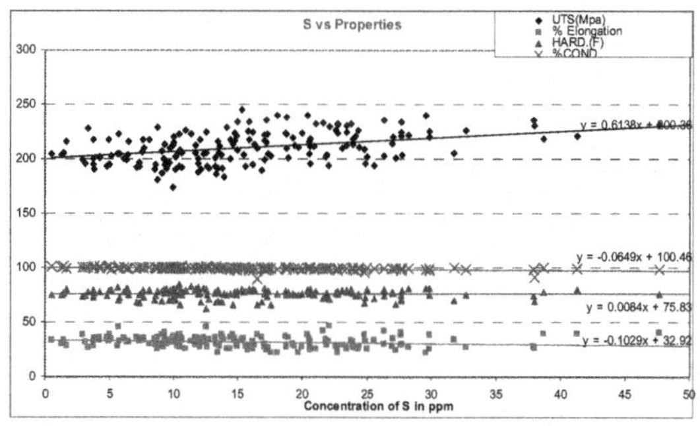

Figure-7: Relationship between S vs Properties.

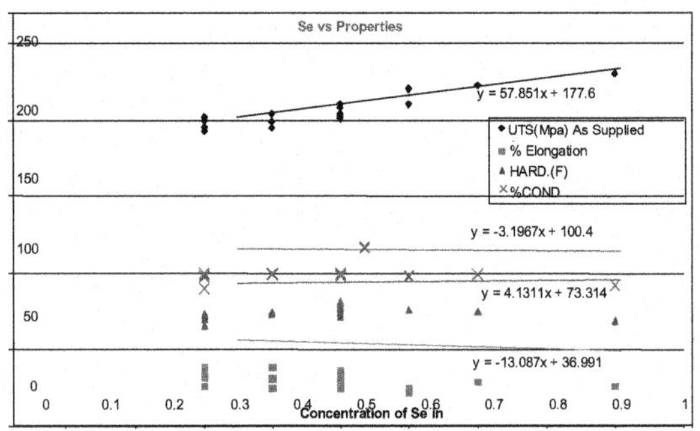

Figure-8: Relationship between Se vs Properties.

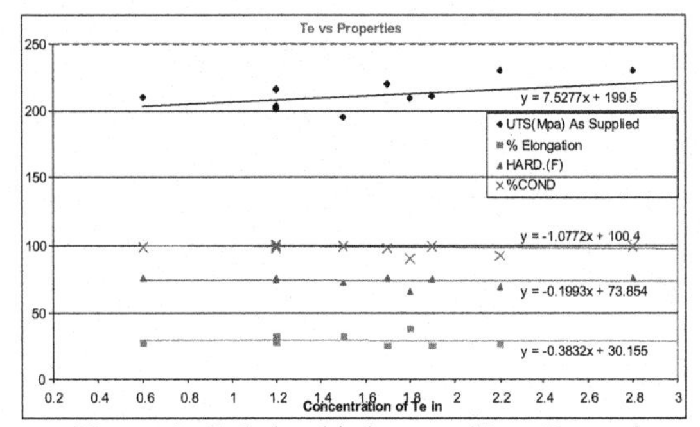

Figure-9: Relationship between Te vs Properties.

This kind of effects must have been predominant reasons of failure of copper anode bars in the present case, particularly by lead as it is found at very high concentrations before revising specifications. To support this view, the tensile property was compared with samples having high concentrations of lead at normal and elevated temperatures and found that the tensile properties are very significantly low at elevated temperature. The stress-strain curve plotted for two samples of copper bar having 20 and 300 ppm of lead at elevated temperatures below supports this view (Figures-10a & 10b).

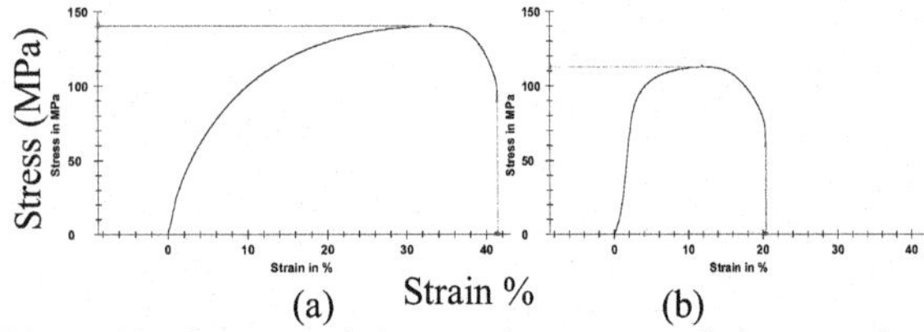

Figure-10: Stress-strain curve of copper anode bar samples with low concentration of lead (a) and high concentration of lead (b) at 450° C temperature.

This indicates that the elements that segregate in the grain boundary help in early failure during hot loaded condition[2].

Sulphur, Selenium and Tellurium are considered together as being generally harmful to copper in affecting the mechanical properties[5]. Selenium and tellurium form brittle sulphides and oxides, and in the presence of these compounds copper seriously degrades in these properties.

All the above five elements showed a negative tendency towards electrical conductivity showing that with increasing concentrations of these elements, the electrical conductivity values decrease within the concentration ranges of these elements studied.

Oxygen is present in all commercial copper except in the deoxidized grades. In castings it exists as a Cu-Cu_2O eutectic and in mechanically worked copper in the form of small globules of Cu_2O. The eutectic consists of spheroidal Cu_2O in a matrix of copper-rich solid solution. Its effect on the mechanical properties is not great, slowly increasing the tensile strength and reducing the ductility as the amount of oxygen increases[5]. In small amounts oxygen actually increases the electrical conductivity of commercial copper, very likely by oxidizing other impurities and removing these from solid solution. Large amounts of oxygen, however, slowly reduce the conductivity by introducing undue amounts of Cu_2O, which reduce the effective cross section of the metal. Further, oxygen, present as Cu_2O in tough pitch and undeoxidised coppers, restricts grain growth at high annealing temperatures.

In the present study, the concentration of elemental oxygen was found from 200 ppm to 1300 ppm and in some samples as high as 3000 ppm also recorded. The Cu-Cu_2O eutectic was estimated from 1 % to more than 7% in the samples studied. The established measurements of eutectic of tough pitch copper containing from 0.03 to 0.05 per cent oxygen having as much as 12 per cent eutectic constituent in its make-up[6]. The deviations observed in the present case may be attributed to the interference of other oxides. The properties tensile strength and elongation were studied in as received condition and hardness and electrical conductivity both in as received and annealed condition. The relationship of oxygen with these parameters is presented in Figure-11. This figure shows that oxygen content has a positive relationship with tensile strength as stated earlier where as the other properties except electrical conductivity showed no such significant relation as the much of the data plotted is at low levels of oxygen.

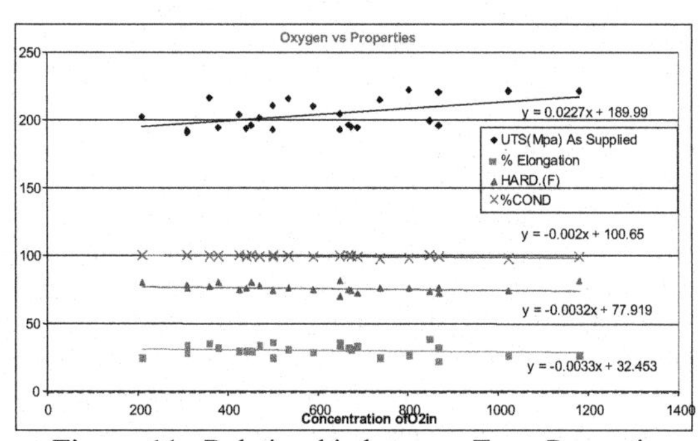

Figure-11: Relationship between Te vs Properties.

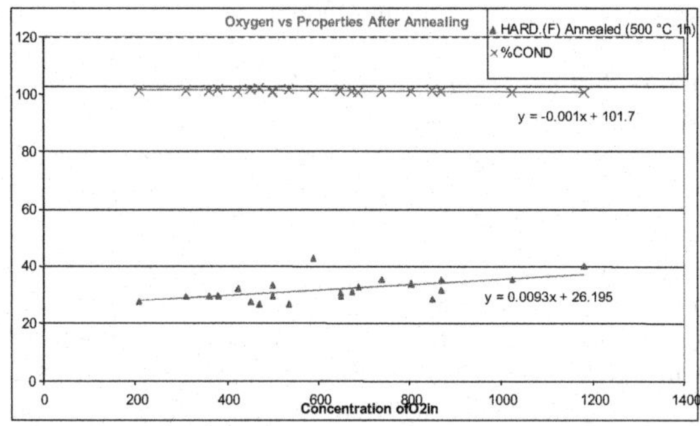

Figure-12: Relationship between Te vs Properties.

The Figure-12 shows the same properties after annealing for the same samples. A comparison of Figure-11 with Figure-12 shows that the anticipated decrease in hardness and increase in electrical conductivity are not proportionate after annealing as these properties are hampered with increasing concentrations of oxygen.

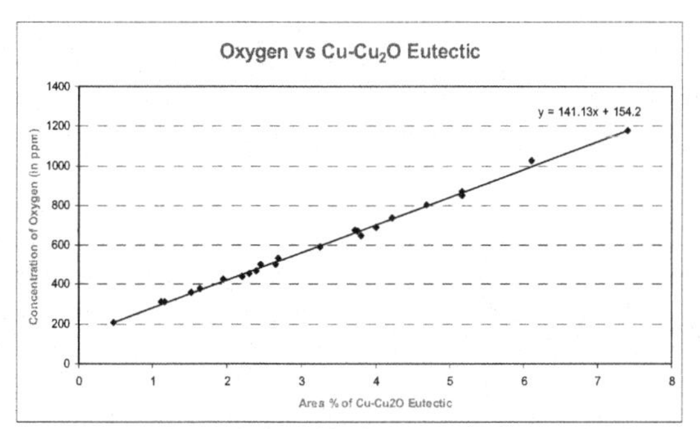

Figure-13: Relationship between O vs $Cu\text{-}Cu_2O$.

The Figure-13 shows that there is a strong relationship with oxygen concentration with area % of $Cu\text{-}Cu_2O$ eutectic confirming that the elemental oxygen levels are proportionate to the $Cu\text{-}Cu_2O$ eutectic formation. Electrical conductivity, though not very significant in the study range of oxygen (200 to 1200 ppm), showed a negative relationship indicating that higher the oxygen levels, there is scope for fall in electrical conductivity values.

Oxygen present in the copper is a source of injury to the metal when it is annealed in a reducing atmosphere as in the present anode environment where the ETP copper bars are exposed to elevated temperatures during the operation. Reducing gases diffuse into the metal and react with the globules of Cu_2O: hydrogen traces available in the anode operating environment generated from dissociation of moisture from freshly added alumina to the pot, for example will diffuse far more rapidly into the metal and water vapor is formed by the reaction[4, 9].

$$Cu_2O + H_2 \rightarrow 2\,Cu + H_2O \quad \text{------ (1)}$$

This is accompanied by an increase in volume. Considerable pressure may thus be developed and extensive cracks or grain boundary separation is produced in the metal[2, 8]. In the present study as there is ample scope for presence of traces of moisture[10], this kind of failure which is also called gassing[9] cannot be ruled out if oxygen is present in larger proportions in copper as observed before revising the specifications. A comparison of oxygen content in copper metal before and after revising specifications is presented in the Figures-14 and 15.

Metallographic observations: During the period of study, all the samples analysed for spectrochemical and physical properties were simultaneously subjected for close surface verification and metallographic examination at different lengths all along the bar. This study facilitated for identification of cracks, inclusions (Figure-16), voids (Figure-17), oxide accumulations, micro-segregation and also estimation of grain size. For example, segregation of $Cu\text{-}Cu_2O$ eutectic causes formation of cracks as shown in Figures-18 and 19. This kind of defects in the bar goes unnoticed if microstructural studies of the bar are not conducted and these defects can reduce the life of copper bar.

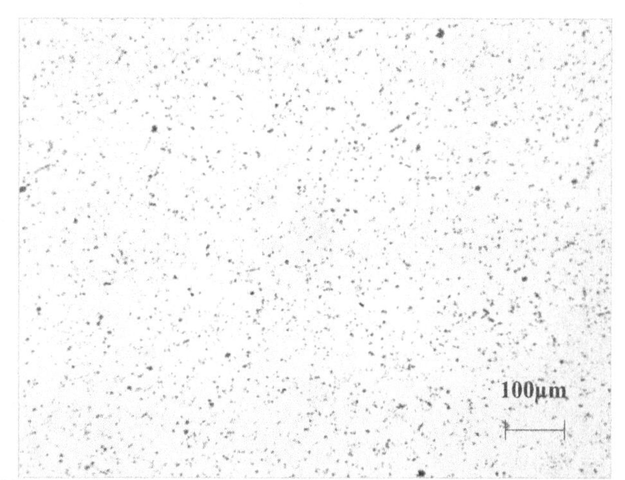

Figure-14: Distribution of $Cu\text{-}Cu_2O$ eutectic phases before revising the specification.

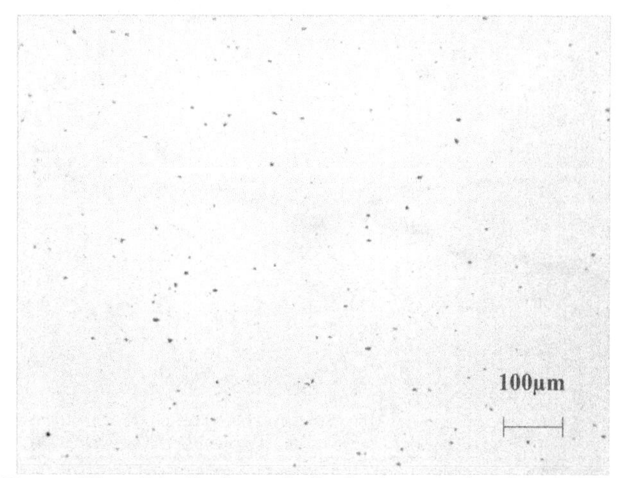

Figure-15: Distribution of $Cu\text{-}Cu_2O$ eutectic phases after revising the specification.

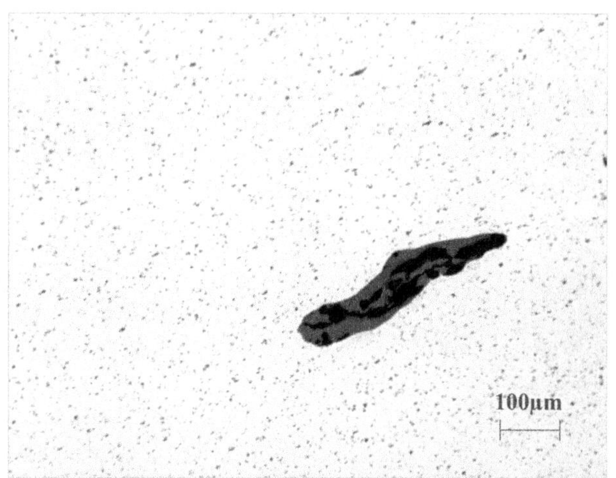

Figure-16: Oxide inclusion inside the bar.

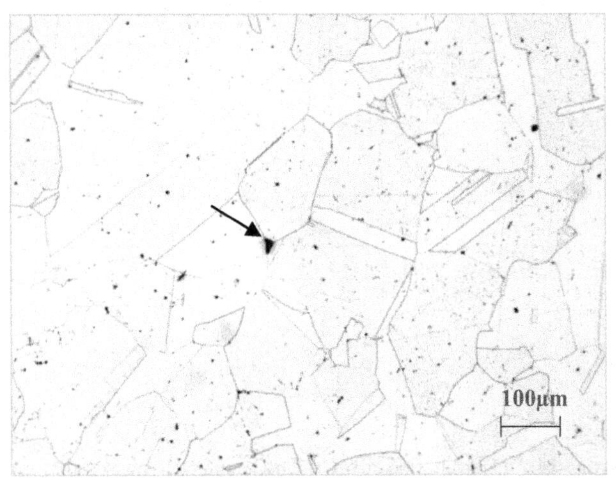

Figure-17: Grain boundary voids shown by arrow marks

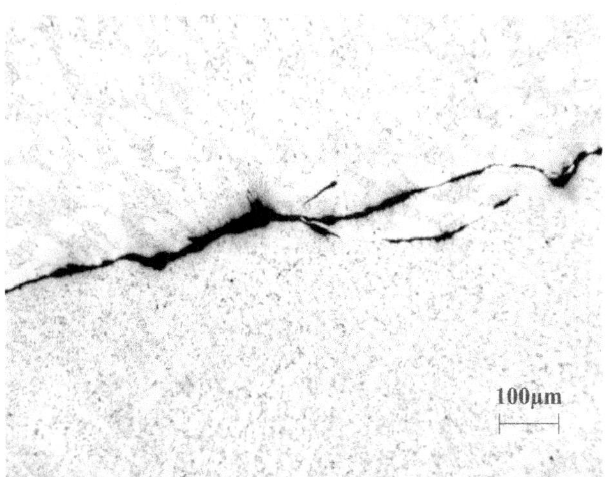

Figure-18: Segregation of $Cu\text{-}Cu_2O$ eutectic causing formation of crack.

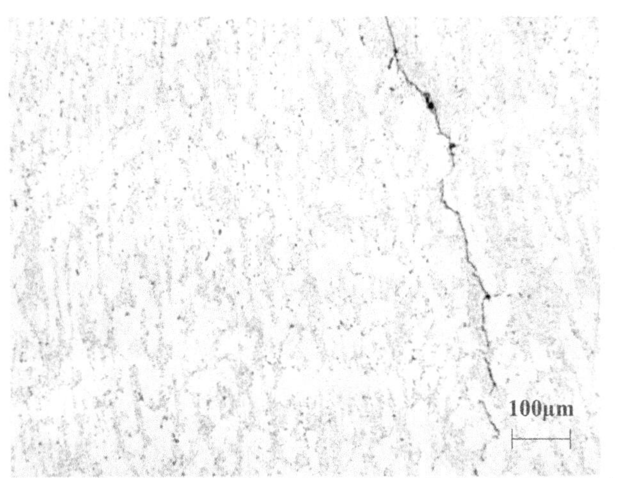

Figure-19. Interdendritic segregation of $Cu\text{-}Cu_2O$ eutectic causing intergranular crack formation.

Conclusions

Copper bars used in the alumina reduction process are subject to different physico-chemical processes and complex conditions of the pot environment. Hence, the normal specifications of ETP copper bars are not sufficient to have a satisfactory life cycle time.

Reducing the impurity levels of particularly low melting elements to near zero concentrations will have great effect on improving the life cycle time of copper bars.

Oxygen in copper bar has a very significant role in controlling the conductivity levels and causing failure of bars in reducing atmosphere by facilitating crack formation and propagation. Reducing the oxygen levels to lowest minimum helps to increase the life of the bar.

Comprehensive study of physical, mechanical, chemical and metallographic properties gives a better assessment than studying individual parameter for complex environmental application like alumina reduction process.

Acknowledgements

The authors acknowledge the management of Hindalco Industries Limited for the infrastructure and support provided for conducting this work.

References

1. Indian Standard - IS: 191 (Part V) – 1980, pages: 19-21
2. A Case Study on Life Cycle Analysis of Copper Anode Bars, by J. Mukhopadhyay, Y. V. Ramana, R. Kumar- Light Metals 2005, TMS Publication, pages 347-351.
3. Engineering Metallurgy, A textbook for users of metals. By Bradley Stoughton, Allison Butts and Ardrey M. Bounds, Mc Graw-Hill Book Company, 1953, Pages 243-327.
4. Engineering Physical Metallurgy and Heat Treatment. Yu M Lakhtin, Mir Publishers, Moscow, 1977, Pages: 401-424.
5. Non-Ferrous Production Metallurgy by The Late John L. Bray, John Wiley & Sons, Inc., New York, Second edition 1959, pages 121-244.
6. Nonferrous Physical Metallurgy by Robert J Raudebaugh, Pitman Publishing Corporation, Toronto, 1952, Pages 112-188.
7. Practical Microscopical Metallography by: Richard Henry Greaves & Harold Wrighton, 4th Edition. Chapman & Hall Ltd., London, 1960, Pages 151-157.
8. Copper, Its Trade, Manufacture, Use, and Environmental Status by Gunter Joseph, Edited by Konard J. A. Kundig, ASM International, 1999, Pages: 67 – 76.
9. Metallurgy for Engineers by E. C. Rollason, Edward Arnold (Publishers) Ltd, 2nd Edition, London, 1949, Pages 235-267.
10. Handbook of Aluminum, Vol. 2, Alloy Production and Materials Manufacturing by George E. Totten and D. Scott ManKenzie, Marcel Dekker, Inc. Publication 2003, pages 1-79.

CARBON TECHNOLOGY

Anode Baking

SESSION CHAIR

Trygve Foosnæs

Norwegian University of Science & Technology

Trondheim, Norway

THERMAL DILATION OF GREEN ANODES DURING BAKING

Juraj Chmelar[1], Trygve Foosnæs[2], Harald A. Øye[2]
[1] SINTEF Materials and Chemistry, NO-7465 Trondheim, Norway
[2] Norwegian University of Science and Technology, Department of Materials Science and Engineering, NO-7491 Trondheim Norway

Keywords: Petrol Coke, Thermal Dilation, Characterisation

Abstract

Pilot scale anodes were made using three single source and one blended coke to determine how the green anode properties affect the final baked anode. Testing was performed in an improved vertical dilatometer using samples 50 mm in diameter and 50 mm long to determine the effect of different heating rates. No sample support was used, however the effect of packing material was evaluated. The sample shrinkage was calculated from the dilatometer data as the difference between the expansion at 550 °C and 950 °C. Dilatometric data for anodes prepared identically and with the same composition show differences due to varying mechanical properties of the cokes.

Introduction

After forming, the anode is baked to temperatures in excess of 1100°C. During this thermal treatment the anode can be subjected to significant stress, mainly due to thermal expansion, shrinkage and inherent temperature gradients, which can affect critical properties of the anode. Understanding these physical and chemical changes is important for both the economic viability of the process and optimal process control.

Dilatometric equipment was developed to study the dimensional changes during baking of green anodes, showing that these changes are influenced by sample orientation parameters (both parallel and perpendicular to the direction of forming), type of filler, specimen dimensions and heating rate [1].

Two studies [2, 3] point out that grain size distribution, pitch content, heating rate and the use of packing mass have significant influence on the dimensional changes during baking.

Description of thermal dilation

Figure 1 presents an example of thermal dilation of a green core sample 50 mm in diameter and 50 mm long, containing 15 % pitch. No packing material was used to support the sample. The sample was heated at 60 °C/h to 950 °C, then held for three hours under nitrogen.

The initial expansion at about 100 °C is due to release of residual stresses from the forming and cooling process. The sample expands further (up to 0.55 % of its initial height) because of trapped volatiles. A further increase in the temperature leads to slumping caused by the plasticity of the sample. The sample expands slightly at about 300 °C due to trapped light binder volatiles. There is a sudden height increase at about 400 °C which most probably is a consequence of the residual pitch expulsion from the coke pores. The sample height is then reduced after the release of volatiles and at about 450 °C the transition from a plastic to a solid matrix starts. The major release of non-coking volatiles takes place up to about 600 °C. The post coking process starts at about 600 °C and continues until 900 °C, followed by the release of cracked volatiles (methane, hydrogen) and sample height reduction. [4]

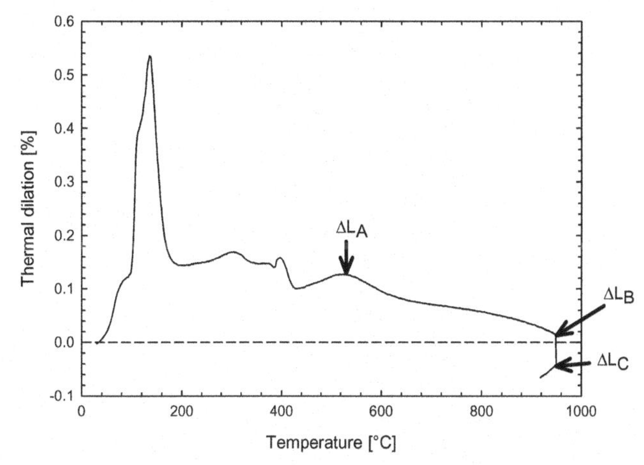

Figure 1. An example of thermal dilation measurement of a green sample versus temperature.

Standard method for determination of shrinkage during baking

The ISO 14428 standard describes the determination of expansion and shrinkage of ramming pastes during baking [5]. The standard describes the experimental procedure, equipment used for testing and how to calculate and interpret results.

In the current study the sample was packed in a mass of carbon particles (electro-calcined anthracite or graphite), then heated in nitrogen at 180 °C/h to 950 °C, where it was soaked for three hours.

The maximum thermal shrinkage before and after soaking at 950 °C were calculated from the following equations (Figure 1):

$$\Delta L_{before} = \Delta L_A - \Delta L_B \quad (1)$$

$$\Delta L_{after} = \Delta L_A - \Delta L_C \quad (2)$$

The distance measurements were corrected for the movement of the sample base by calibrating the equipment. However, the ISO 14428 method does not discuss the significant changes that occur before the sample reaches 450 °C.

Equipment

This investigation was performed in an improved vertical dilatometer, with a water cooled cylindrical furnace (Figure 2).

Two type S thermocouples were installed, one in the middle of the furnace to control the furnace temperature and the other from the bottom to measure the sample temperature. A core sample of 50 mm in diameter and 50 mm long was placed on a graphite washer inside the cylindrical quartz tube. A quartz push rod rested on top of the sample and followed the sample movements, which were then transferred to a linear transducer.

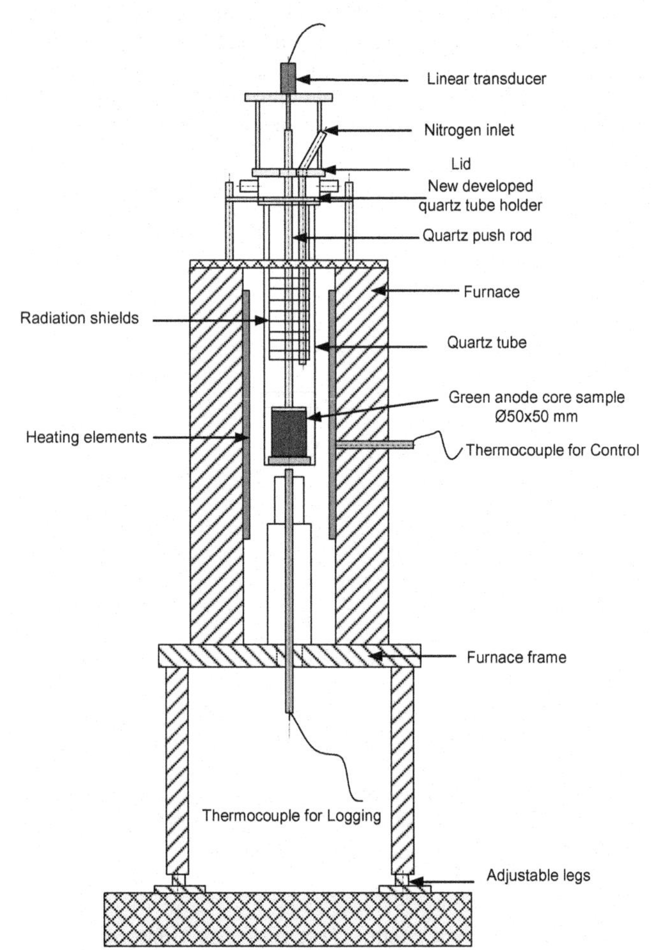

Figure 2. Schematic drawing of the dilatometer.

A quartz tube was suspended in the oven and kept in position with a holder suspended on four long bolts with nuts on the top of the oven. The top of the quartz tube was attached to a holder with two o-rings compressed between the quartz tube and the metal holder. This arrangement allowed the quartz tube to be positioned firmly in the furnace.

The new holder was additionally equipped with two ducts placed on opposites sides, but only one of them was used during the experiments for the outlet of the off-gasses. A rubber hose was installed on this duct and led to a vent pipe installed above the furnace. The under-pressure created by the rubber hose was sufficient to draw out the off-gasses, preventing condensation of volatiles on the push rod and linear transducer.

An enhanced nitrogen purging system was developed where a quartz tube of internal diameter 5 mm was mounted into the lid and through the radiation shields. The tube ended about 8 cm over the sample. This solution ensured nitrogen purging of the sample, prevented oxidation and also provided helpful circulation of the off-gasses from the green carbon sample. The nitrogen flow rate was approximately 40 l/h.

The furnace operation and thermal dilation measurement was controlled with LabView 7.1, a graphical programming language. The Human Machine Interface (HMI) allows the user to set the heating profile of the furnace as well as logging frequency, nitrogen purging and start/end of the water cooling. The sampling time can be defined between 0.5 seconds and 10 minutes.

Equipment calibration and dilation corrections

Prior to experimental testing the dilatometer was calibrated. The position of the quartz tube was adjusted so that the sample temperature was isothermal (± 2 °C). Subsequently, the quartz tube with the push rod was calibrated with a quartz reference sample (50 mm in diameter and 50 mm long; R&D Carbon).

Experimental and Results

Method optimisation

The aim of this investigation was to find an effect of the heating rate and the sampling position on the thermal dilation results. The shrinkage for each investigation is also presented.

The anodes were made from one coke type, one composition (coarse 17 %, medium 23.8 %, fines 34 % and dust 25.2 % with 63 % of -63 µm) and 15 % pitch content. The paste was mixed in a sigma mixer and formed in a vibration compactor. Core samples 160 mm in diameter and about 180 mm high were taken from the pilot scale anodes. Two core samples of 50 mm diameter were drilled from one green anode. The outer 15 mm of the samples was cut off in order to remove the outer inhomogeneous layer.

Influence of the heating rate

Four heating rates were investigated; 20, 60, 120 and 180 °C/h. The samples were free standing in the dilatometer and purged with nitrogen. The average for two samples is presented.

Figure 3 presents the thermal dilation curves for each heating rate. The maximum thermal expansion for the lowest heating rate (20 °C/h) was at about 120 °C, while the highest heating rate (180 °C/h) gives the highest maximum at 140 °C. The heating rates of 60 and 120 °C/h have a very similar curve profile. At the lower heating rates the dilation changes occur earlier than with the higher heating rates, except the initial expansion at a heating rate of 180 °C/h.

Calculated shrinkage ΔL_{before} and ΔL_{after} are presented in Figure 4. The highest shrinkage occurred for the heating rate of 180 °C/h, while the lowest shrinkage was found at 20 °C/h.

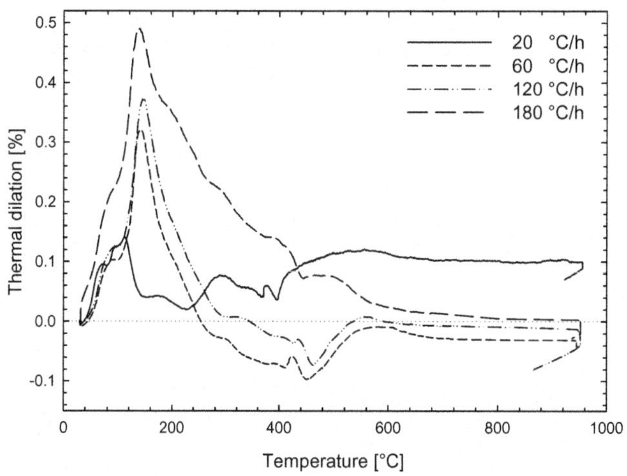

Figure 3. Thermal dilation of identical samples at different heating rates.

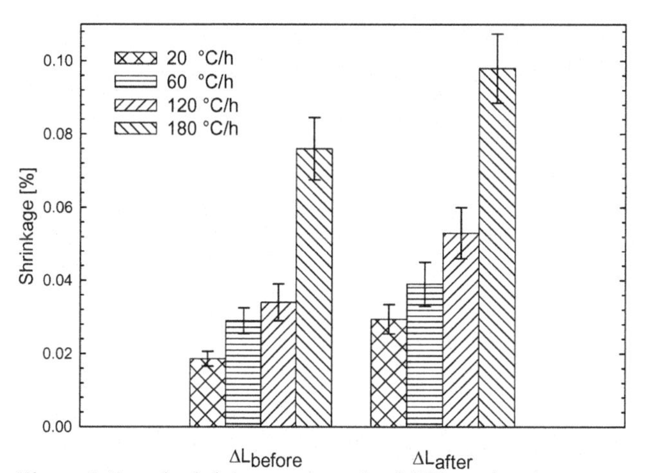

Figure 4. Sample shrinkage as a result of different heating rates.

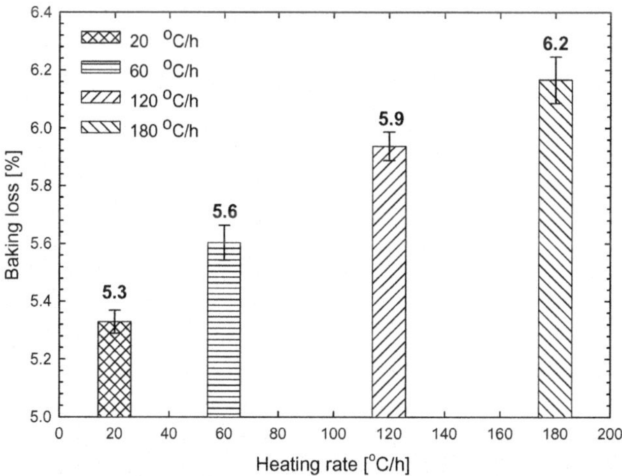

Figure 5. Baking loss of the samples at different heating rates.

The shrinkage results show how the heating rate can affect the final sample shrinkage and the tension in the samples thus created. This may have a critical effect on the final properties (air permeability, specific electrical resistance and Young's modulus) of the baked anodes. Measurements were only performed in the vertical direction (parallel to the vibration direction), thus there is limited information about the horizontal expansion/shrinkage of the sample.

Finally, the baking loss of the samples was measured and is presented in Figure 5. The baking loss increases with increasing heating rate.

Influence of the sampling position and packing mass

It was found that the sampling position in the anode was of critical importance for the behaviour of the sample during the thermal dilation investigation [1].

During forming the compaction force decreases with the sample depth, due to friction between the coke particles and between the paste and the mould. The force distribution corresponds to the thickness of the bridge layer of pitch between the particles. At the top of the sample the bridge layer will be thinnest while at the bottom the thickest. This can be seen after the forming process of the green anodes as a dark area at the top on the periphery due to pitch pressed out from the anode as a consequence of the pressure in this area.

Eight samples were tested in the dilatometer with a heating rate of 60 °C/h. Four samples were run with a graphite packing mass (graphite particles +0.5 mm to -1.0 mm) and the other four were free standing. Figure 6 shows the thermal dilation profile of the samples from the top and bottom positions. The drawn curves are the calculated average of two parallel measurements.

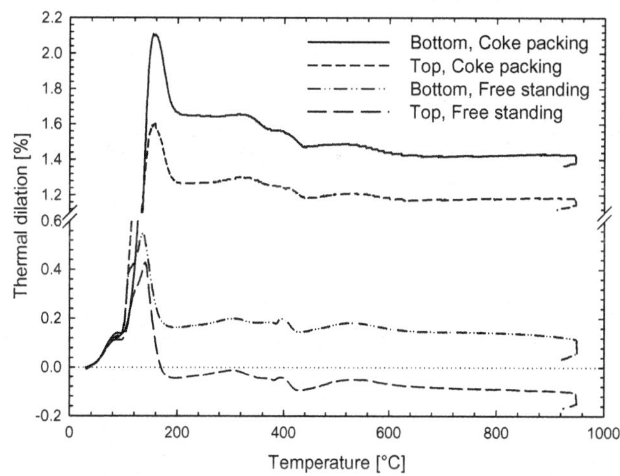

Figure 6. Thermal dilation of the graphite packed and free standing samples from the top and bottom of the green anode.

There is a significant difference in the thermal expansion when the sample is freely standing or packed. Much higher expansions are obtained when the sample is packed as slumping is minimized.

The bottom samples reached a higher maximum expansion than the top samples. The higher expansion maximum was caused by a higher pitch concentration between the particles in the bottom part of the green anode. The bottom samples of the anode have correspondingly slightly higher shrinkage than the samples from the top (Figure 7). Also, there is a difference between free standing and packed samples. The packed samples have lower values for both ΔL_{before} and ΔL_{after} due to the radial thrust on the sample from the packing mass.

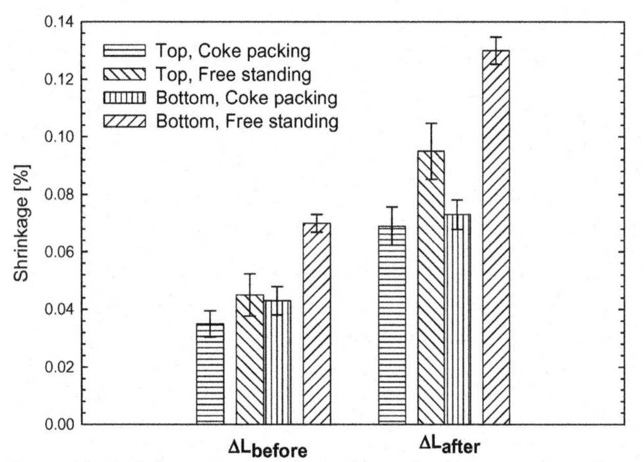

Figure 7. Shrinkage for samples with and without packing from top and bottom position.

Testing of three single source cokes

Anode samples from three single source cokes marked as SSA, SSB and SSC were investigated in the dilatometer. Table 1 presents the coke specifications. The cokes differ in sulphur content where SSC has the highest amount (2.89 wt%), SSA (1.09 wt%) and SSB (1.06 wt%). There is also a difference in the grain stability, where SSA and SSC are the mechanically strongest (87 and 85) while SSB is the weakest (66). Porosity was analysed with an image analysis technique on a 0.5 to 1 mm coke fraction [6]. The porosity was analysed with 3 μm cut-off size. The SSA and SSC cokes have a very similar total porosity of 16.8 and 17.1 %, while the SSB coke has the highest total porosity of 21.1 %.

Table 1. Coke specifications.

Properties	Method	Unit	SSA	SSB	SSC
Grain stability	ISO 10142	%	87	65	85
Total porosity	Image analysis	%	16.8	21.1	17.1
Sulphur	ISO 12980	%	1.09	1.06	2.89
		45*	0.60	0.72	0.62
Dust roundness	Image analysis	63*	0.67	0.76	0.66
		94*	0.74	0.87	0.76

* dust size expressed in % of particles passing a 63 μm sieve

Each coke was sieved separately in specified fractions for further recipe mixing (coarse 17 %, medium 23.8 %, fines 34 % and dust 25.2 %). The dust was produced by grinding the -2 mm fraction in an air swept ball mill [7]. Three different dust sizes were produced; Dust 45, Dust 63, and Dust 94, which corresponds to 45 %, 63 % and 94 % of the particles passing a 63 μm sieve, respectively.

The roundness of the particles was characterized by image analysis (Pharma Vision 830). Roundness is characterized on a scale between 0 and 1[7]. A perfect circle has a roundness of 1.0, while a very narrow elongated object has a roundness close to 0. Table 1 presents the mean dust size values. The differences in the mechanical strengths of the investigated cokes resulted in varying roundness of the produced fine particles. The SSB coke has higher roundness values compared to the SSA and SSC cokes. The standard deviation of particle roundness was about 5 % for Dust 45, 4.1 % for Dust 63 and 3.2 % for Dust 94. The differences in roundness can influence the powder flowability and internal friction [8]. The binder matrix using SSB coke is expected to flow more easily while in the SSA and SSC cokes the flowability was reduced due to irregularly shaped interlocking particles. However, the SSB coke has a similar roundness factor in Dust 45 as SSA and SSC have in Dust 94.

The anode paste (15 % pitch) was compacted in a vibrator at 170 °C. Core samples were drilled from identical positions in order to maintain the same testing conditions. The dilatometer was heated at 60 °C/h to 950 °C where it was soaked for three hours before cooling. All samples were run without a packing mass. Two samples were run for each test and an average of these is presented in Figures 8, 9 and 10. Figures 11, 12 and 13 show the measured shrinkage. Figure 14 presents four graphs where the first three (SSA, SSB and SSC) are plotted correlations between the dust size and the residual sample dilation (thermal dilation after the initial expansion when the height flattens out). The fourth graph presents the green apparent densities for each coke and dust size, and indicates that 15 % pitch is optimum for SSB coke with Dust 45, while the optimum for SSA and SSC cokes is Dust 63.

Dust 45 produced samples with a very similar shrinkage (Figure 11) despite a large expansion of the SSB coke, first at about 110 °C then strongly at 240 °C followed by slumping at about 350 °C (Figure 8). The expansion at 350 °C is probably caused by a denser matrix where binder volatiles were trapped and rounder particles produced higher flowability in the paste. At this dust size the SSB coke had the lowest shrinkage for both ΔL_{before} and ΔL_{after} and it seems that 15 % pitch is close to optimum for SSB at this grain composition (Figure 14). During the mixing process more fine particles were generated in the mechanically weak SSB coke, thus generating a higher surface area. It is expected that some of the largest grains in the coarse and medium fractions were crushed during mixing, thus reducing the samples overall strength.

The SSB coke with Dust 63 has a similar dilation profile (as with Dust 45); however both the initial expansion at 110 °C and the expansion at 350 °C are lower, although it reached the highest shrinkage of all cokes for this dust size. The lower expansion is considered by the larger surface area of the particles and thus a thinner binder layer with lower adhesion strength. The SSA coke initially expanded more to 160 °C with this size fraction after a significant residual stress release at 70 °C. The higher expansion is caused by a more dense structure with a higher volume of trapped volatiles in the voids and pores. The SSC coke with Dust 63 has nearly identical initial expansion (110 °C) as with Dust 45, however the whole thermal dilation profile is higher. Dust 63 produced the best composition with 15 % pitch content, considering green apparent densities (Figure 14) for both SSA and SSC.

The largest fine dust fraction of 94 % changed the thermal dilation profile of the SSB coke dramatically. The secondary expansion at about 350 °C (also observed with Dust 45 and Dust 63) was diminished and the sample shrunk extensively from 500 to 950 °C. For this composition visible cracks were observed at the periphery, which might contribute to the sample shrinkage. The sample became under-pitched with Dust 94 and thus no significant expansion occurred because of a limited amount of trapped volatiles in the voids and pores. The highest initial expansion is observed for the SSC coke at about the same temperature (160 °C) as measured with Dust 63. The higher thermal dilation might be caused by more trapped volatiles in a denser structure. However, the SSA coke does not have a large initial expansion, but instead three subsequent peaks from about 70 to 190 °C. The highest shrinkage was found for the SSB coke, while the lowest for the SSC coke (Figure 13).

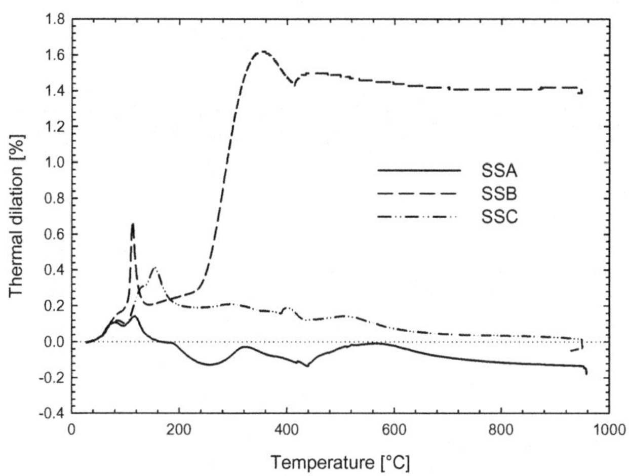

Figure 8. Thermal dilation of three single source cokes made with Dust 45 and 15 % pitch.

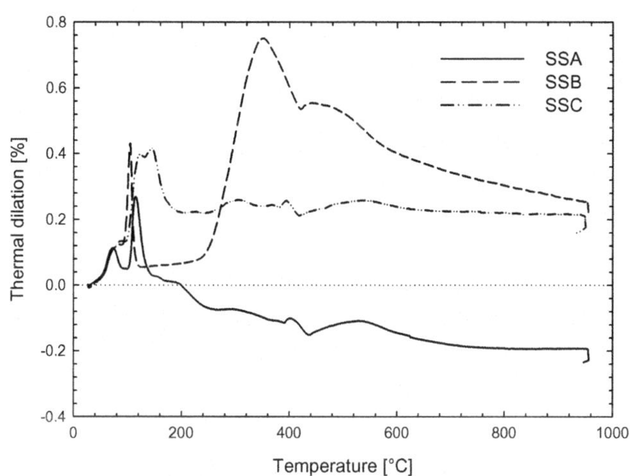

Figure 9. Thermal dilation of three single source cokes made with Dust 63 and 15 % pitch.

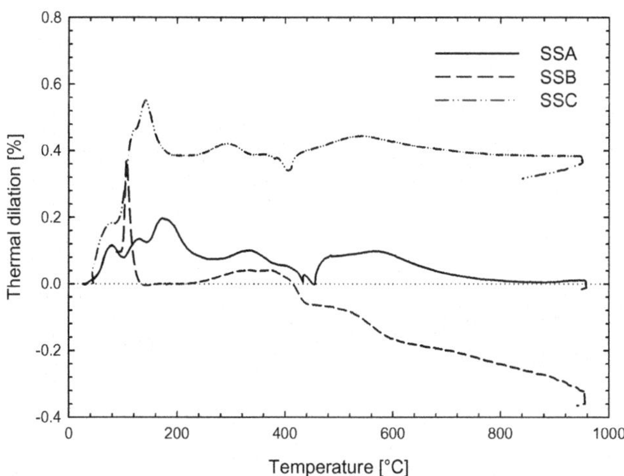

Figure 10. Thermal dilation of three single source cokes made with Dust 94 and 15 % pitch.

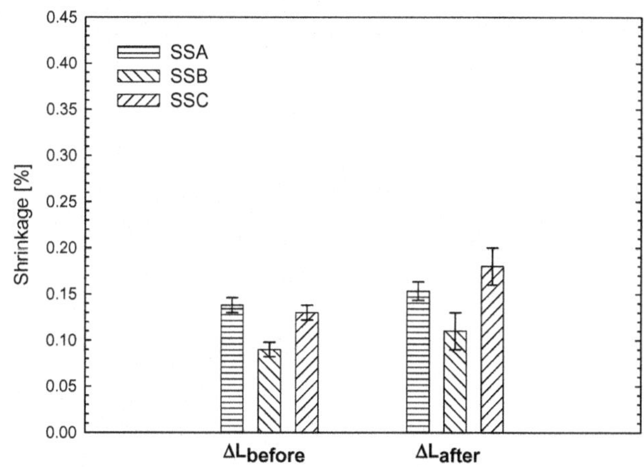

Figure 11. Shrinkage for three single source cokes made with Dust 45 and 15 % pitch.

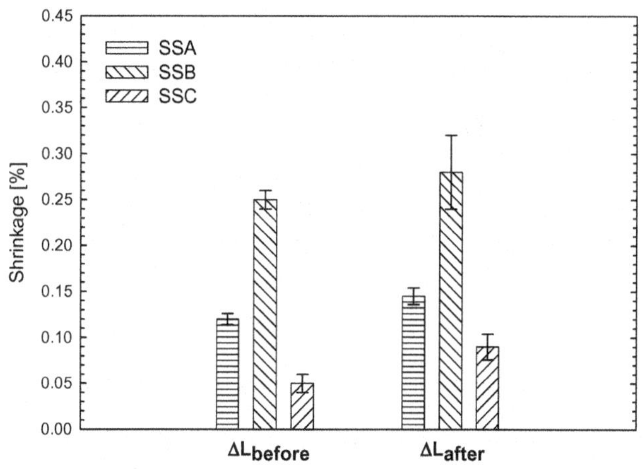

Figure 12. Shrinkage for three single source cokes made with Dust 63 and 15 % pitch.

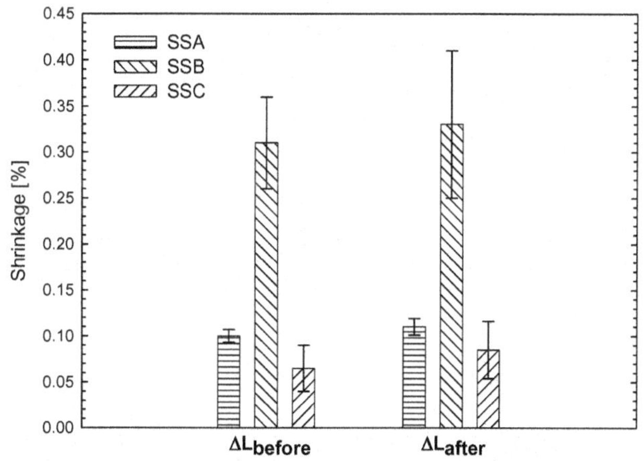

Figure 13. Shrinkage for three single source cokes made with Dust 94 and 15 % pitch.

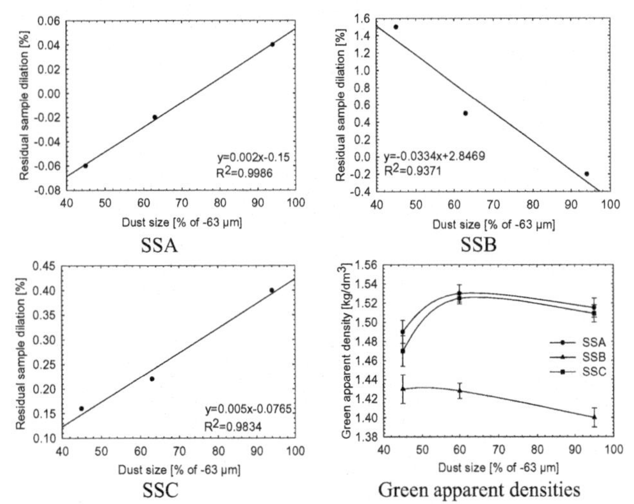

Figure 14. Residual sample dilation for SSA, SSB, SSC and green apparent densities versus three dust sizes.

The correlation between the residual sample dilation (thermal dilation after the initial expansion when the height flattens out) and dust size fraction in Figure 14 shows an increasing trend for the SSA and SSC cokes while it decreases for the SSB coke. The most probable reason why the SSB coke has a decreasing trend and lower green apparent density is the excess fines and lack of large coke grains, due to crushing during mixing, and thus a binder deficit resulting in reduced overall strength and crack formation.

Discussion and Conclusion

The new dilatometer construction showed an improvement in the reproducibility and measurement stability.

Four different heating rates were investigated with green core samples of identical pitch content and granulometry. An increase in the heating rate contributes to a larger initial expansion due to more trapped volatiles (at about 150 °C) and also to a higher final baking loss. The calculated shrinkage showed that the highest heating rate (180 °C/h) produces samples with the highest shrinkage, while the lowest heating rate (20 °C/h) had the lowest shrinkage. Additionally, changes in the early thermal expansion can be observed with a decrease in the heating rate.

The importance of the sampling position was investigated with the samples drilled from the top and bottom position of the green pilot anodes. Also, the difference between samples with graphite packing versus free standing samples in the dilatometer was investigated.

The distribution of the forming force inside the anodes is reduced with the height due to friction between the coke particles and between the paste and the mould. Therefore, an inhomogeneous binder distribution is generated in the anode's height and width. The results show that the bottom samples have a higher expansion due to a thicker pitch bridge layer while the top samples have reduced expansion. The sample packing contributed to a higher initial expansion due to the reduced free surface area for binder volatile release, and the calculated shrinkage was reduced because of the radial thrust from the packing material.

Anodes made from three single source cokes (SSA, SSB and SSC) were investigated in the dilatometer with the same pitch content and identical granulometry, but with three different dust sizes (45, 63 and 94 % of –63 µm). The SSA and SSC coke are the mechanically strongest and produced more irregular particle shapes (particles with lower roundness) during grinding in the ball mill, while the softer SSB coke has rounder particles for each size distribution. There is also a significant porosity difference, which is similar for SSA and SSC cokes (16.8 and 17.1 %, respectively) while SSB coke was 21.1 %.

The measurements show decreasing shrinkage with decreasing dust particle size (only for SSA and SSC cokes), while for the samples made of SSB coke the shrinkage increases.

The mechanically weak SSB coke had the highest green apparent density already with Dust 45, while the SSA and SSC coke had the highest density with Dust 64. This is due to an increase in the fines (surface area) and lack of large grains (coarse and medium fractions) by crushing during the mixing process.

Acknowledgment

Financial support from NFR and the Norwegian aluminium and ferro-alloy industries through the CarboMat program is gratefully acknowledged.

References

1. Fitzer E., Huettinger K.J., Megalopoulos A., "Dilatometric study of the baking behaviour of pitch bonded carbon artifacts". Carbon 1973, p. 621-626
2. Letizia I., Calderone F., "Linear dimensional changes occurring during the baking of pitch-bonded green carbon bodies". Carbon 1980, p. 297-301
3. Martirena H., "Laboratory studies on mixing, forming and calcining bodies". Light metals TMS 1983, p. 749-764
4. Foosnæs T., Naterstad T., in "Introduction to aluminium electrolysis". Second edition, Grjotheim K., Kvande H., Editors, Aluminium Verlag, ISBN 3-87017-233-9, p. 87-138
5. ISO 14428, "Expansion and shrinkage during baking", 2005
6. Rørvik S., Øye H.A., Sørlie M., "Characterisation of porosity in cokes by image analysis". TMS 2001, p. 603-609
7. Chmelar J., Foosnæs T., Øye H. A., Sandvik K.L., "Coke quality effect on the grinding in an air swept ball mill circuit". Light Metals TMS 2005, p. 647-652
8. Bumiller M., Carson J., Prescott J., "A preliminary investigation concerning the effect of particle shape on a powder flow properties". Malvern Instruments, 2005
9. Dell M.B., Peterson R.W., "Dilatometry of pitch-anthracite mixes for carbon". Metallurgical Transaction Volume 4 1973, p. 2077-2080
10. ISO 10142, "Determination of grain stability using a laboratory vibration mill". 1996

Light Metals 2006 *Edited by Travis J. Galloway* ***TMS (The Minerals, Metals & Materials Society), 2006***

NEW PROCESS CONTROL SYSTEM APPLIED ON A CLOSED BAKING FURNACE

Inge Holden[1], Frank Heinke[2], Frank Aune[3], Lorentz Petter Lossius[1]
[1] Hydro Aluminium a.s Technology & Operational Support, P.O.Box. 303, NO-6882 Øvre Årdal, Norway
[2] Innovatherm, Prof. Dr. Leisenberg GmbH+Co.KG. Process Automation Div.; Roter Lohweg 20, D-35510 Butzbach, Germany
[3] Hydro Aluminium a.s Sunndal Carbon, P.O.Box 51, NO-6601 Sunndalsøra, Norway

Keywords: Furnace Retrofit, Capacity Increase, Process Control

Abstract

In 2004 the Årdal Carbon plant increased total baked anode production with 30 % through retrofit of Furnace #3. Using the latest technology of the Hydro baking furnace design, annual capacity of a two fire-zone furnace was increased from 50.000 to 105.000 tons.

The closed top furnace technology has traditionally higher tar content in the flue-gas than furnaces of open type design. At plants with electrostatic precipitators the major part of un-combusted volatiles are collected as tar that has been dealt with either by deposition, recycling or combustion. With this retrofit, flue gas treatment is now based on the technique of regenerative thermal oxidation.

The paper demonstrate the improvements achieved with a new process control system and evaluate the most relevant parameters of the baking process such as combustion of volatiles, energy consumption and quality consistency.

Introduction

The production of prebaked anodes in Årdal started around 1958. In the mid 1990s anodes were produced on 4 baking furnaces of vertical flue design, with an annual production capacity of about 150 000 tons.

Since 1998 retrofit of two baking furnaces using the latest technology of the Hydro baking furnace design has increased the annual production capacity to about 235 000 tons on 3 baking furnaces. The latest expansion was made in 2004 with the retrofit of Furnace #3. Both retrofits were carried out inside existing buildings.

Retrofit of Furnace #3

Main Furnace Data

The available space between building columns of the existing building for Furnace #3 is 29 m (c/c), setting physical limitations for the retrofit of the furnace. The baking furnace is designed for baked anodes with dimensions 1600 (length) x 700 (width) x 600 (height). An optimum utilisation of the available space in the building was achieved by designing the furnace with 7 pits and orienting the anode top and bottom surfaces towards the headwall.

However, with more available space between building columns, as e.g. in a greenfield plant, the baking furnace could have been constructed with 8 pits. Such a furnace would have had the same section load, and oriented the anodes with the top and bottom surface towards the flue wall. The annual capacity for an 8 pit furnace would be 113 000 tons, an additional 8 000 tons.

The principle design and fire zone configuration of the Hydro baking furnace is shown in Figure 1, showing the exhaust bend, additional air fan and the A- and B-part gas burners.

Figure 1. Principle design of Hydro baking furnace.

Stub hole drilling was introduced in Årdal in 1998, with experiences reported in a 2003 Light Metals presentation [1].

Slots in anodes were introduced on regular basis in 2004. Slots with depth up to 34 cm are sawed in the baked anode, maintaining the principle of baking compact anode blocks.

The main characteristics of the retrofit are shown in Table 1.

Table 1. Main characteristics of the retrofit

Baking furnace characteristics	Unit	Old furnace	New furnace
Nos. of sections	Nos	30	30
Nos. of fire-zones	Nos	2	2
Nos. of pits	Nos	5	7
Section capacity	t/section	92,5	168
Cycle time	h	32	28
Annual capacity	t/a	50 500	105 000

The advantage of our baking furnace concept regarding production capacity is connected to the way the energy input is distributed. A certain amount of packing material is used as energy source to reduce the heat load on the refractory. Finally the gas input is divided in two parts, where the combustion chamber either is large (as under the cover) or is constructed with high temperature refractory materials as in the headwall. This concept allows us to operate with a high production rate per pit. Furnace #3 is presently operated at 27 hours fire step, giving a production rate per pit of 0.89 t/h with a pit load of 24 tons baked anodes.

Process Safety

The total energy consumption in modern baking furnaces amounts to typically 4.8 – 5.2 GJ/t baked anodes, which are covered by the following entries:

- Oil or gas: 40 – 50 %
- Pitch volatiles: 40 %
- Packing coke: 10 – 20 %

Particular safety aspects of baking furnace operations are related to the following specific features:

- Only 40 - 50 % of the total energy input is controllable by immediate actions.
- Strict control of the air to fuel ratio is in practice hampered by air ingress into the furnace atmosphere.

European and IEC safety standards give useful guidelines for requirements and solutions for risk reduction. By definition the Directive of Machinery (law in the European Union) applies to the Anode Baking System, including burner system, baking furnace and fume treatment plant with emergency ventilation system. The Directive of Machinery outlines basic requirements, guidelines and standards related to safety, health and environmental requirements of baking furnace design and operation.

International safety recommendations, i.e. European and IEC safety standards, prescribe essential design and operational characteristics to be fulfilled for safe operation of similar units.

Operational safety levels are by international standards graded in terms of Safety Integrity Level (SIL) Figures, which are linked to the severity of hazardous consequences and the probability of occurrence.

Since no standards are specific to the anode baking process, the standard EN 1539 "Dryers and ovens, in which flammable substances are released – Safety requirements" provides useful guidelines for evaluation of the process. Two main process requirements are targeted to safeguard against fires or explosions in the ring main or fume treatment plant:

- An upper concentration of combustible substances as percent of the Lower Explosion Limit (LEL) downstream the combustion region (e.g. 25 %), dependent on which level of safety requirements that is implemented.
- A Limiting Temperature for the flue gas, in practical terms defined to be the temperature at the exhaust manifold.

Furnace #3 is furnished with a Safety Supervision System in order to comply with the essential health and safety requirements in the Directive of Machinery related to risks of fires and explosions. The Safety System is divided into the following categories:

1. A programmable electronic system (PES); which includes elements such as sensors, data communication, the programmable electronic controller etc.

   The PES (hardware and embedded software) fulfils the requirements of IEC 61508. The application software fulfils the requirements of IEC 61511-1. The safety supervision and control PLC system for the actual purpose fulfils the requirements of SIL 2.

   The PES is supervising the process status in the fire zones as well as in the ring main for flue gas, and controls the energy supply to the furnace and the flue gas flow outlet of the furnace. The status of the emergency power system is supervised by the PES.

   The first priority of the PES is to maintain sufficient flue gas rates within defined time limits. The flue gas draft system includes an emergency ventilation alternative and automatic emergency power supply switch.

   Incidental loss of the furnace draft may anyhow occur. Immediate action by the PES is to close down the energy supply of the furnace, and to shut down the draft by closing outlet dampers of the flue gas ring main system. This furnace modus is defined as Safe State. A procedure of flaring of volatiles until air surplus is established in all fire zones is carried out prior to establishing normal operation.

   Fail Safe functions are defined by interlocked sensors signal level for draft supervision, draft and external energy supply control. By critical failure of sensors or equipment, the Fail Safe functions set the furnace to the Safe State defined.

2. Non-programmable safety functions where interlocks are included on specific equipment, e.g. flame supervision of start up burners.

3. Utility system:
   i) Un-interrupted power supply (UPS), e.g. for the PES.
   ii) Emergency power generator, e.g. for the emergency ventilation system.

4. Mitigative functions to reduce the consequences of a hazardous event, e.g. pressure relief panels on the ring main and protection of personnel working area.

Process Control System

The main challenge of draft control of baking furnaces is to control the energy released from the pitch volatiles, which cover approximately 40 % of the total energy consumption.

The overall process response time to transients during anode baking is generally slow, due to the large mass and thermal inertia of anodes, packing material and refractory materials per section. Changes of the furnace under-pressure will change the heat-up gradients after a relative short process time. The actual effect on the pitch volatiles release rate arises, however, gradually over a very long time period. Such a system behavior is described by a so-called PDTn behavior with trailing phases [2].

<u>Vaporization and Pyrolysis of Pitch</u>
Vaporization and pyrolysis of pitch includes a large number of components throughout the ordinary heat treatment range of the anodes. Hence the internal fuel generation varies. In our study of the energy supply from pitch volatiles, the following average composition was chosen [3, 4]:

- Tar: 91.3 weight%
- Hydrogen 6.5 weight%
- Methane 2.2 weight%

A comprehensive anode-temperature logging program was carried out on Furnace #3 in order to establish average heat up rates in different anode positions in the different pits. Based on curves presented by Charette et al [4], giving the relation between anode temperature heat up rate and pitch volatiles evaporation, an estimate of the energy supply could be established as shown in Figure 2.

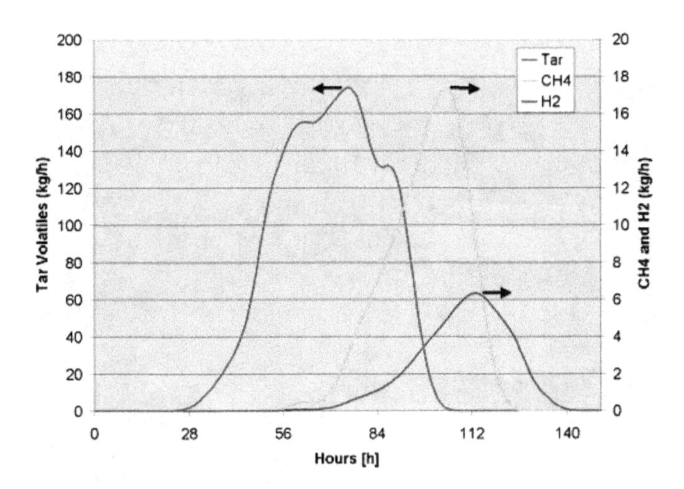

Figure 2. Pitch volatiles release rates per section. Tar at left Y-axis. CH_4 and H_2 at right Y-axis. Fire step 28 hours.

The tar evaporation starts shortly after the first fire advance period in the 2nd section. The highest energy generation rate by combustion of pitch components is achieved in the 3rd section. Hydrogen and methane are released mainly in the 4th section.

Draft Control System

The primary characteristics affected by the efficiency of the draft control system are:

- Anode quality consistency
- Fuel consumption
- Emissions
- Flue gas rates

The draft control system shall ensure repeatable behavior of the pre-heating and tar burn section, achieving high tar burn efficiency in combination with low flue gas rates. The tasks of the draft control system can be stated as:

1. The heat up zone shall reach the tar burn ignition temperature at a defined time despite possible differences in start conditions (e.g. thermal history, air leaks, refractory wear).
2. The front burners for the A- and B-part shall start when the energy supply from the pitch volatiles is near fully utilized. The spread of this start time is a measure for the regularity and repeatability of the control of the heating up zone.

Control Structure of Furnace #3

The principle of the control structure of Furnace #3 can be exemplified by the structure of A-part temperature control for the gas burners as shown in Figure 3.

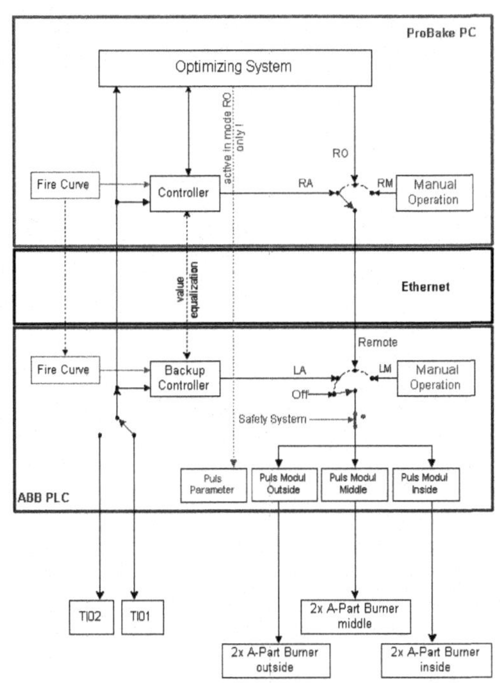

Figure 3. Structure of A-part temperature control.

The Remote Control system is PC based with 2 PCs whereof one serves as a back-up. The following control modes are available:

Remote Manual (RM) provides manual control of:

- Exhaust bend damper position
- Burner capacity

Remote Automatic (RA) provides classic automatic control of:

- Furnace draft by set value of draft level
- Burner capacity by fire curves

Remote Optimised (RO) provisions of automatic control:

- Furnace draft by calculated set value of temperature gradient
- Burner capacity by fire curves and temperature distribution across the sections

A back-up PLC control system is located around the furnace, consisting of 10 PLC cabinets where each PLC controls 3 sections. This Local Control system includes manual control modes and classic automatic burner and draft control. The back-up system is used only in case of hardware or software failures affecting the communication or superior PC control system. The local cabinets of the PLC system serve as an interface towards the furnaces Safety System.

The control structure of the Optimised Draft Control is described in the next section.

Principle of Optimised Draft Control

The draft is controlled by the temperature gradient in the 1st section in the fire zone. The set value for the temperature gradient is continuously calculated, where the target point for the calculation is a defined temperature after e.g. 45 hours in cycle. The gradient control is thus aiming for a process status between 17 and 45 hours onwards in time. The set value for the temperature gradient is allowed to vary within a certain predetermined range. The automatic draft control includes restrictions in allowed operational limits and features as automatic draft control during fire step.

The front position of the tar burning in the fire zone is in a practical sense limited by the maximum allowed outlet flue gas temperature.

Defined target values of the PI-regulator has to reflect the impact of pitch volatiles release rate by variable flue gas temperature gradients. All changes in the draft level are made very slowly, i.e. by low amplification factor (Kp) and high integral time constant (Tn).

The success of the new draft control algorithms are closely related to the design of the fire curves for the section with significant release of hydrogen and methane, i.e. the 4th section in the fire zone. Temperature control according to pre-defined fire curves is only used for sections with gas burners. The temperature curves in the 4th section define the starting point of the A- and B-part burner. Excessive fuel injection prior to the pitch volatiles combustion may accelerate the heat front towards the exhaust outlet end, and thereby impose oxygen deficit.

In Figure 4 the pitch volatiles release rate per section is separated into the sections A- and B-part, and recalculated to MW to show the thermal effect released by pitch volatiles combustion. The starting point of the A- and B-part burner groups are shown, as well as the Fire Curves. The typical start of the A- and B-part burners are when the internal energy supply from the pitch volatiles to the corresponding section part are reduced to approximately 0.2 MW.

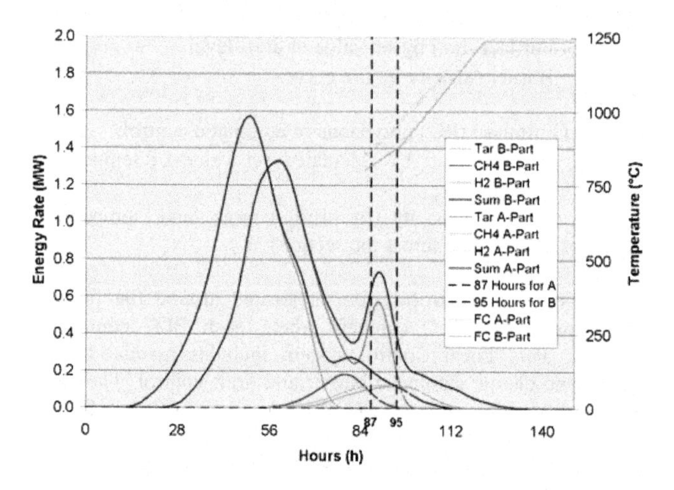

Figure 4. Typical heat generation (MW) of A- and B-part by pitch volatiles combustion versus time. Start of A-part burners at 87 hours and start of B-part burners at 95 hours.

Additional Air for Tar Burn

Additional air is supplied to the 3rd section in the fire zone by a fan system. The air distribution system is an integrated part of the section cover as shown by the foto in Figure 5. The air fan provides a stable basic level of air for combustion of tar, which is independent of the furnace draft level set by the draft control algorithm. The main purpose of the system is to ensure stable high tar burn efficiency. Secondary positive effects are:

- Reduction of the total flue gas rate outlet of the fire zone.
- Reduction of LPG consumption by a reduction of the flue gas rate through sections with gas burners.

Figure 5. Additional air system

Results

After approximately 7 months of operation of Furnace #3 a comprehensive performance test was carried out with measurements of anode quality, energy consumption and process characteristics.

The performance test was carried out at production of anodes with baked dimensions 1580 (length) x 700 (width) x 600 (height). The baked anode weight after drilling of stub holes and sawing of slots was 962 kg.

Production capacity

The performance test was carried out at 28 hours fire step, corresponding to an annual production capacity of 101 125 tons.

Energy consumption

The energy consumption is listed in Table 2.

Table 2. Energy consumption

	Unit	Value
LPG	GJ/t	1.56
Packing coke	kg/t	31.2
Pitch volatiles	kg/t	48.6

Exhaust gas composition

Flue gas characteristics at the outlet of the fire zone are listed in Table 3.

Table 3. Flue gas characteristics

	Unit	Average	StDev
Flue gas rate	Nm^3/t	3360	594
Emission NS 16 PAH [5]	mg/Nm^3	59,8	41,7
Tar burn efficiency	%	97,7	

Anode calcining level

The equivalent temperature method [6] is used to express the overall baking level and distribution of a section or a furnace. Figure 6 shows results achieved at measurements of all anodes in a section at the performance test.

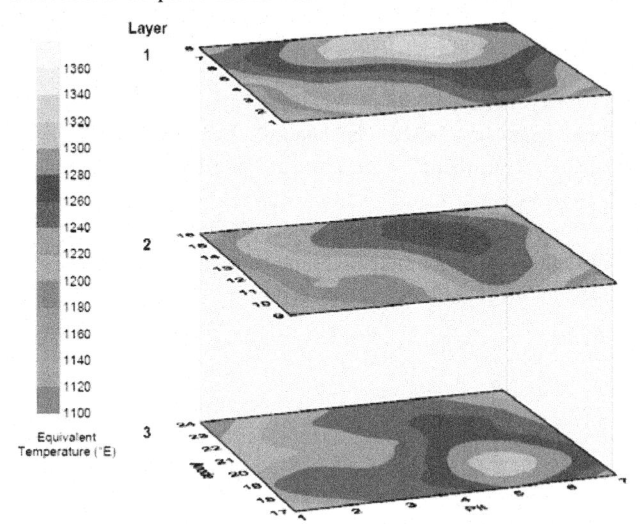

Figure 6. Baking level distribution by anode layers for one section, $T_{eq} = 1236 \pm 41$ °E.

As part of regular anode quality follow-up, the equivalent temperature is measured on 1.6% of the produced anodes. In Figure 7 the baking level distribution of one section is compared with rolling average of the regular anode quality follow-up for anodes delivered to the Hydro Aluminium Årdal Metal Plant (AAM).

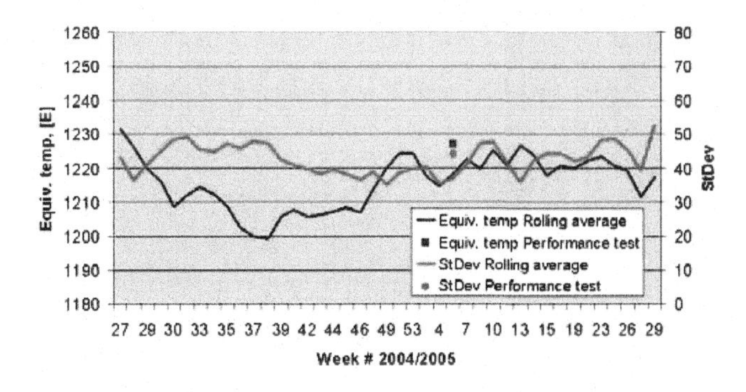

Figure 7. Rolling average of last 75 anodes to AAM with Performance Test results indicated.

The calcining level rolling average and StDev are fairly constant and close to equal with the test sections, showing that the repeatability of the process control system is excellent.

Anode properties

The performance test also included analysis of anode properties of all 168 anodes in the section. Figures 8-11 show 3D plots of anode property distributions.

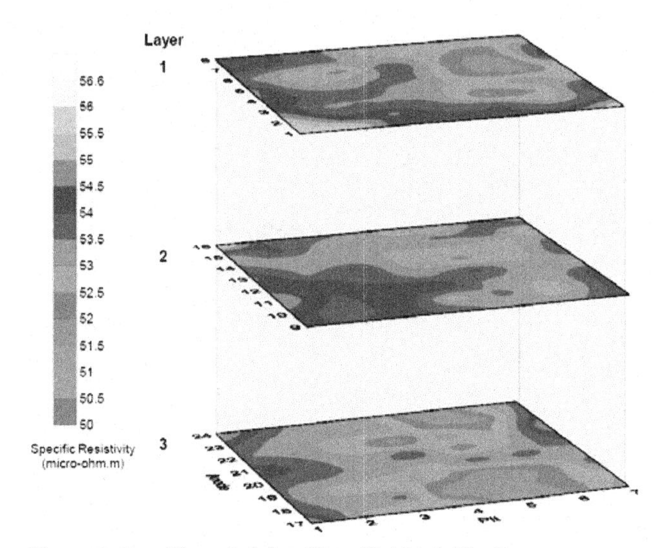

Figure 8. Specific resistivity, SR = 53.45 ± 0.95 μΩm.

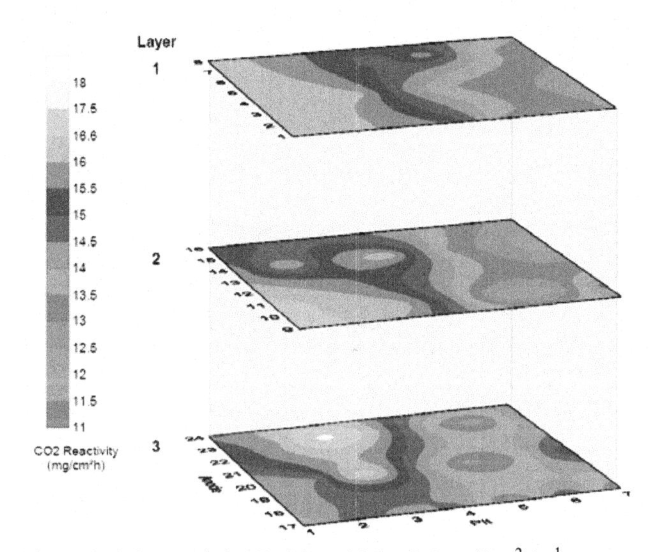

Figure 9. CO_2-reactivity, R.$CO_2 = 14.6 \pm 1.5$ mg/(cm^{2}h). [1]

[1] Reactivities were determined using high-sensitivity Hydro Aluminium methods, which are proportional with the corresponding R&D Carbon methods.

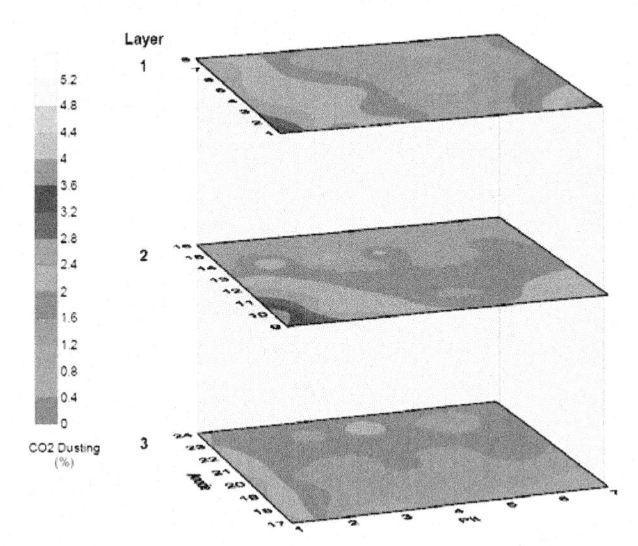

Figure 10. CO_2-dusting, CO_2. Dust = 1.8 ± 0.7 %.

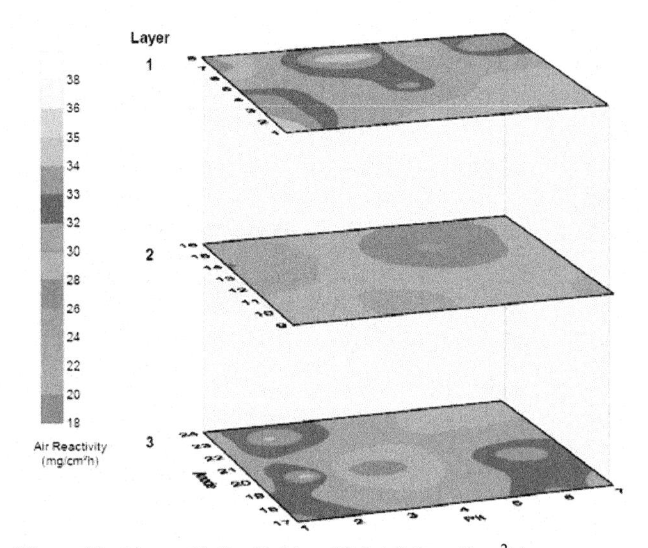

Figure 11. Air reactivity, R.Air = 30.8 ± 2.9 mg/(cm^2h).

The spread in physical properties shown in the figures should be as low as possible. All the data shown are well below the desired specifications set for anodes in Hydro.

References

[1]. Bjørn Erik Aga, Inge Holden, Hogne Linga, Knut Solbu, "Drilling of stub holes in prebaked anodes" TMS Light Metals 2003, 541-545.

[2]. F. Heinke, "Experimentelle Untersuchung des instationären Verhaltens brennstoffbeheizter Glasschmelzwannen", - Dissertation A, HAB Weimar 1994

[3]. Øyvind Gundersen, "Modeling of structure and properties of soft carbon with application to carbon anode baking" (Dr. ing. Thesis, Department of Engineering Cybernetics, Norwegian University of Science and Technology, 1998) 172-173.

[4]. Francois Tremblay and André Charette, "Cinetique de degagement des matieres volatiles lors de la pyrolyse d´electrodes de carbone industrielle" (The Canadian Journal of Chemical Engineering, Volume 66, February 1988), 93.

[5]. Norsk Standard NS 9815. "Water and air analysis. Gas chromatograpic analysis for the determination of polycyclic aromatic hydrocarbons". 1. utgave september 1995.

[6]. Lorentz Petter Lossius, Inge Holden, Hogne Linga, "The equivalent temperature method for measuring the baking level of anodes". TMS Light Metals 2006, in print.

Light Metals 2006 *Edited by Travis J. Galloway* ***TMS (The Minerals, Metals & Materials Society), 2006***

THE EQUIVALENT TEMPERATURE METHOD FOR MEASURING THE BAKING LEVEL OF ANODES

Lorentz Petter Lossius, Inge Holden, Hogne Linga
Hydro Aluminium a.s Technology & Operational Support, P.O.Box. 303, NO-6882 Øvre Årdal, Norway

Keywords: Anode Quality, Baking Level, Analysis Method

Abstract

The equivalent temperature is a measure of baking level using a temperature scale (°E) based on heat treatment temperatures.

The method has been established as an ISO method, ISO 17499: "Carbonaceous materials used in the production of aluminium — Determination of the baking level expressed by the equivalent temperature". The background for the ISO method is described, including the 2003 round robin for determining the precision which resulted in a repeatability of 9 °E and reproducibility (between laboratories) of 14 °E in the range 1050 to 1400 °E.

The main use is for monitoring anode quality in routine production and for performance testing. Examples will be given from Hydro Aluminium baking furnaces where the method has been in use since 1982 [1,2].

Introduction

The equivalent temperature is determined by placing a test portion of a reference coke in a graphite holder with a small hole in the lid to secure gas outlet. The holder is sent with the anode through the baking furnace. Following baking the holder is recovered and the now-calcined reference coke is analysed with regard to L_c using ISO [3] or ASTM [4] methods.

A pre-determined calibration curve linking equivalent temperature with the L_c crystallite height is used to determine the equivalent temperature from the measured L_c-value.

Figure 1. Storage of used and new 100 mm high holders.

The ISO Standard Method

In 1998 ISO Technical Committee TC226 "Materials for the production of primary aluminium" decided to establish the principle of the equivalent temperature as a standard method [5].

The precision statement is a critical part of a standard method. The ISO meeting in Rigi, Switzerland, June 2002 initiated the round robin (RR), which was run by Hydro Aluminium in 2002-3. The results were presented to ISO at the meeting in Tromsø, Norway, September 2003.

The standard proposal went through Committee Draft and Draft International Standard votes and was accepted for publication at the ISO[1] meeting in Jeju, Korea, April 2005 as method ISO 17499: "Carbonaceous materials used in the production of aluminium — Determination of the baking level expressed by the equivalent temperature".

Definition and Calibration

Equivalence

The term equivalent temperature (T_{eq}) comes from "level of calcination equivalent with a two hour heat treatment at the temperature". A small (20-gram) portion of a reference coke is calcined in a fast heating and cooling furnace by soaking for 2-hours. The heat treatment temperature is the equivalent temperature. After calcining, the crystallite height measurement (L_c) is made, and the (L_c, T_{eq}) pairs are used to determine a calibration curve, see Figure 2 and Eq. 1 [6].

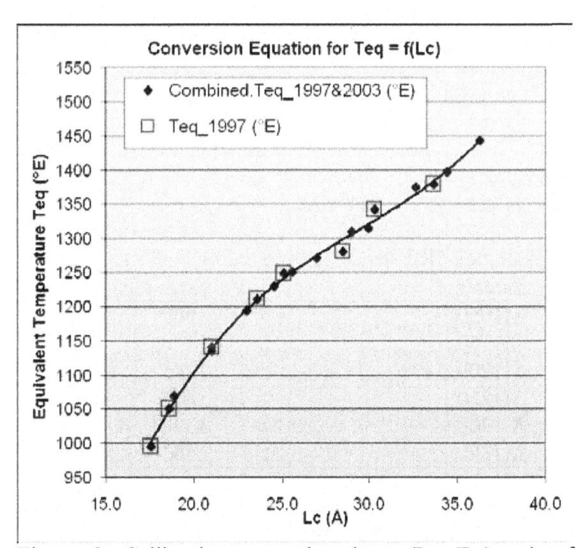

Figure 2. Calibration curve based on (L_c, T_{eq}) pairs for the reference coke. Two series of 2-hour heat treatments (1997, 2003) have been combined to give the calibration curve.

$$T_{eq} = 0.0812 * L_c^3 - 7.1537 * L_c^2 + 225.16 * L_c - 1187.5 \qquad Eq.1$$

[1] Until 2004 Technical Committee TC226 was TC47, Sub-Committee 7 for "Aluminium oxide, cryolite, aluminium fluoride, sodium fluoride, carbonaceous products for the aluminium industry"

A third order expression fitted the data best. Colin P. Hughes reported a similar behaviour in the 1996 TMS presentation [7].

Figure 3. Tray with samples for x-ray diffraction analysis for L_c.

ISO Round Robin

RR Procedure

The participants are listed in Table 1.

Table 1. Participants in the ISO RR for the precision statement.

Metal Plant	Contact Person
Albras Aluminio	Max Heilgendorff
Angelsey Aluminium Metal Ltd.	Donna W. Evans
Comalco Research and Tech. Support	Dr Alan Tomsett
Dubai Aluminium Co Ltd	S. C. Tripathi
Elkem Aluminium ANS Research	Turid Vidvei
Hydro Aluminium Kurri Kurri Pty Ltd.	Darrell Harman
Hydro Aluminium Neuss	Marco Nierfeld
Hydro Aluminium Sunndal	Frank Aune
Hydro Aluminium Årdal	Kirsti Gulbrandsen
R&D Carbon	Stefan Meichtry
Slovalco	Jozef Lovcican

The RR samples consisted of two green petrol cokes, ISO249 (Statoil Mongstad) and ISO250 (Conoco), each heat-treated to eleven baking levels in a large laboratory furnace. Six of the eleven were used as RR standards for the calibration curve; see Table 2, and five as RR unknowns. The equivalent temperature was determined at Hydro Aluminium LSS Årdal using separate $T_{eq} = f(L_c)$ calibrations for each coke.

Table 2. Example of RR standards from the ISO249 sample set.

Sample	Teq Value (°E)
M1050	1069
M1150	1190
M1200	1245
M1250	1311
M1300	1398
M1350	1502

The L_c analysis was performed according to the standard laboratory practice of each participant. The following requirement was made regarding precision: If one parallel of three was more than ±0.5 Å different from the average it was considered an outlier and a new portion should be milled and analysed. If again one parallel was more than ±0.5 Å different from the average it was considered an outlier and discarded. Then the average of the remaining two was used. All participants used specially distributed, tailored Excel-spreadsheets for calibration and return of results. This simplified the final data-treatment considerably.

Precision Calculation

The precision was determined using ASTM E691 "Interlaboratory Study of Precision Program". With ten participants, ten materials and three parallels for each unknown the round robin more than met the minimum requirements prescribed by ASTM E691 [8].

Result

The precision was

- *Repeatability: r = 9 °E*
- *Reproducibility (between-labs): R = 14 °E*

in the range 1050 °E to 1400 °E. The precision figures were independent of the measured equivalent temperature values as can be inferred from Figure 4 [9].

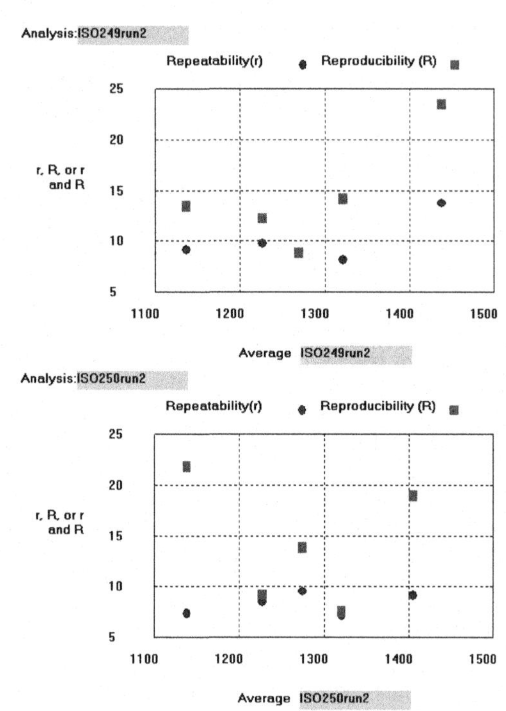

Figure 4. Precision for each of the RR "unknowns" (from E691).

Usage

Repeatability: Given a number of anodes (or test portions) all with equivalent temperature 1200 °E, if measured by the same operator at the same laboratory, the repeatability expresses that 95 out of 100 measurements will be within the range [1191, 1209] °E.

Reproducibility: Given a number of anodes (or test portions) all with equivalent temperature 1200 °E, if measured by different laboratories, the reproducibility expresses that 95 out of 100 measurements will be within the range [1186, 1214] °E.

Discussion

The span of equivalent temperatures tested in the RR was 1050 to 1400 °E. This encompasses the span of normal variation observed for anodes in baking furnaces. The Repeatability (within-lab) was 9°E, and relative to the span of 350°E this is as low as 2.6 %rel, at the 95 % confidence level. This is a very promising result both as regards monitoring anode quality and seeing differences when mapping the baking level in anode furnace sections.

Using the Equivalent Temperature

The Green Reference Coke

A calibration is unique for a chosen green petrol coke. A substantial amount of green reference coke should be stored. For use in Hydro Aluminium baking furnaces, 12 tons of green 1982 Mongstad coke has lasted 20 years and still 2-3 tons are left.

The reference coke should be a single source petrol coke. Before determining the calibration curve the coke should be dried at 540 °C, crushed to ±4 mm particle size and thoroughly mixed.

Reference Coke Standard Sets

The sets of reference coke are available with Certificate of Analysis; please contact the authors for information. The sets are useful for

- Making it easier to recalibrate
- Starting using analysis of equivalent temperature if a fast heating and cooling furnace is not available
- Starting a new reference coke

Table 3. Reference material for a calibration consisting of eleven samples spanning a wide range of equivalent temperatures.

Standard	Equivalent Temperature, T_{eq} (°E)
C1050	1061
C1100	1138
C1150	1195
C1180	1230
C1200	1250
C1220	1275
C1250	1306
C1260	1321
C1300	1364
C1325	1399
C1350	1443

Starting Using the Equivalent Temperature

A reference coke standard set can be used to establish a new reference coke. The following procedure avoids the complex heat treatments with a fast heating and cooling furnace to get the (T_{eq}, L_c) data pairs. The procedure requires green coke of both the old and new reference cokes.

To get a good calibration twelve pairs of graphite sample holders with old and new green reference coke should be used. The holders are sent pairwise through a baking cycle positioned in a section so as to reflect the entire span of equivalent temperatures expected. Ideally, some pairs should be calcined under controlled conditions to have one low result with T_{eq} of approximately 1050 °E and two high results with T_{eq} in the range 1300-1400 °E, to stabilize the curve fit.

Analysing the pairs of test portions with the calibration based on the reference materials yields (L_c, T_{eq}) points for the new reference coke via the T_{eq} values determined fro the old reference coke. The calibration equation for the new coke can then be determined.

Example: Quality Control

As part of routine anode quality follow-up, the equivalent temperature is determined for 1.6 % of the produced anodes. A quality specification is in use, e.g. baking level minimum 1200 °E.

In routine analysis the equivalent temperature calibration is programmed into the x-ray diffractometer application and the result is calculated and uploaded to the product quality databasc automatically.

Example: Mapping a Section

After approximately 7 months of operation of the new Årdal Carbon Furnace #3 a comprehensive performance test was carried out. Furnace #3 is a 2-fire, 30-section vertical flue furnace with annual production 105 000 tons of anodes [10]. The sections are 7-pit with totally 168 anodes stacked vertically in three layers.

The baking level distribution was determined for several sections by the equivalent temperature method. Figure 5 and 6 illustrate the baking level in one section vertically by pit and horizontally by anode layer. Pit 1 is the outer pit in the section.

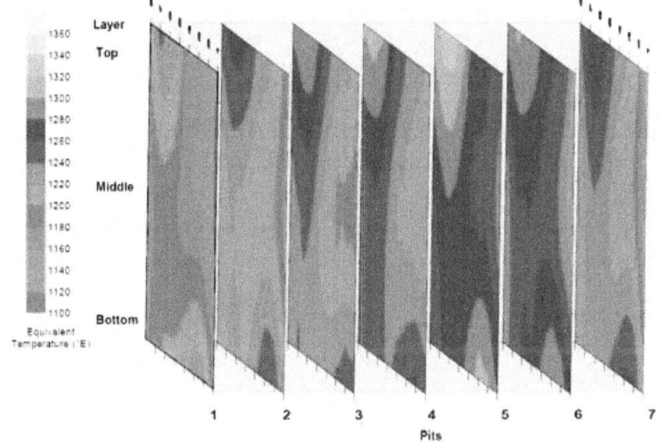

Figure 5. Baking level distribution vertically by pits. The average and standard deviation of the section was T_{eq} = 1236±41 °E.

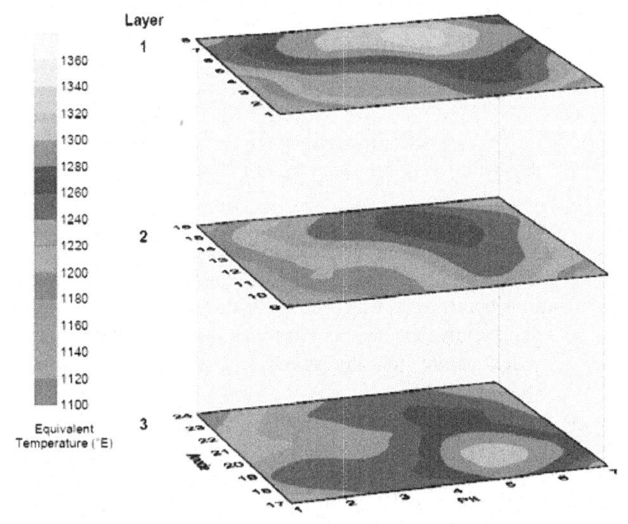

Figure 6. Equivalent temperature distribution by anode layers. Each layer represents 7x8 = 56 anodes.

The plots show that

- The sensitivity of the equivalent temperature method was sufficient for a detailed plot of the balking level distribution
- The baking level along pit five was somewhat higher and the result was useful for tuning the baking through the entire heat treatment including the fire bridge control system

Example: Position of Sample

The equivalent temperature sample can be placed in a stub hole or a specially made small depression on the anode. In Årdal Carbon furnaces #3 and #4 anodes are stacked so that the sample is in the centre of the 8-anode pack. A test was run to see if a sample placed toward the pit wall packing coke would show different equivalent temperature level than in the centre of the pit.

In the test an additional reference coke sample was sent with the anode placed in a hole in the anode side. The test comprised 121 anodes. Positioning was random as regards furnace #3 or #4, section, pit or anode layer. Table 4 shows a statistical comparison, and Figure 7 the distribution of values.

Table 4. Mean and standard deviation from 121 anodes with equivalent temperature samples in two positions.

Position	Mean T_{eq} (°E)	StDev T_{eq} (°E)
Centre of anode stack	1221.1	52.1
Side toward pit wall	1221.5	53.2

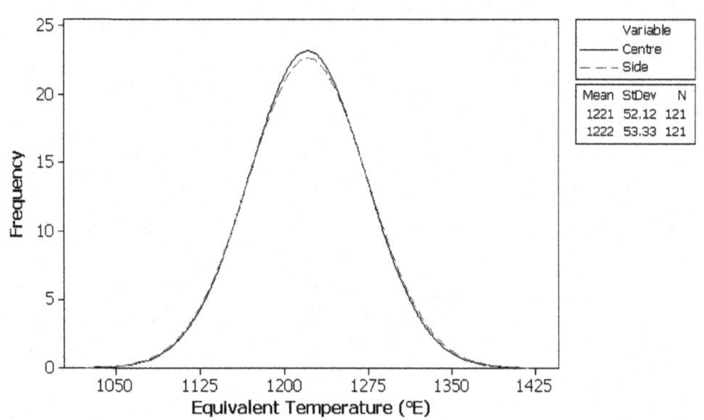

Figure 7. Distribution of two samples in 121 anodes.

A paired t-Test using MiniTab Release 14.13 gave a difference of 0.40 and the confidence interval at 95% confidence level for the two positions being equal was (-4.69; 3.90). This meant

- The equivalent temperature was the same in the anode pack centre and the anode side toward the pit wall packing coke
- The anodes are baked equally all through

Although the anode side toward the pit wall will be heated earlier than the centre, the soaking time and the slower cooling of the centre is sufficient to allow the anode centre to catch up and to equalize the baking all through the anode.

Example: Laboratory Scale Anodes

At the Hydro Aluminium testing facility in Årdal laboratory scale pilot anodes are made for petrol coke and anode paste test purposes. Making pilot anodes with properties equal to full scale industrial anodes requires a chain of equal treatments from aggregate handling and recipe through mixing, vibroforming and baking. The control of baking is critical. Full-scale heat treatment has been successfully reproduced in a laboratory furnace using the equivalent temperature method to define the baking level.

Figures 8-11 are included to illustrate this. The results are from a study of the effect of several factors including pitch level on properties of anodes with mixed normal and high sulfur cokes.

In the plots, the pilot anodes (blue points) are compared to industrial scale Årdal Carbon anodes made with petrol coke from three different coke shipments (I, II, III). The production property variation is shown by the average centre point and one standard deviation spread. The pilot anodes were made from coke II.

The baking level of the pilot anodes was 1200 °E, which is the target for the full-scale anodes. Of the properties shown, specific resistivity is most influenced by baking level. The other three are only weakly influenced within the baking level applied.

Observations

- Density – pitch level 14.0 to 14.7 % most similar
- Specific resistivity – pitch level 14.7 %
- Carboxy reactivity – the 15.2 % anodes' deviation due single outlier

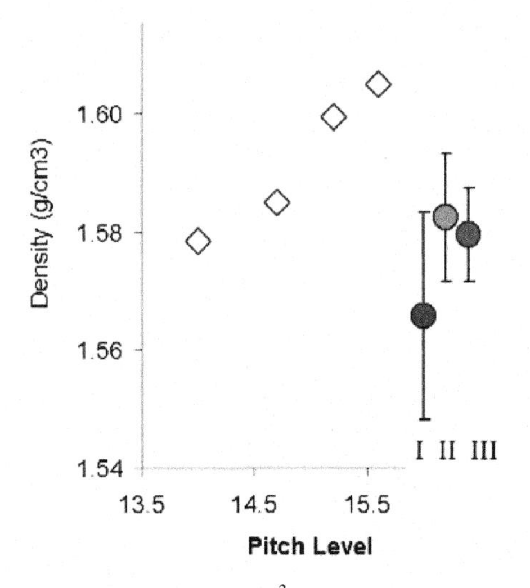

Figure 8. Density (g/cm^3) of pilot anode compared to full-scale anodes. Production anode pitch level is not given.

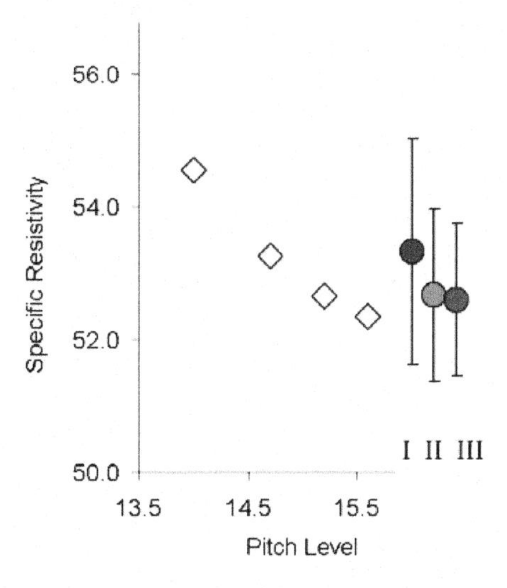

Figure 9. Pilot anode comparison – SR (μΩm).

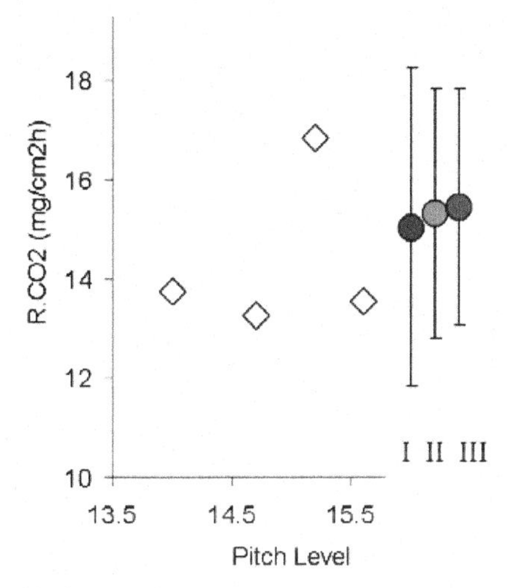

Figure 10. Pilot anode comparison – CO_2-reactivity ($mg/(cm^2h)$).

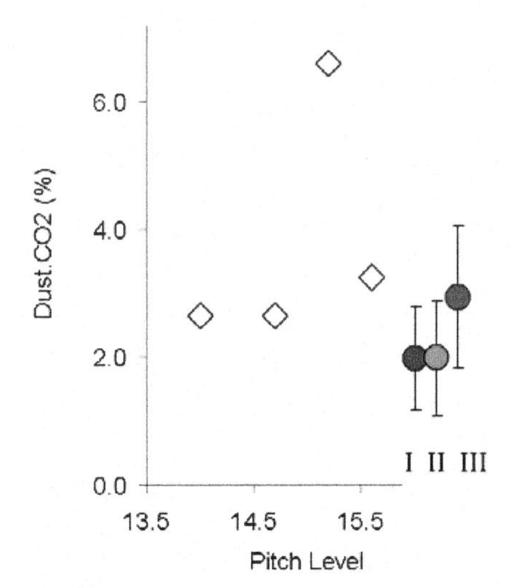

Figure 11. Pilot anode comparison – CO_2 dusting (%).

Conclusions

The equivalent temperature method has been established as an ISO method, ISO 17499:

- The within-lab precision is 9 °E at 95% confidence level, which is small compared to the normal variation range allowing good monitoring of baking level
- The between lab precision is 14 °E, also fairly narrow, which allow comparison of baking furnaces

Examples have been shown of usage in Hydro Aluminium for

- Anode quality control
- Mapping of the baking level in entire sections of a baking furnace
- Test for equal baking all through anodes
- Control of laboratory scale baking to ensure equal baking level with industrial scale anodes

Acknowledgements

The authors wish to thank Dr. Wolfgang Schmidt-Hatting for help with the precision calculations and Mr. Audun Bosdal and Mrs. Kirsti Gulbrandsen for help with organizing the round robin.

References

[1]. Foosnæs, T.; Jarek, S.; Linga, H., "Vertical flue baking furnace rebuilding concepts and anode quality", Light Metals 1989, ed. Paul G. Campbell, pp.569-574.

[2]. Foosnæs, T., Kulset, N., Linga, H., Næumann, G.R. and Werge-Olsen, A., "Measurements and Control of the Calcining Level in Anode Baking Furnaces", Light Metals 1995, ed. J. Evans, pp.649-652.

[3]. ISO/FDIS 20203 "Carbonaceous materials used in the production of aluminium — Calcined coke — Determination of Crystallite Size (L_c) of Calcined Petroleum Coke by X-Ray Diffraction" is at final draft voting stage.

[4]. ASTM D5187-91 (2002) "Standard Test Method for Determination of Crystallite Size (L_c of Calcined Petroleum Coke by X-Ray Diffraction".

[5]. Original proposal reference ISO/TC47/SC7 N1186, Work Item WI 7.4.13, 1998-03-10 (Arne Werge-Olsen, Hydro Aluminium, Norway).

[6]. The calibration curve was introduced by Tormod Naterstad and H. Berg in "Metode for bestemmelse av ekvivalenttemperatur i brennovner", RAPÅ 82/030 ÅSV 1982-05-06.

[7]. Hughes, Colin P., "Methods for determining the degree of baking in carbon anodes", Light Metals 1996, ed. W. Hale, Proc. of 125th TMS Annual Meeting, Anaheim, California, p.521-527.

[8]. ASTM E691 for Windows, Prepared by ASTM E11 Quality and Statistical Committee, Version 2, 1996.

[9]. Precision statement for the Equivalent Temperature, Ref. ISO/TC47/SC7 N1458 WI 7.4.13, PWI 17499, 2003-08-21.

[10]. Inge Holden, Frank Heinke, Frank Aune, Lorentz Petter Lossius, "New Process Control System Applied on a Closed Baking Furnace", Light Metals 2006, in print.

Light Metals 2006 *Edited by Travis J. Galloway* ***TMS (The Minerals, Metals & Materials Society), 2006***

NEW REQUIREMENTS AND SOLUTIONS FOR THE FUME TREATMENT AT PASTE MIXING AND ANODE BAKING PLANTS

Matthias Hagen,

LTB Lufttechnik Bayreuth GmbH & Co. KG,
95474 Goldkronach, Germany

Keywords: Emission, Fume treatment, VOC, PAH, RTO, benzene, HF

Abstract

The new IPPC regulation (1) as well as local legislation forces manufacturers of anodes to improve their existing fume treatment center . The most critical emissions are Benzene and polycyclic aromatic hydrocarbons (PAH). Generally a regenerative thermal oxidiser (RTO) would be the best applicable solution. But the first systems installed had huge problems with heavy tar and soot. The new RTO-based technology should avoid problems due to condensates like heavy tar and soot and use the calorific value of the tar. For this task a new test plant was installed at a closed type baking furnace. The results of this test confirmed the basic design of the RTO but also caused a totally new design of the inlet valves. Furthermore a ceramic prefilter is required to avoid condensations at the inlet of the RTO. The result of the test and the new concept will be shown.

Type of emissions

A general information about the occurring emissions is quite difficult, as the conditions of each baking process are different. The most important factors are:

- Type of furnace (open or closed type baking)
- Type of fuel
- Raw materials for anodes (e.g. reuse of butts; sulphur content of coke 0.5- 4%)

Emissions are the result of a weight loss of the produced anodes of about 5 % during baking. In the tables the range of possible emissions at furnaces without a fume treatment is shown. The main difference is the type of furnace. In an open type furnace more and especially hydrocarbons with a low boiling point get burnt inside the furnace. This effect will also reduce the energy consumption of the furnace.

The firing of fuel creates sulphur dioxide. So the difference in emission is depending on the sulphur content of the fuel, which can vary between low contents in natural gas and higher contents in fuel oil or even heavy oil.

Fluorides are a result of the use of spent anodes (butts) in the raw material of anodes. These butts contain fluorides, which result in some hydrogene fluoride (HF) release during baking.

		open type furnace	close type furnace
Anode specific flue gas volume	m^3/t	5000	3500
CO	mg/Nm^3	<1200	<1500
Total carbon	mg/Nm^3	100 -300	<1000
Tar/condensates	mg/Nm^3	200 - 400	800 - 1200
PAH	mg/Nm^3	20 - 200	40 - 500
Soot/dust	mg/Nm^3	100 -200	50 -100
SO_x	mg/Nm^3	100 - 800	100 - 800
HF	mg/Nm^3	5- 300	5 - 300

Table 1: typical emissions during anode baking

The most important pollutants are the aromatic hydrocarbons. Among these we can find substances like benzene, which has one benzoic ring, as well as the polycyclic aromatic hydrocarbons (PAH), which consist of more than two benzoic rings. Most of them are a poison, some are even carcinogenic. The list of PAH starts with Naphthalene, which has two benzoic rings, a molecular weight of 128 and a boiling point of 218°C. A typical example in the middle of the PAH list is benzo(a)pyrene, which has 5 rings, a molecular weight of 252 and a boiling point of 496°C.

As the substances with a lower molecular weight are under the conditions of a furnace mostly gaseous, they will be burnt in the furnace (especially in open type). The focus for emissions is therefore on the heavy PAH. The typical example is benzo(a)pyrene.

Measurement methods and emission limits

In Germany the VDI- regulation 3467 (2) started in 1998 to describe the process and the emission control methods in production of carbon and electro graphite materials. For the PAH two packages (VDI I and VDI II) were created to measure and report the PAH emissions. Meanwhile a draft of the VDI 3874 (3)for a new measurement method exists, where the testing phase is reduced to 30 minutes and the identification of the PAH is done by a GC-MS, which works via mass-identification. Also the OSPAR members of northern Europe, who try to protect the northern Atlantic sea, have created an own package of PAH. More and more states and organisations like the Environmental Protection Agency (USA) with the EPA 16, have created own packages.

PAH Compounds	VDI-I	VDI-II	VDI-I+II	OSPAR II	EPA
Naphthalene					X
Acenaphthylene					X
Acenaphthene					X
Fluorene					X
Phenanthrene				X	X
Anthracene				X	X
Fluoranthene				X	X
Pyrene					X
Benzo (a) pyrene	X		X	X	X
Dibenzo (a.h) anthracene	X		X	X	X
Benzo (a) anthracene		X	X	X	X
Benzo (b) fluoranthene		X	X	X	X
Benzo (j) fluoranthene		X	X		
Benzo (k) fluoranthene		X	X	X	X
Chrysene		X	X	X	X
Indeno (1,2,3,-cd) pyrene		X	X	X	X
Benzo (ghi) perylene				X	X
Benzo (b) naptho (2,1-d) thiophene		X	X		

Table 2: PAH compounds

As different methods of measurement are used all over the world, it is very difficult to compare their results. The best solution would be a general analysis on all known PAH.
As this would create an enormous effort many countries use just selected PAH as an emission limit. So the focus is set in many countries on Benzo(a)pyrene, which is used as a reference PAH.

All heavy PAH usually have a boiling point of more than 200°C. Due to the temperature at the outlet of the furnace of approximately less than 200°C they are all condensed. These tar particles create an increasing sticky mass in the fume treatment system, which is difficult to handle. This is even more important when we consider that the 16 PAH substances measured in the EPA 16 test will just show 15 – 30 % of the total tars. The measured Benzo(a)pyrene concentration will show only 0.5 – 0.8 % of the total tar emission.

In Germany the only legal limits for existing plants with a fume treatment system are fixed in the TA-Luft 5.4.4.7 for total hydrocarbons <150 (20) mg/m³ and for benzene < 3 (1) mg/m³. In case of using a thermal treatment system to reach the limit of benzene the lower limits in brackets are applicable.

The Reference Document on Best Available Techniques in the Non Ferrous Metals Industries, known as IPPC (Integrated Pollution Prevention Control), summarised informations about existing plants in Europe. Also recommendations on the best available techniques (BAT) are included. For the purpose of PAH treatment the only method mentioned is an afterburner or oxidiser. This paper could be a guideline for authorities to receive an overview on technical options. It is not a law.

Historical background

The first fume treatment systems for baking furnaces were electrostatic precipitators, which were able to reduce the particle emissions. To improve the reduction efficiency lots of them were equipped upstream with a cooler. Due to this the gas temperature dropped to approximately 70°C, which was sufficient to condense most of the tar aerosols. Most of these plants still exist due to a good corrosion protection by the tar inside.

Based on more strict requirements for the total carbon as well as the HF and SO_x - emissions during the late 80ies, a new system was developed. The so-called fume treatment centre (FTC) uses a dry scrubbing process with alumina as an ab- and adsorbent to eliminate the pollutants.

During the last years baking furnaces were in the focus of environmental agencies again. The reason was the more strict limits for polycyclic aromatic hydrocarbons (PAH) especially benzene, which is supposed to be carcinogenic. The only known technology to reduce these PAH as well as total carbon below the required limits is a thermal oxidation, which is equipped with a regenerative heat exchanger to recover most of the energy the process requires.

Already in 1994 the German Environmental Agency started a test plant at SGL in Meitingen, Germany to reduce PAH emissions downstream of an electrostatic filter (4). This RTO was manufactured by Reeco-Stroem, a Danish company, which since 2004 has gone out of business.

In the late 1990's two plants were built for the use at a baking furnace for special graphite without the upstream use of an electrostatic precipitator (ESP). They worked quite well except some problems at the rotary valve system. Based on this experience the manufacturer installed 2 systems at baking furnaces in Hamburg, Germany and in France. The size of the plants was four to ten times bigger. This effected upsizing problems at the rotary valve system. By clogging with condensables a regular bake-out process was required, which caused temperature distortion of the rotary valve system. To avoid this effect and the resulting maintenance standstill one plant used the existing ESP as a prefilter. So it was clear that a RTO needs to be equipped with a prefilter to avoid clogging of the valve system.

Based on this a new plant was built in Norway and it works well. In 2005 another plant was built for an open furnace type in the Netherlands. This system was the first combination of prefilter, RTO and packed bed filter for HF reduction.

Test plant

As tests with a rotary valve equipped RTO failed at site it was necessary to use a specific valve design at a RTO test plant. At the same time LTB had finished the manufacturing of a brand new 3-chamber RTO test plant (figure 1), designed with special valves, which are able to withstand a bake-out temperature of 500 °C.

Figure 1: LTB RTO test plant

The plant is equipped with a Siemens PLC S7, which is connected with a visualisation (figure 2). Via modem the plant can be operated from the LTB office.

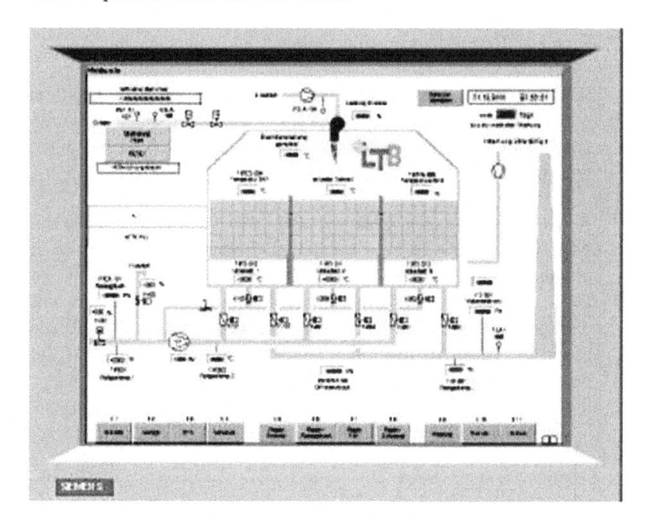

Figure 2: RTO visualisation

In August 2004 the test run started at a Riedhammer closed type-baking furnace.
A stream of 1500 m³/h was diverted from the ring duct. The plant operated for several weeks with different tests. To determine the optimal temperature for the treatment of hydrocarbons a temperature range of 800 to 1000°C was applied. In order to reach sufficiently low emissions the temperature should be above 850°C.

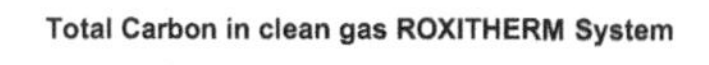

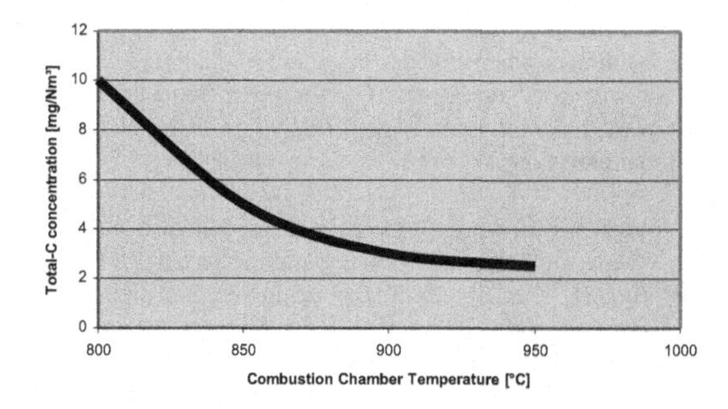

Figure 3: Emission of total carbon

The situation for PAH was more critical. As expected the required energy for breaking the benzene rings is much higher and a temperature of more than 950°C is required.

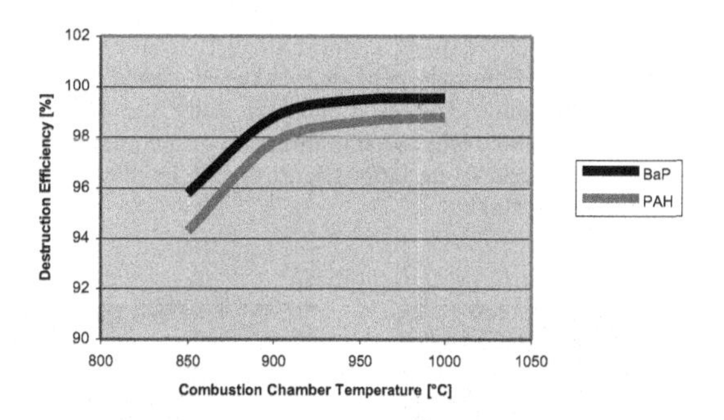

Figure 4: Emission of BaP and PAH

One of the main concerns of the customer towards every fume treatment is the possibility of clogging due to dust and condensates from tar and soot. To create a huge amount of deposits inside the test plant the whole system was not insulated. The effect could be seen on the photos (figure 5 and 6), showing one of the inlet valves after a running time of two weeks.

Figure 5:Tar condensates

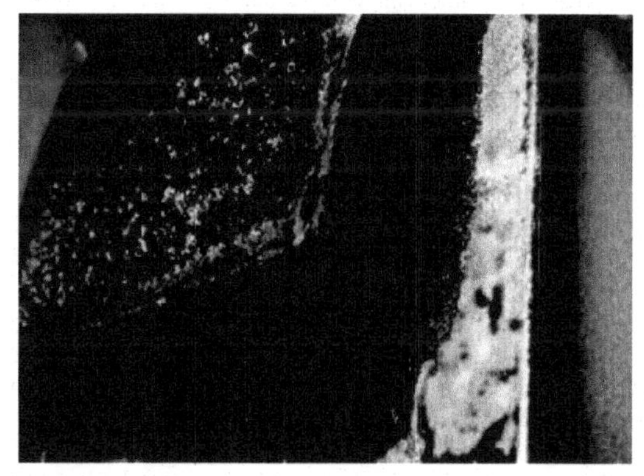

Figure 6: Condensates at the valve

Despite of the rough operating conditions the clean gas emissions during the whole test were constantly low.
The special LTB valves with outside actuator and a shaft-free seal avoided any leakage and ensured the low emission values.

A similar situation with deposits was found at the ceramic heat exchanger inlet. The pressure drop was expected to increase due to the deposits. But it kept stable over more than 2 weeks, as the special honeycomb structure of the ceramic is responsible for a laminar flow, which avoids deposits inside the ceramic media.

After 2 weeks of normal operation a bake-out should verify the cleaning efficiency. To check the efficiency the plant was opened, and a photo was taken in direction of the inlet ceramic bed before (figure 7) and after (figure 8) the bake-out, which was operated at 500°C.

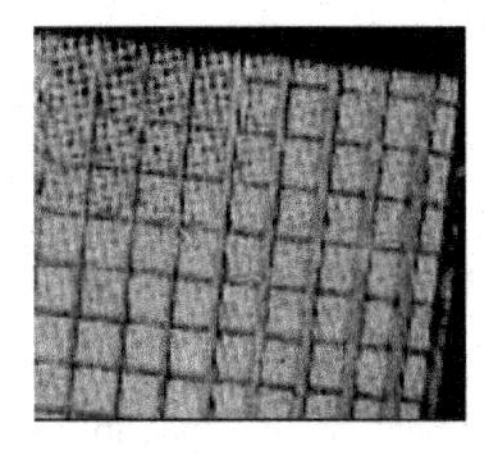

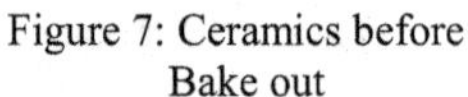

Figure 7: Ceramics before Bake out

Figure 8: Ceramics after Bake out

The result was that all deposits on the ceramic inlet were removed.
During the increasing temperature at the ceramic bed inlet, most of the deposits liquefied and dropped out of the bed down onto the valves and into the valve box. Even though these deposits disappeared at the end of the bake-out, the fouling of the valves should be avoided for future plants in general.

During the bake-out process, which is carried out in one of the three chambers, the remaining two chambers are still treating the flue gas. The bake-out gases are transported towards the stack without a treatment. This causes an increase in emissions.

New plant design

To offer the customer a perfect solution it was necessary to improve the current RTO- design. The following items should be realised:

- Extended time without bake-out
- Valves outside the dropping area
- Continuous emission-reduced bake-out
- Use of the energy of condensates
- Discharge of non-combustibles to waste

To avoid clogging of the RTO it is necessary to separate most of the condensables prior to the RTO in a pre-filter system. The small, condensed droplets have a diameter between 0.5 to 4 microns. To separate them, a filter needs high turbulence.
When we consider the amount of condensate, which could reach up to 50 kg/h, the filter needs the possibility to store several tons of condensate. Finally the filter should withstand a temperature of 500°C during purification.

These requirements were fulfilled with a horizontal filter bed made of ceramic saddles, which are installed in a metal casing. To run the plant in continuous operation two redundant filters have to be installed (figure 9). In this way it is possible to clean the fume gases in one filter during purification of the other pre-filter. For this purpose hot air enters the filter vertically, increasing the temperature of the ceramics and the condensates. The liquefied phase will drop down, can be collected at the bottom of the casing, where it can be discharged easily. Due to this discharge most of the non-combustible content, like dust and ash, will not stay inside the system. The pressure drop will hence not increase for a long time. An exchange of the ceramics could be avoided.

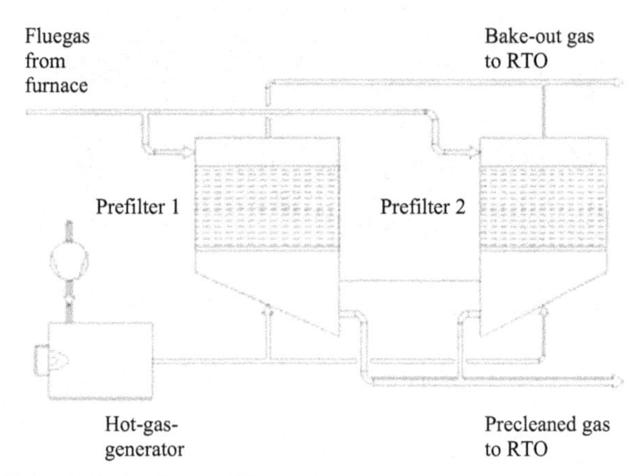

Figure 9: Redundant prefilter system

With increasing temperature the heavy tars will get gaseous and mixed with bake-out air. This polluted air is transported into the combustion chamber of the RTO, where it will be oxidised completely.

A similar bake-out process is used in the RTO. Additionally the valves there have to be protected. During the Drop-out lots of condensates fall down from the ceramic bed of the RTO. To avoid a clogging of the valves the design of the valve-box was changed (figure 10). The valves are now located outside the dropping area. All dropping condensates are collected in the heated cone below, from where they could be discharged easily.

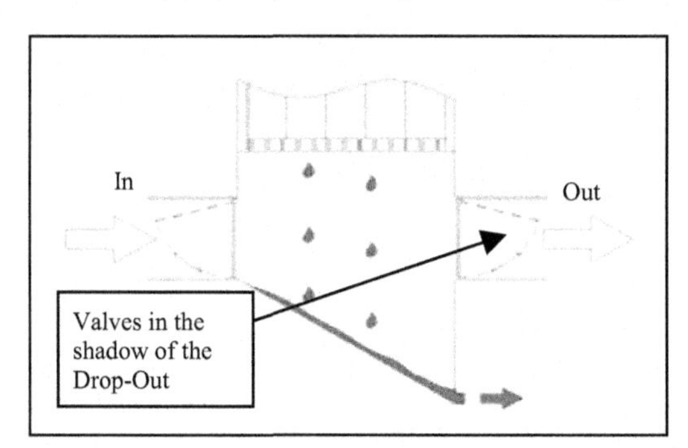

Figure 10: Valve box design

A standard RTO, working according to the 3-chamber method (also 5- or 7- chamber systems) fall back to a 2-chamber mode during the bake-out of one chamber. Therefore two effects occur. On the one hand the typical 2-chamber peak will happen every 2 minutes when the gas direction is changing. This will result in increased emissions. On the other hand the polluted gases of the bake-out usually go directly into the stack, creating additional emissions.

To avoid both effects the system is extended to a 4-chamber RTO, which allows the use of a separate chamber for bake-out whilst running the system in the peak-free 3-chamber mode with a purging cycle. In Figure 11 the flue gas is entering the system in chamber 1, where it is heated by the energy stored in the ceramics. After the oxidation of the organic pollutants in the combustion chamber the clean gases exit the plant via chamber

3, leaving the energy in the ceramic bed. Chamber two, which was the inlet of the prior phase, is now purged with clean gas. Chamber 4 is heated by a gas flow from the combustion chamber up to a maximum temperature of 500°C. The gaseous condensates are transported by a fan directly to the burning chamber, where they are oxidised. The energy content is used for running the RTO and reduces the secondary energy demand, required for the two redundant burners.

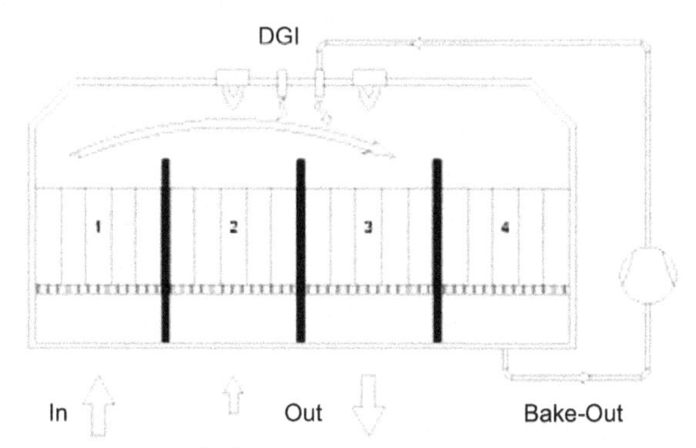

Figure 11: RTO design showing one phase of working

To reduce the energy consumption of the plant, it is additionally equipped with a direct gas injection (DGI) shown in figure 11. This system starts injecting natural gas after reaching the required minimum temperature in the burning chamber by heating with the burners.
The advantage is the saving of combustion air, which would require higher energy to heat it. Figure 12 shows a view into the burning chamber with the DGI.

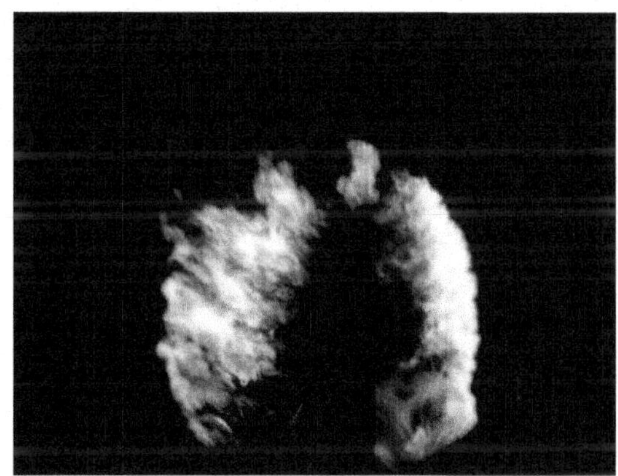
Figure 12: Direct gas injection

As a final stage of the plant a HF treatment system could be added. As wet scrubbing systems would create waste water for this application a dry scrubber is preferred. This technology was developed by LTB in the 1980's and used in several installations of the ceramic industry. Inside the casing there are several cascades, filled with grainy limestone or alumina. These absorption zones insure a good contact of the HF with the absorbent, which will react spontaneously to CaF_2 or AlF_3.

Figure 13: LTB HF absorbers

The complete system to treat the flue gases from anode production will use the following stages:

- Redundant pre-filter with heater
- 4-chamber RTO with special valves
- Redundant fans
- Optional HF- absorption dry scrubbing

An example is shown in figure 14.

Figure 14: Layout with furnace

Conclusion

The RTO based technology will be the best available technology to be used in the anode baking industry, based on the necessity to reduce carcinogenic emissions of PAH.

The first customer has ordered a full size plant of this type in autumn 2005. The installation will be ready for operation in summer 2006. The results of the plant in operation as well as the achieved emissions will be reported later, as they will be the basis of subsequent plants.

References

(1) Reference Document on Best Available Technology in the non ferrous Metals Industry, May 2000, European IPPC Bureau, Sevilla http://eippcb.jrc.es
(2) VDI 3467 Emission Control Production of Carbon and Electrographite Materials; March 1998
(3) VDI 3874 (draft) determination of polycyclic aromatic hydrocarbons (PAH) – GC/MC method; August 2005
(4) UBA-FB AP 2058; Verminderung von PAH-Emissionen durch Errichtung und Betrieb einer thermischen Nachverbrennung mit Wärmerückgewinnung für Ringöfen zur Herstellung von Elektrographit; Hammer, Wolfgang e.a.- SGL Carbon GmbH; August 1994

Light Metals 2006 *Edited by Travis J. Galloway* ***TMS (The Minerals, Metals & Materials Society), 2006***

INTEGRATED TECHNOLOGY FOR BAKING FURNACE AND FUME TREATMENT PLANT

Wolfgang Leisenberg
Innovatherm Prof. Dr. Leisenberg GmbH+Co. KG, Am Hetgesborn 20, D-35510 Butzbach
Giessen-Friedberg University of Applied Sciences, Wilhelm-Leuschner Strasse 13, D-61169 Friedberg, Germany
Rosenstrasse 19, D-61231 Bad Nauheim, Germany

Keywords: Baking, Pitch burn, FTP, Coke, Heating, Emissions, Environment protection

Abstract

In the state of art anode plants baking furnace and fume treatment are regarded as isolated stand alone plants. But fume treatment begins in the baking furnace. The flue gas starts as preheated air in the cooling area of the baking furnace and ends at the stack. Regarding this process as an integrated one, we will find hidden potentials for both plants. Special situations as moving procedure or pitch burn can be anticipated in the FTP as feed forward information, while flue gas parameters can help to improve the baking process as feed back information. The paper also deals with the energy situation of the process including the use of free heat content of the flue gases. Last but not least a concept of the optimisation of the flue gas cleaning is presented, using the furnace for thermal oxidation of VOC and condensable tar emissions.

Introduction

Former papers [1, 2, 3] have shown some results that can be achieved using modern firing and control technologies on open pit (horizontal flue) baking furnaces. Theses results are affecting significantly the efficiency of the baking process. They also have a high impact on the pitch volatile load and on the temperature of the fume gases. This paper opens new possibilities for the anode plant, regarding the baking furnace and the fume treatment plant as an integral process.

Pitch Burn Technology and its Consequences

State of the art pitch burn technology automatically ensures that when starting the pitch burn, the temperature is sufficiently high to ignite the volatiles. It also controls the draft and the fuel capacity of the burners in a way that sufficient oxygen always is available for complete volatiles combustion. Advanced technologies [1, 2] allow tar pitch combustion of 99.84 % on average.

Energy

The fact, that practically all energy of the pitch is used inside the furnace, leads to excellent fuel consumption which typically ranges around 1.8 GJ/t of anode.

Baking Level

As a first side effect long term investigations have proven that the baking level increases using an identical firing profile.

The reason for that phenomenon is that the preheat temperature level is higher due to the volatile combustion and owing to the higher heat transfer to the anode during pitch burn. Many anode plants use the difference of real density between calcined coke and baked anode as an indicator for the baking level. For example in [2] has been found that according to the plant standard the real density of the baked anode should not be lower than that of the calcined coke. After introducing pitch burn technology the real density of the baked anodes has been found significantly higher using identical firing profiles.

Preheat Temperature Gradient

With the pitch burn technology a second side effect appears: Fig. 1 shows a typical pitch burn anode temperature profile compared to a conventional one. Due to the higher heat transfer to the anode during pitch burn the temperature rise starts earlier. Therefore the maximum gradient drops from 14 °C/h to 12 °C/h with identical top temperature and cycle time. This side effect may help to avoid cracking of the anodes reducing thermal tensions during the preheat phase.

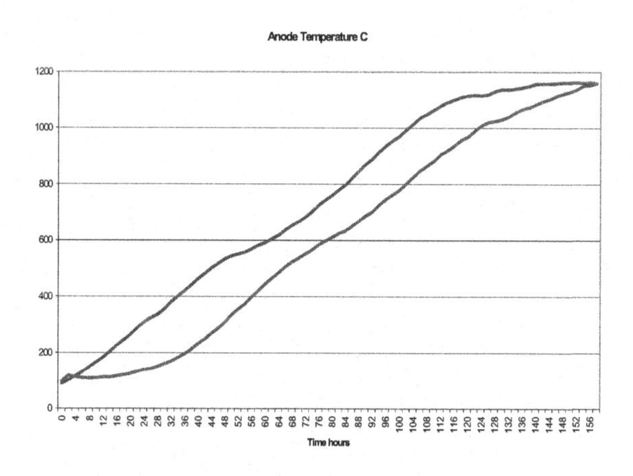

Fig. 1 Temperature profile (blue curve with pitch burn)

Emissions

The VOC in the flue gas subsequently are very low since most of the volatiles are burnt in the process. Measurements show that modern firing and control systems achieve total unburned tar pitch content in the waste gas of less than 25 mg/Nm³ at the outlet of the furnace (Table 1).

Emissions	Unit	Conventional System	ProBake (ALBA)
Total Tar	mg/Nm³	350-700	22
Cond. HC	mg/Nm³	250-500	14
VOC	mg/Nm³	65-200	8
PAH 16	mg/Nm³	6.5-20*	0.8*
Benzpyrene	µg/Nm³	65-200*	8*
		* estimated	

Table. 1: VOC Load of raw flue gas

Flue Gas Temperature

The heat balance of a furnace with complete pitch burn shows, that the heat content of the volatiles cannot be utilized completely in the baking process (Fig. 2). Therefore the flue temperatures at the exhaust ramp are roughly 100°C higher than without complete pitch burn.

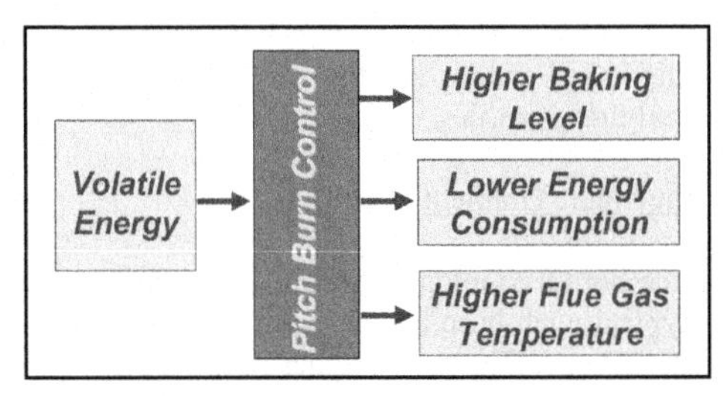

Fig. 2: Schematic energy balance of volatile heat

Integrated Flue Gas System

Fume Treatment Plant

The VOC load of a furnace performing complete pitch burn is less than what some conventional plants achieve after fume treatment. The fume treatment plant (Fig. 3) can be simplified or modified with this low VOC and condensable load.

Fig. 3: Fume Treatment Plant

The first fact is that the furnace acts as a conditioner for the flue gas: The volatile and VOC load can be influenced by the firing control. So the data from the raw gas and the clean gas can be used to modify the operation of the firing system, and vice versa the operation mode of the furnace can be used for feed forward control of the fume treatment plant.

In practice, if some methane appears in the fume treatment plant, the reason must be not completely burnt fuel in the furnace. Since methane cannot be removed by adsorption materials as alumina, the firing control has to react by increasing oxygen content by draft or by decreasing fuel capacity. If the firing is not in move procedure or in corner situation, the recirculation of alumina can be much higher than in move or corner situations. This means that the fume treatment plant can be operated using a minimum of adsorption matter. Fig. 5 shows the scheme of a compact fume treatment plant.

These facts may illustrate that it makes highly sense to regard the furnace and the fume treatment as conditioners of the clean waste gas. Therefore both plants should be controlled not isolated but as an integrated system exchanging data and reacting in best and most economic way to achieve the parameters given by the environment authorities.

Heat Transfer to the Paste Plant

The second challenge is to use the heat of the flue gases within the plant. There is just one relevant heat sink in the anode plant to heat the coke before it is contaminated with the hot fluid tar pitch for better wetting to minimize the pitch content of the anode.

Coke heating in the paste plant is generally performed using a heated screw conveyor using heat transfer oil. Because the average difference between flue gas temperature and the desired heat transfer oil temperature is not big enough, the flue gas cannot replace the firing system of the heat transfer oil.

Trials with direct heat transfer from flue gas to coke using a moving bed process have shown very promising results. With an ingoing flue gas temperature of 200°C, the outgoing coke temperature can be brought easily to 150°C. This temperature is sufficient for most applications, since the hot fluid pitch transfers additional heat to the coke. As shown later, if higher coke temperature is needed, the flue gas temperature can be increased easily.

The specific heats of flue gas and coke are in the same level, while in average the flue gas mass flow is, due to the inleaked air, three to six times more than the mass flow of the coke. As a first result it can be stated, that more than enough heat is available at sufficient temperature level to heat the coke to the desired temperature.

Since the flue gas volume available is rather too large for heating the coke, a part of the flue gas bypasses the heat exchange unit.

Nevertheless, depending on the number of fires connected, the flue gas temperature of the ring main may fluctuate up to 100°C. It is low after fire move, and increases steadily until the next fire move. So it may happen, that for a period of time the flue gas temperature falls below 200°C.

In general the cooling zone has a big potential for heat recovery. A small portion of the heat content of the baked anodes is used as preheated air in the firing zone. The larger portion of the heat and of the refractory is blown into the furnace hall.

Installing exhaust ramps in the cooling area which feed hot air into the flue gas duct, additional heat can be used without installation of additional hot air ducts. The hot air temperature depends on the position of the ramp within the cooling area. The most suitable exhaust temperature is in the range of 400°C allowing the use of identical exhaust ramps all over the furnace. Adding a controlled portion of this hot air into the ring main, a minimum temperature of 200°C can be achieved easily. In the worst case of two fires on the furnace, after fire move a decreasing volume may last for about 6 hours averaging in roughly 20% of the flue gas volume.

Including Pitch Fume Treatment

Since the flue gas has to be lead into the paste plant for heat exchange it makes sense to feed the pitch fumes of the paste plant into the flue gas duct. A calculation of the temperature after mixing shows values far below 200°C, which excludes an ignition of the fumes coming from the paste plant. One fume treat-ment plant is sufficient to clean all fume gases of the anode plant.

This FTP can be a conventional dry adsorption unit, or, as an alternative, a multiple fluidized bed reactor (Fig. 4). This reactor consists of fluid beds for several absorption materials, which are passed by the flue gas.

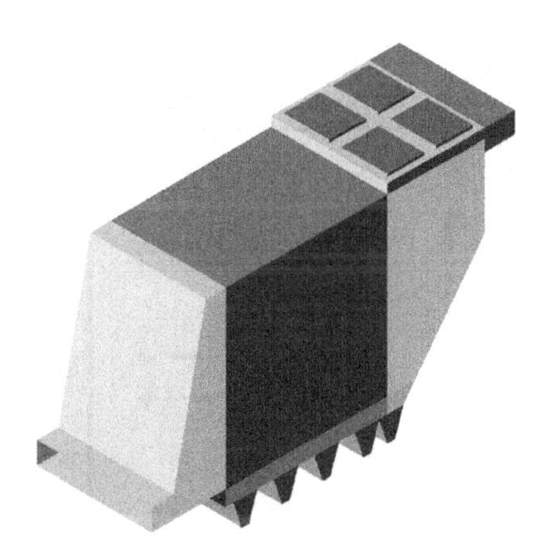

Fig. 4: Multiple fluidized bed reactor

As a further alternative, similar to the alumina reactor, fine fraction of the coke can act as an adsorption matter. Since tar pitch fume treatment plants in paste plants, meeting high emission standards, are using this adsorption matter, the fine fraction of the coke should be suitable for flue gas as well.

Fig. 6 shows a complete system of this technology. The coke, split in different grain size fractions (3, 4) is lead through the heat exchange unit (1), where the green scrap (5) is fed directly to the mixers or kneaders in the paste plant (6). The flue gas (2a) with mixed with hot (11) or cold (18) air from additional exhaust ramps in the cooling zone is split in two portions, the one going into the heat exchanger (1) and the other (2e) bypassing the heat exchanger. During disturbances at the furnace, hot air can be generated by an additional burner (12). Both volumes (2a) and (2e) are joined (2c) and the pitch fumes (14) of the paste plant (6) are added. Depending on the FTP-technology the flue gas (2d) passes an adsorption reactor (15), optionally a limestone grit bed and finally a fabric filter before it leaves the plant via the chimney. To minimize the electrical energy of the fans, the bypass volume (2e) can be lead separately (19) through the fume treatment plant.

Financial Aspects

Running cost

To give a rough overview, the heat recovery for the process is roughly 150 MJ per ton of green anode. Taking this into account, the effective specific energy consumption of the baking process may drop to about 1.65 GJ per ton of baked anode. Total heat savings of 20,000 GJ are possible for a smelter of 250,000 tons of aluminium per year. This significantly affects the running cost, especially in view of increasing energy prices. It goes without saying that this also has a positive impact on the environment.

Investment

Since the flue gas duct has to be lead to the paste plant, gases from the green material production can easily be given to the flue gas duct. There is no need for a separate fume treatment plant for the green anode plant. The heating mixers and heat transfer oil equipment are not required as well. Taking all this in account, the investment for the new technology should not be higher than that of a conventional plant. In combination of an compact FTP or a multiple fluidized bed reactor fume treatment, this technology for new plants will be most cost effective.

Summary

The control strategy for complete pitch burn leads to lower energy consumption, higher anode quality, lower preheat gradients and less emission. Since most of the volatiles are burnt in the process, the unburned tar pitch content in the flue gas is very low. Therefore the fume treatment plant can be built more compactly to meet the environmental standards. Since the heat content of the volatiles cannot be utilized completely in the baking process, the additional heat can be used in the paste plant for heat the coke by direct heat exchange in a fluid bed reactor. Especially for stand alone anode plants, a multiple fluid bed reactor can be considered, using limestone and activated carbon or the fine fraction of coke for adsorption. The new technology at same investment cost as conventional plants will reduce running cost for fuel and contribute to environment protection.

Literature

[1] W. Leisenberg, *Flue Gas Management*, Light Metals (1999) 579-584

[2] J. Ameeri; K. M. Khaji; W. K. Leisenberg, *The Impact of the Firing and Control System on the Efficiency of the Baking Process*, Light Metals (2003) 589-594.

[3] W. Leisenberg, *Firing and Control Technology for Complete Pitch Burn and its Consequences for Anode Quality, Energy Efficiency and Fume Treatment Plant*, 9th international Conference on non ferrous metals, July 8-9, 2005, Pune, India.

Fig. 5 Compact Fume Treatment Plant

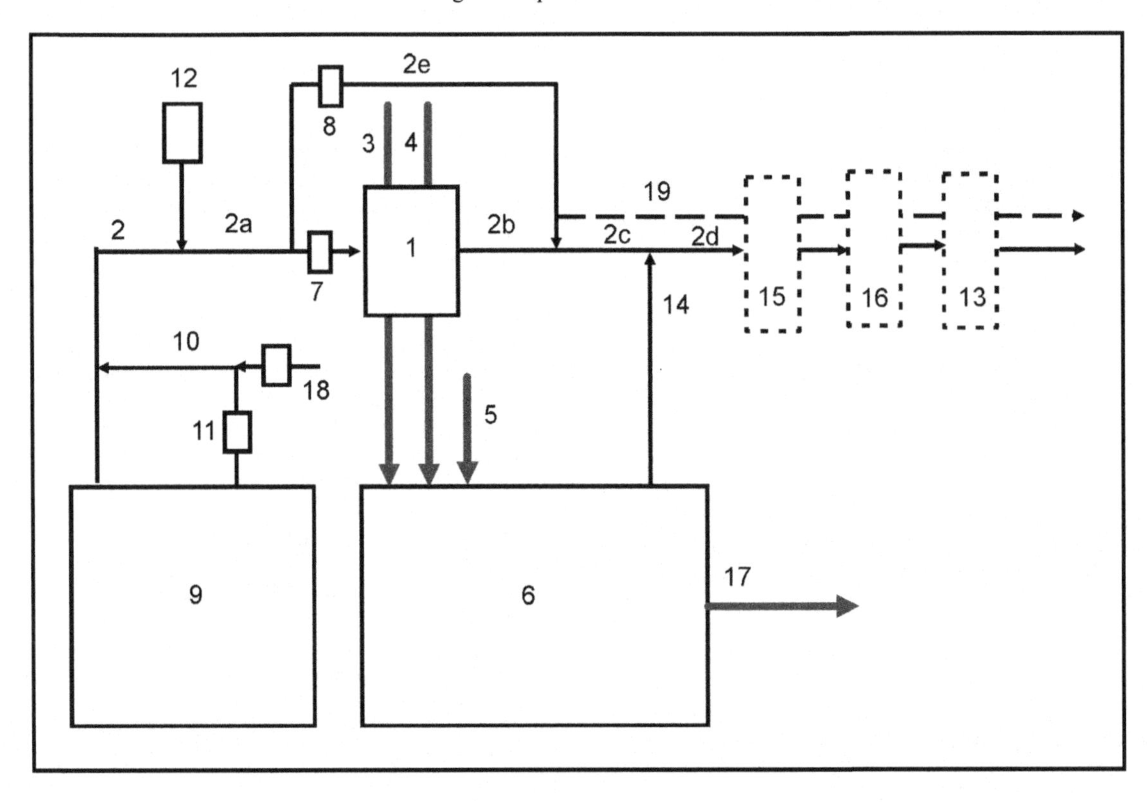

Fig. 6 Scheme of the complete System

Light Metals 2006 *Edited by Travis J. Galloway* ***TMS (The Minerals, Metals & Materials Society), 2006***

FLUE CONDITION INDEX – A NEW CHALLENGE TO INCREASE FLUE LIFETIME, OPERATIONAL SAFETY AND FUEL EFFICIENCY IN OPEN PIT ANODE BAKING FURNACES

Detlef Maiwald, Wolfgang Leisenberg
Innovatherm Prof. Dr. Leisenberg GmbH + Co. KG; Am Hetgesborn 20, D-35510 Butzbach, Germany

Abstract

The condition of the flue walls in open pit anode baking furnaces is an important factor in terms of production efficiency. The flue walls change their physical properties which change the flow resistance, produce air leakages and even affect the mechanical stability due to the periodical heating/cooling cycle. Therefore the flue walls have to be observed regularly by maintenance staff and exchanged in average after a lifetime of 150 fire cycles.

With the introduction of a flue condition index, each flue in a furnace is evaluated continuously. An on-line mathematical model detects the actual condition of each flue by correlation of the relevant process data available in the firing system. As a consequence the firing properties like the maximum fuel input or the draft can be adapted or limited to the actual condition of the flue. This prevents critical situations, avoids hot spots and increases operational safety, flue wall lifetime and fuel efficiency.

Introduction

The anode baking process is running basically as a two convection heat exchanger with a firing zone in between, thus forming three areas. These areas are the preheat, firing and cooling area. Therefore, the basic requirement of the Firing Control System is to control each flue of a fire in these three areas.

The baking process is determined by a temperature-versus-time function relating to the anodes. The heat is mainly produced by combustion of primary fuel in form of natural gas or heavy fuel oil inside the flues, introduced by a set of equipment called a "fire" which is moved around the furnace.

Figure 1 shows a typical arrangement of a fire with the related temperature profile.

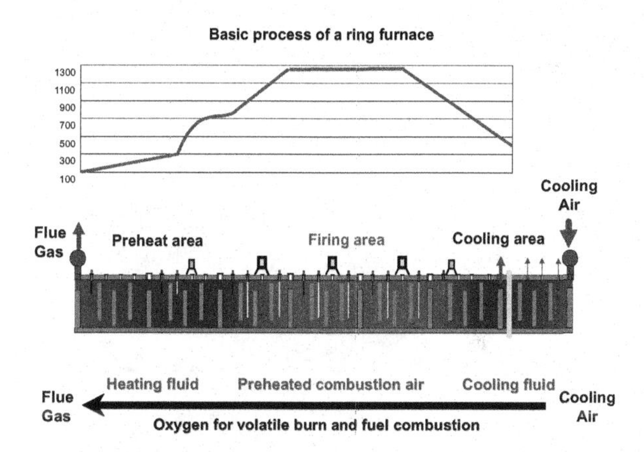

Figure 1: Baking furnace principle

For the production of baked anodes, green anodes have to be loaded into the baking furnace. They have to be heated up to 1080 °C and cooled down afterwards to reach unloading temperature. The cycle time is about 160 h to reach the maximum temperature and another 160 h to get back to unpacking temperature of about 230 °C.

So in average every 14 days the furnace is operated in an alternating temperature range of more than 1000 °C.

This cyclic and non continuous procedure performs thermal stress to the furnace structure and especially to the flue walls made of high temperature resistant refractory materials. As a result the flue walls start to deteriorate from day one of production, and after some time of operation visual changes can be observed like bending of the flue walls, vertical cracks along the flue wall or expansion of the joints to gaps. Figures 2-4 are typical spots through a life time of a flue wall.

Figure 2: Flue wall condition of a new furnace

Figure 3: Flue wall condition after years of operation

Figure 4: Typical cracks along a flue wall

Production consequences

Since the anode baking furnace structure mainly consists of flue walls (a typical furnace contains more than 324 flue walls), special attention has to be laid on the changes of the flue wall condition. Consequences can be as follows:

- Bending of the flue walls may change the physical properties of the flow resistance
- Bending of the flue walls lead to gaps at the joints, where air leakages will occur
- Cracks and gaps at the flue walls may lead to ingress of packing coke materials, which will block the flue partially or totally
- Total collapse of a flue wall may lead to serious operational hazards

Therefore in most of the anode baking facilities a "refractory maintenance team" is established to observe the flue wall condition after each fire cycle, plan necessary repairs or partial changes of flue walls after a specific life time.

But this maintenance team is checking mechanical damages only, negative effects in the thermal process are not observed.

Evaluation of a Flue Condition

The basic idea for the evaluation of a Flue Index is the utilization of the existing data collected by the firing system and the integration into a fuzzy logic system. A typical condition of an undisturbed flue situation can be discussed by Figure 5.

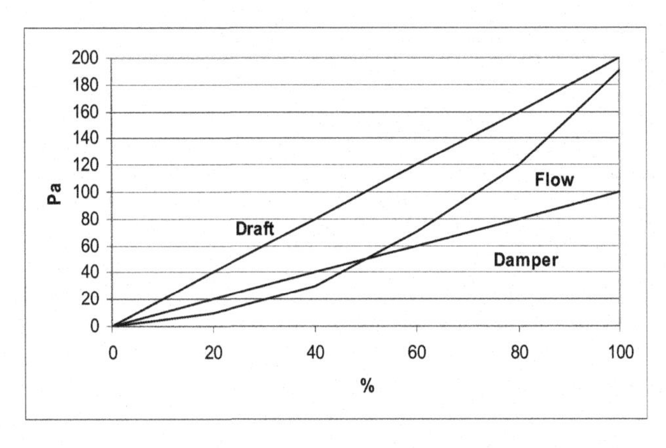

Figure 5: Condition in an undisturbed flue

This Figure illustrates the damper position in per cent at the exhaust manifold for one flue versus the actual draft value and indicates the resulting flue gas flow rate of an "undisturbed" flue in good or new condition.

Figure 6 shows the same values for a disturbed flue, or a flue in a poor condition.

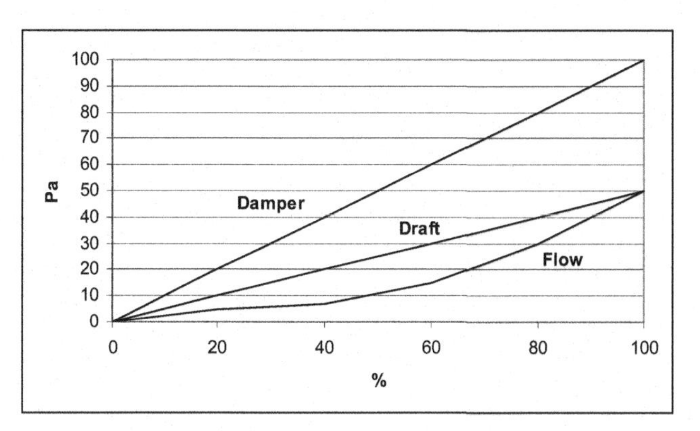

Figure 6: Condition in a disturbed flue

In comparison to Figure 5 for the same damper position, the draft value is lower and also the corresponding flue gas flow is lower. Taking this phenomena into account, the flue must have changed its physical properties, i.e. the flow resistance has increased.

Similar dependencies can also be found in the firing zone at the burner ramps. The target temperature can be achieved with different burner capacities, depending on the actual flow resistance of the flue. If the flue is in poor condition, for the same target temperature a higher burner load can be found in relation to other flues. And this situation even worsens the condition of the flue, resulting in higher hot spot temperatures underneath the burners and higher fuel input into the furnace.

The Flue condition index

As described earlier, certain conditions and values of the firing system give an indication of the current flue condition. These are mainly:

- The damper position
- The temperature deviation
- The burner capacity
- The draft deviation

Therefore a fuzzy logic model is chosen to perform a Flue Condition Index by introducing these data as an input data base. The principle is shown in Figure 7.

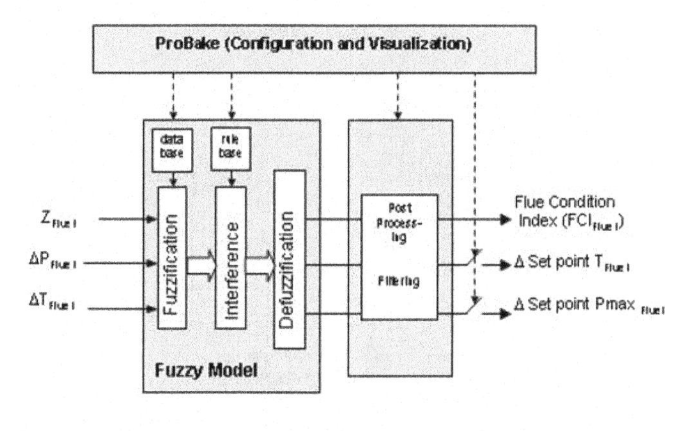

Figure 7: Flue condition index fuzzy model

Fuzzy technology contains theoretical methods und technical realizations based on fuzzy logic algorithms. While within the crisp logic only two values of degrees of truth exists (true=1 and false=0), the fuzzy logic is a multivalent logic. Within the fuzzy logic the degree of truth can be defined as function of weighing factors. Therefore values between 0 and 1 for degrees of truth are possible. Although the fuzzy technology uses humans experience, vague knowledges and heuristic methods, it performs on a defined scientific base.

The input signals are correlated, introduced into the fuzzy logic model, where the dependencies and rules are set and processed. As a result a Flue Condition Index is continuously calculated, indicating a Flue Condition in a range of 0 – 100 per cent.

Survey of the Furnace by Flue Condition Index

Having this Flue Condition Index as a base, the furnace is now continuously supervised in terms of an actual condition of each flue. The operator is informed about the status of every flue, the same applies for the refractory maintenance staff. Degradation of the flues can be observed in an early stage, before a situation becomes serious. Additionally the flue life time of each flue in terms of fire cycles will be observed automatically. A typical result of the flue condition index is shown in Figure 8.

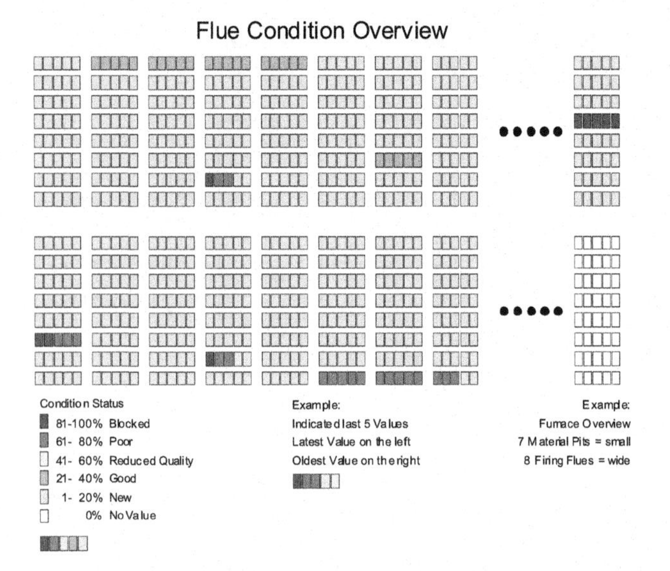

Figure 8: Flue condition furnace overview

Each flue contains 5 indication bars of the last 5 evaluations of a flue condition index each time a fire passes across this flue. Grey and green colours of this bar indicate flues in nearly new condition, yellow colour indicates ongoing degradation where as a red colour shows up bad condition or even a blocked flue. Now it is possible to select one specific flue and get details data which are shown in Figure 9.

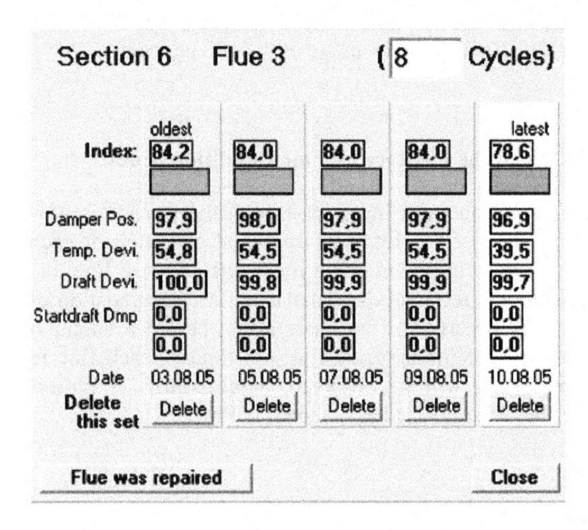

Figure 9: Flue condition index details

The top line focuses the selected flue and indicates the actual life time in terms of fire cycles. The last five calculated Flue Condition Index (FCI) Values are shown which reflect directly the status of the flue. Also the input data of the fuzzy logic model are shown in correlation to the FCI found.

Now it becomes obvious to integrate these results into the firing control strategy in order to adapt and optimize the baking process.

Adaptive Control

The standard firing and control system does not recognize any variations in the condition of the flue. It even acts in the wrong direction when flues become older and the flow resistance is increasing. The standard control of the burner ramp can be discussed as a typical example. It introduces a certain amount of fuel at the first peepholes of the flue to reach the target setpoint temperature, which is measured at the end of the flue. If the flue becomes older, a higher amount of fuel is necessary to reach the same target temperature. This leads to higher hot spot temperatures at the burner entries and consequently higher temperature stress to the refractory materials. Carbon plants will recognize, that after some years of operation the specific fuel consumption of a baking furnace is slightly increasing.

An adaptive control algorithm using the FCI prevents partial overheating by limiting the maximum fuel input dynamically. The control strategy is shown in Figure 10.

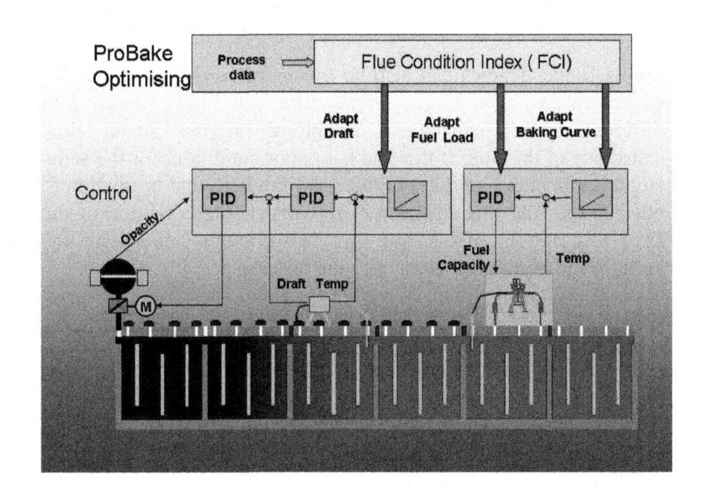

Figure 10: Adaptive control strategy

Further adaptive controls which can be executed through the FCI are the adaptation of the draft and the adaptation of the baking curve. If a flue is detected in a poor condition, immediately a higher draft setpoint is activated to reach the same flow condition compared to an undisturbed flue. The standard control would rely on the same draft setpoint, which may lead to bad combustion condition in terms of available oxygen in the flue gas.

Finally, also the baking curve can be adapted dynamically to the condition of the flue. If a flue is in bad condition, the heating up gradient will be lowered and the final soaking temperature adapted at a later stage to reach a similar heat transfer to the anode.

As a consequence, older flues will be treated with lower thermal stress which leads to an additional increase of the flue lifetime in terms of number of fire cycles.

Operational safety

Each flue will be supervised on-line and in every phase of production. Therefore hazardous situations are minimized. Bad flues are detected and announced before the situation becomes serious. Even blocked flues will be detected and consequently the fuel input will be stopped immediately. Also repairs and flue changes can be planned and executed on the base of real data and statistics on top of visual inspections before a collapse occurs. This improves operational safety and prevents hazardous situations utmost.

Summary

For the optimization of the control of an anode baking furnace, a Flue Condition Index (FCI) Module was invented. Using the conventional data from a firing and control system, the condition of each flue is evaluated by a fuzzy logic model. This additional information is a basis for further improvements of the furnace operation. The integration of the FCI into the firing strategy increases the lifetime of the flues, improves the fuel efficiency especially on older furnaces and maximizes the operational safety.

References

[1] W. Leisenberg, "Flue Gas Management", Light Metals 1999, 579-584

[2] David Wilburn, Noranda Aluminium, "Test Results of Flue Condition Index" , Internal Report 2005

[3] P. Mnikoleiski, Innovatherm, Investigation of various Flue Conditions at ALCAN Sebree Works, Internal report 2005

[4] A. Himmelreich, Innovatherm, Flue Condition Module, Fuzzy Logic Fundamentals, Development Report 2005

Cathode Properties/ Refractory Materials

SESSION CHAIR
Glen Goeres
Alcoa Aluminum
Rockdale, TX, USA

Light Metals 2006 *Edited by Travis J. Galloway* ***TMS (The Minerals, Metals & Materials Society), 2006***

INFLUENCE OF INTERNAL CATHODE STRUCTURE ON BEHAVIOR DURING ELECTROLYSIS PART III: WEAR BEHAVIOR IN GRAPHITIC MATERIALS

Pretesh Patel[1], Margaret Hyland[1], Frank Hiltmann[2]

[1]Department of Chemical and Materials Engineering, University of Auckland, Private bag 92109, Auckland, New Zealand
[2]SGL Carbon GmbH, Griesheim Plant, Stroofstrasse 27, 65933 Frankfurt, Germany

Keywords: Porosity, graphitic cathode, aluminium carbide formation, wear

Abstract

Graphitic cathode samples with varying granulometries were tested under a number of controlled laboratory electrolysis conditions to determine the effect of internal cathode structure and operating conditions on wear behavior and rates.

The graphitic samples produced were all relatively dense materials with low open porosities and in some cases a narrow pore size range which helped to reduce bath penetration into the cathode and thus reducing the ability for internal aluminium carbide formation to occur. However, though bath penetration was minimal, wear rates were still significant and it was thought that accelerated and preferential aluminium carbide (Al_4C_3) formation was occurring with the disordered carbon in the binder phase material. This wear behavior was accelerated with increasing current density and excess AlF_3 content indicating that the main wear process was electrochemical.

Introduction

Graphitic and graphitized cathode blocks have become more prevalent in modern aluminium smelters as they have proven to have superior thermal and electrical properties than previously used anthracite based cathode blocks. With lower electrical resistivity, smelters have been able to increase production through continual increases in cell current density; however this positive increase in production does have an associated negative in terms of cathode life. Accompanied by the shift from anthracitic cathodes to cathode blocks with higher degrees of graphitization cell life has dropped considerably in most smelters.

It has been widely accepted that the increase in wear rate and non-uniform wear corresponds to an increase in current density. This increase in current density is believed to be the driving force behind aluminium carbide formation which is one of the main degradation reactions occurring at the cathode. Al_4C_3 can form through a number of means, firstly through direct contact between molten aluminium and the carbon material [1].

$$4Al_{(l)} + 3C_{(s)} \rightarrow Al_4C_3$$

Though this reaction is thermodynamically favoured at electrolysis temperatures, the amount of aluminium carbide actually produced this way is well below that predicted by the reactions kinetics and thermodynamics. Therefore, other phenomena must be occurring to account for the actual degree of aluminium carbide formed [1]. The second proposed method for aluminium carbide formation is a chemical reaction between intercalated sodium, cryolite and carbon [2].

$$4Na_3AlF_{6(l)} + 12Na_{(in\ C)} + 3C_{(s)} \rightarrow Al_4C_3(s) + 24NaF_{(l)}$$

This reaction relies on significant amounts of intercalated sodium in order for it to be the dominant method for aluminium carbide formation. This is unlikely to be the case in highly graphitized cathode blocks where sodium intercalation is minimal within the bulk material [2].

Increasing wear rate with increasing current density suggests that an electrochemical reaction may be the dominant aluminium carbide forming mechanism.

$$4Al^{3+} + 3C + 12e^- \rightarrow Al_4C_3$$

This reaction relies on a supply of Al^{3+} ions which is formed through the anodic dissolution of molten aluminium at the metal/electrolyte interface. Increasing the current within the system will increase the driving force for the anodic dissolution of Al^{3+} and will increase the driving force for the transport of Al^{3+} ions to and into the cathode material, thus increasing the chance of aluminium carbide being produced [3,4].

The formation of aluminium carbide is not the sole mechanism that contributes to the cathode wear. The aluminium carbide formed will dissolve into the bath if exposed to it through the following suggested dissolution reaction [5].

$$Al_4C_{3(S)} + 5AlF_{3(diss)} + 9F^-_{(diss)} \rightarrow 3Al_3CF_8{}^{3-}{}_{(diss)}$$

The bath however has a limited solubility of aluminium carbide which when reached will cause the aluminium carbide layer on the cathode surface to grow. This phenomenon is also believed to occur within the pores of the cathode and can lead to internal stresses in the material leading to particle detachment and uneven wear [3,6].

This paper is an extension of previous papers presented at TMS 2005 [6-7]. Part one of the previous papers discussed comparisons in mechanical properties between graphitized and graphitic materials with similar granulometries. For each case a range of materials were prepared with each having varying degrees of porosity. Generally it was found that isotropic coke based graphitized material were more sensitive to changes in formulation than the graphite based graphitic material. This was shown with the large spread in open porosity between samples with varying composition and also in the large difference in the mechanical properties between high and low porosity samples. Greater sensitivity to formulation changes was attributed to the granulometry of the isotropic coke particles [7].

Isotropic coke material as its name suggests means the grains are similar in size and shape in all directions, therefore during extrusion there will be no preferred orientation for grain alignment and this could lead to inadequate packing of the aggregate material. Further restrictions to grain alignment and packing could also occur due to the hard nature of the isotropic grains. This would restrict the movement during extrusion and will lead to a less than ideal-packed product [7].

Graphitic material on the other hand proved to have a much higher tolerance to granulometry and formulation changes. Graphitic material produced much denser material with more consistent mechanical properties. This was attributed to the alignment of the anisotropic graphite grains during extrusion. During extrusion graphite grains will preferentially align their major axis with the direction of extrusion. This coupled with the soft, self lubricating characteristics of the graphite grains allowed greater aggregate movement and alignment to take place and create a denser final product [7].

In terms of mechanical properties the following conclusions were made for isotropic coke samples versus graphite samples.

- Graphitized isotropic coke materials tended to form products with higher porosity and less grain orientation than the graphitic samples under the same extrusion conditions and formulation.
- Graphitic material had superior density, flexural and compressive strength than isotropic coke samples with the same grain size formulations. This was found to be due to the superior packing arrangements found in the graphitic material.
- Isotropic coke based samples had lower electrical resistivity than graphitic samples. This was expected due the difference in processing of the two types of samples. Isotopic coke samples were fully graphitized while graphitic samples were baked to 1100°C thus giving the graphitized material higher structural order and thus lower resistivity.
- Cut direction had a significant effect on the properties in graphitic material, indicating definite grain orientation and alignment. The same phenomenon was practically not found with the isotropic coke samples.

Experimental

Experimental work was based around the design of a 2nd order factorial matrix, which defined the compositions of the cathode samples and the operating parameters that would be used during the laboratory electrolysis experiments. The objective of this investigation was to investigate the effect of porosity and pore size distribution in graphitized and graphitic material during electrolysis, therefore the matrix was used to define the compositions needed to achieve a range of porosities and pore size distributions. Table 1 shows the variables which were investigated in this work.

Table 1: Key variables assessed

Variable	Variance 1	Variance 2
Filler type	Isotropic coke	Graphite
Granulometry	Coarse	Fine
Flour content	Low	High
Cut direction	Parallel	Perpendicular
Bath acidity	5% excess AlF_3	10% excess AlF_3
Current density	0.6 A/cm^2	1 A/cm^2

From this matrix four core formulations for both isotropic coke and graphitic material were formulated. These formulations defined the granulometry of the samples.

Table 2: Core formulations for sample production

ICL	Isotropic coke, Coarse granulometry, low porosity
ICH	Isotropic coke, Coarse granulometry, high porosity
IFL	Isotropic coke, Fine granulometry, low porosity
IFH	Isotropic coke, Fine granulometry, high porosity
GCL	Graphite, Coarse granulometry, low porosity
GCH	Graphite, Coarse granulometry, high porosity
GFL	Graphite, Fine granulometry, low porosity
GFH	Graphite, Fine granulometry, high porosity

All samples were extruded and initially baked to 800°C to allow carbonization of the binder pitch to occur. Isotropic coke based samples were then graphitized at 2700°C to increase structural order within the material (will be referred to as graphitized isotropic coke material). Anisotropic graphite based material was re-baked to 1100 °C. Re-baking the graphitic material at 1100 °C left the binder phase in a relatively disordered state compared to the graphitized material (will be referred to as graphitic material). An in-depth account of the processing and mechanical properties of the produced samples can be found in previous work [7].

All samples were tested in an inverted cell configuration as shown in figure 1.

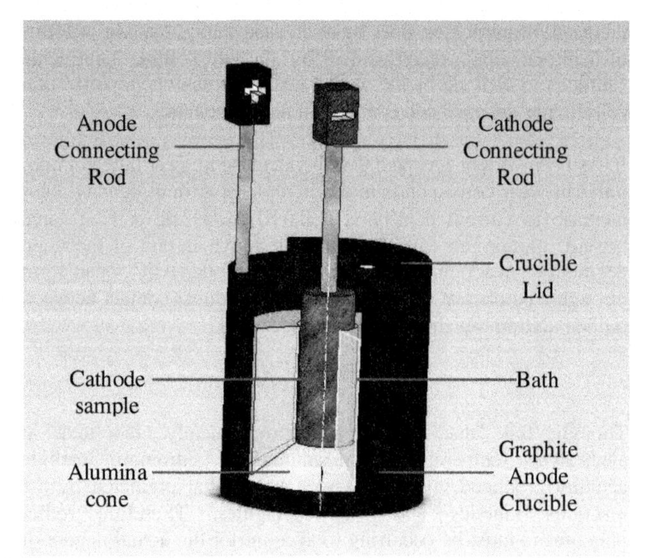

Figure 1: Schematic showing the laboratory electrolysis configuration

Samples, 30mm diameter x 110mm length were subjected to 96 hour electrolysis trials; this length of time produced measurable wear and allowed enough time for wear phenomena such as aluminium carbide formation and dissolution to take place and be observed.

Following electrolysis, samples were cleaned of excess bath and the volume loss measured through mercury displacement and dimensional analysis techniques.

Wear rates were then analyzed to determine which of the variables outlined in Table 1 were important contributors to wear. Analysis of the effects of variables on wear rates were conducted using a second order factorial analysis method which allowed individual effects and combinations of effects evaluated.

Results and Discussion

The results reported in this paper mainly concentrate on the wear phenomenon observed during laboratory electrolysis trials of graphitic cathode materials with varying granulometries and open porosities. Samples were tested under constant current density which was either 0.6 or 1 A/cm^2 with bath compositions of 5% or 10% excess AlF_3 content. Configuration of experimental conditions and samples were determined by the experimental matrix outlined earlier.

In part II of this series of papers it was found that the graphitized isotropic coke materials showed distinct wear behavior as open porosity was increased [6]. As porosity was increased it was found that the uniformity of the wear declined. With greater open porosity, a higher degree of bath penetration took place and led to a greater degree of internal degradation reactions occurring such as aluminium carbide formation. Internal aluminium carbide formation, dissolution and growth coupled with the already weak internal structure of the cathode due to high porosity lead to the phenomenon of particle detachment. This was the cause of the pitting and uneven topography on the high porosity graphitized cathode material [6].

With the graphitic material however, much denser materials were formed even with formulations designed to generate high porosity samples. This increased density coupled with the processing steps used to produce the graphitic cathodes gave rise to quite different wear patterns and behavior than what was observed with the graphitized isotropic coke. Figure 2 shows the open porosity and pore size distributions achieved for graphitic samples.

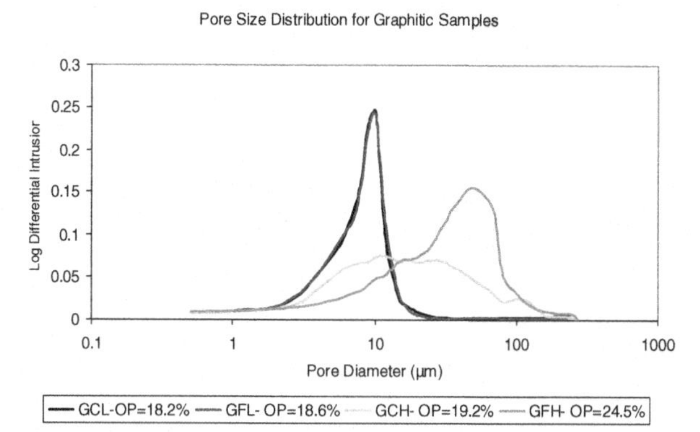

Figure 2: Pore size distribution of graphitic samples

The main contrasting difference between the graphitized and graphitic material tested was that all graphitic materials showed uniform wear behavior with very little evidence of pitting and particle detachment. This was generally expected as previous testing of graphitized material lead to the conclusion that the denser a material is the more uniform the wear profile will be. Figure 3 shows the typical wear pattern of low and high porosity graphitic material.

(a)

(b)

(c)

Figure 3: Typical wear pattern of **(a)** low porosity graphitic material **(b)** high porosity graphitic material **(c)** high porosity graphitized sample for comparison

Though Figure 3a-b show even wear along the electrolyzed length (in terms of no great evidence of pitting or particle detachment as shown in figure 3c), there was however a strong indication of preferential wear occurring at the graphitic cathode surface (fig 3a-b). In many cases a grainy, coarse appearance was found on the cathode surface. This is believed to be due to the preferential aluminium carbide formation and dissolution of the disordered carbons in the binder phase material [3]. Figure 4 shows a close up of a worn graphitic cathode sample.

Figure 4: Course, grainy topography of electrolyzed graphitic cathode material

The graphitic cathode material used in these trials had only been heat treated to 1100°C which while still allowing carbonization to occur will only allow low levels of structural re-alignment to occur, thus leaving the binder phase in a relatively disordered state compared to the highly ordered graphite filler material. Thus as mentioned during electrolysis this binder phase will facilitate aluminium carbide formation and dissolution preferentially over the graphite filler leading to the binder phase having a higher wear rate. Therefore resulting in the course grainy appearance of the cathode material as the graphite filler has worn but not to the same extent as the binder phase.

Bath Penetration and Internal Aluminium Carbide Formation

In theory the preferential degradation of the binder phase over the filler material could lead to possible pitting and particle detachment due to bath penetration and internal aluminium carbide formation. However, as mentioned no definitive evidence of particle detachment was found in any of the graphitic cathode material.

Cross-sections of high and low porosity graphitic material provided a number of explanations to why pitting caused by particle detachment was not a major contributor to wear in the tested graphitic material. Figure 5a-c shows various cross-sections of low and high porosity graphitic cathode material.

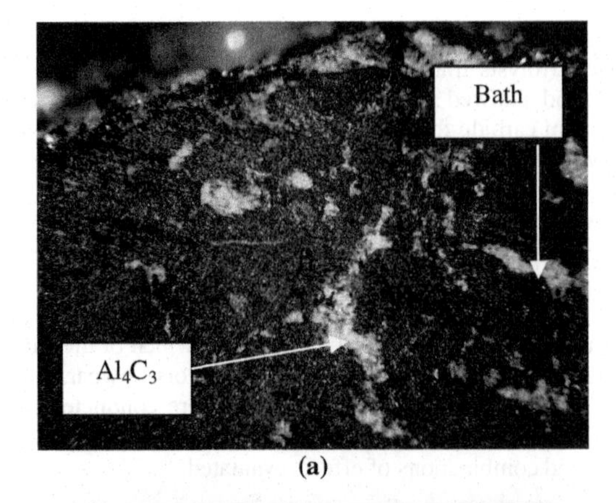

(a)

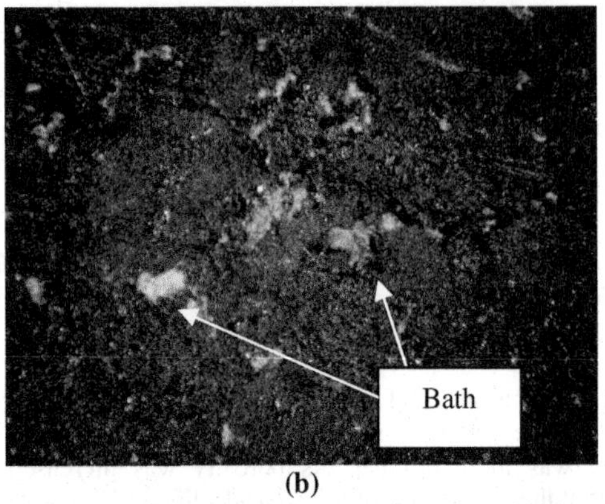

(b)

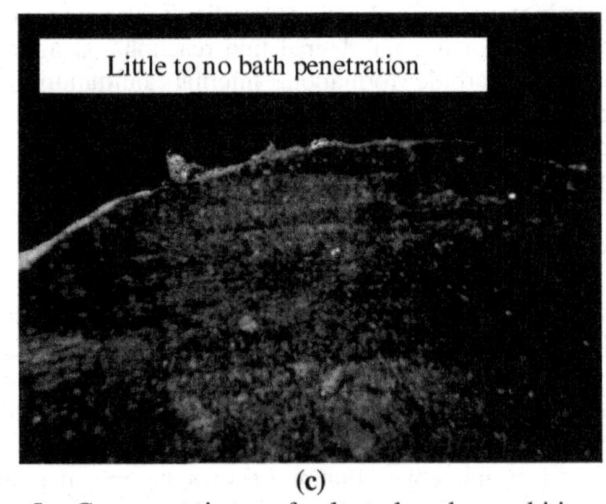

(c)

Figure 5: Cross sections of electrolyzed graphitic cathode material. **(a)** Fine granulometry, high porosity, edge of sample **(b)** Fine granulometry, high porosity, center of sample **(c)** coarse granulometry, low porosity, edge of sample

Figure 5a shows the edge of a cross sectioned graphite, fine granulometry, high porosity sample after 96 hours of electrolysis. The key point of interest is the presence of a yellow compound within the pores of the cathode. This yellow compound was established to be aluminium carbide and could be found mainly situated close to the sample/bath interface of the cathode material. Figure 5b shows a section from the center of the same sample where yellow aluminium carbide is not present. Here a greater degree of white areas were found which indicates bath penetration without carbide formation.

It is thought that the internal aluminium carbide formation is mainly situated close to the bath/cathode interface as in this region there will be a higher concentration of available Al^{3+} ions which are needed for the electrochemical formation of aluminium carbide.

$$4Al^{3+} + 3C + 12e^{-} \rightarrow Al_4C_3$$

Al^{3+} ions are generated by the anodic dissolution of the aluminium metal when it comes in contact with the bath through the following mechanism. [3,4]

$$4Al_{(metal)} \rightarrow 4Al^{3+} + 12e^{-}$$

This dissolution reaction allows sustained current flow through the penetrated bath in the cathode which will be further enhanced with increasing potential gradient in the system. Thus, leading to the theory that increasing current density will lead to increased aluminium carbide formation due to a larger potential gradient in the system which will increase the driving force for Al^{3+} and Al_4C_3 formation. This theory has been confirmed by previous trials with graphitized cathode materials and also by recent work carried out by Wilkening et al [8].

Though in high porosity graphite samples there is evidence of internal aluminium carbide formation there was still no evidence of particle detachment which was apparent in high porosity graphitized isotropic coke samples [6]. This is due to the greater structural integrity of the graphite samples due to the low open porosities and high densities. The GFH sample shown in figure 5a has an open porosity of 24.5% which will allow bath to penetrate into the material but due to grain alignment and high packing efficiencies of the aggregate material the internal strength of the material will be high enough to resist particle detachment. Figure 2 shows the pore size distributions of the various graphitic materials.

Low porosity graphite samples however showed little to no penetration of bath into the cathode. This lack of bath penetration can be seen in Figure 5c were there are no large white or yellow regions indicating the presence of bath and aluminium carbide. This is due to the superior density of the low porosity samples and the small range of pore sizes present in the material. As can be seen from figure 2, GCL and GFL samples had very low open porosities of approximately 18% and the pores were generally less than 20μm in diameter. With the apparent lack of bath penetration in the low porosity samples it does suggest that there is possibly a critical pore diameter that will start facilitating bath penetration and subsequent internal reactions.

With the absence of wear phenomena such as particle detachment and pitting it can be said that for graphitic samples the dominant wear mechanism was not internal aluminium carbide formation and degradation but surface aluminium carbide formation and dissolution. On most samples there was clear evidence of aluminium carbide on the cathode surface as shown by the yellow colouring on the cathode surface in Figure 3a.

Wear Rate Analysis

As mentioned earlier, a 2nd order factorial matrix was used to define the testing regime in this investigation. The matrix allowed the analysis of the effects and interactions of five variables that were considered to be important to cathode wear. These were:

1. Bath chemistry
2. Granulometry
3. Porosity (Flour content)
4. Cut direction
5. Current Density

Note: Table 1 defines the variance for each variable used in the 2nd order factorial design.

The evaluation of the matrix results showed some indifferent results which were due to the fact that the matrix formulations used to produce the cathode samples produced samples with similar densities, open porosities and properties. This meant that during electrolysis different materials under similar operating conditions exhibited quite similar wear behavior and wear rates. Therefore making it difficult to determine the effects and interactions of porosity and granulometry on the wear behavior.

This was not the case for the graphitized isotropic coke material where different formulations gave rise to distinct properties and wear behavior which allowed effects and interactions to be clearly seen [6].

Though it was difficult to establish the clear effect of porosity and granulometry on the wear behavior the effect of operating parameters such as current density and bath chemistry could be seen.

Table 3: Relative impact of variable interactions on wear

Variable effect/ Interaction	Effect
Bath	4.55*
Granulometry	-1.09
Porosity	4.76*
Cut Direction	3.32
Current Density	8.25*
Granulometry, Current Density	1.20
Cut Direction, Current Density	1.41
Bath, Granulometry, Current Density	1.97
Bath, Cut Direction, Current Density	2.31

(* indicates key variables found)

Table 3 shows some of the 1st, 2nd and 3rd order interactions found during this investigation. The variable effect/interaction column lists the specific effects and interactions being looked at and the effect column shows the relative increase/decrease on cathode wear that changing the variables from their variance 1 state to their variance 2 state, whilst all other variables are kept constant at the variance 1 state.

The key effects in Table 3 are those highlighted. These seem to have the greatest effect on the cathode wear. Variable 1 and 5 are bath chemistry and current density respectively. The results show that changing the current density from $0.6A/cm^2$ to $1.0A/cm^2$ would increase the wear rate by 8.25cm/year. This is a considerable increase and does support the theory that cathode wear in graphitic cathodes is current density driven with increasing current density increasing the rate of electrolytic aluminium carbide formation.

Increasing the excess AlF_3 content (variable 1) in the bath does have an adverse effect on the wear rate as shown in Table 3. This is due to the aluminium carbide solubility in the bath increasing with increasing AlF_3 content, thus allowing more carbon material to be removed from the cathode and into the bath as dissolved aluminium carbide. However, in industrial applications the effect of AlF_3 may not be as pronounced as these results suggest. This is because as in industrial cells the molten aluminium pad on the cathode surface will protect the cathode from large amounts of bath contact, thus hindering the aluminium carbide dissolution reaction.

Though it is difficult to compare wear rates generated through the use of a factorial experimental design due to the uniformity in the samples produced; approximate comparisons between operating conditions can be made. Table 4 shows some of the conditions and wear rates found in this investigation. This table clearly shows the effect of current density and bath chemistry on the wear rate of the cathode samples.

Table 4: Wear rate comparison between GCL samples

Sample	Bath Chemistry (excess AlF3)	Current Density A/cm^2	Extrapolated Wear Rate cm/year
GCL Perpen	5	0.6	19.45
GCL Parallel	5	1	24.56
GCL Parallel	10	1	35.9

In summary the results from the graphitic samples showed that denser materials produced uniform wear over the electrolyzed area, with little evidence of pitting or particle detachment. However, though the samples were dense it was found that wear rates were still quite high (when comparing them to the graphitized cathodes) especially at high current densities and high excess AlF_3 content. This was probably due to the ease of the aluminium carbide formation reaction with the relatively disordered binder phase material in the graphitic cathode.

Conclusions

Graphitic Material

- Graphitic material produced dense, low porosity cathode samples even with varying grain formulations. This indicates that graphite is quite tolerant to changes in grain formulation when producing extruded cathode samples.
- Due to the superior internal structure of the cathode material, very little evidence of cathode degradation by pitting and particle detachment was found.
- Low porosity graphitic cathode samples were found to have little to no evidence of bath penetration. These results could indicate that there is a critical pore size at which bath penetration will occur.
- Though dense samples were produced the wear rates observed were still significant. High wear rates were attributed to the preferential aluminium carbide formation with disordered carbons in the non-graphitized binder phase.
- Due to the relative similarity in the various cathode properties between samples, it was difficult to conclude on the effect of porosity and granulometry on wear rate. However results showed a strong dependence of wear rate on current density and bath chemistry. This supports the theory that the main cause of wear is the electrochemical formation of aluminium carbide.

Comparisons between Graphitized and Graphitic Cathodes

The following conclusions are made based on the results reported in part II and III of this series of papers.

- Graphite filler material was less sensitive to formulation changes than isotropic coke material. Therefore accurate formulations will be needed to produce isotropic coke based samples with the same density as the graphitic based material.
- However, based on wear rate analysis on both graphitic and graphitized samples it was found that the wear rates were high for the dense graphitic material. Thus, indicating that density is not the only performance determining parameter.
- The results from this investigation suggest the best material regarding electrochemical behavior would be a combination of both these materials; a dense graphitized cathode material.

References

1. M. Sørlie and H.A Øye, *Cathodes in Aluminium Electrolysis*, 2nd Edition, (Düsseldorf, Aluminium-Verlag GmbH 1994)
2. H.A Øye and B.J Welch, "Cathode Performance: The Influence of Design, Operation and Operating Conditions", *JOM*, 1988, 18-23
3. P. Rafiei, F. Hiltmann, M.M Hyland, B. James and B. Welch, "Electrolytic Degradation within Cathode Materials", *Light Metals 2001*, 747-752
4. R. Keller, J.W Burgman and P.J Sides, "Electrochemical Reactions in the Hall-Heroult Cathode", *Light Metals 1988*, 629-631
5. R. Ødegard, A. Sterten, and J. Thonstad, "On the Solubility of Aluminium Carbide in Cryolitic Melts", *Light Metals 1987*, 295-302
6. P.M Patel, M.M Hyland and F.Hiltmann, "Influence of Internal Cathode Structure on Behavior during Electrolysis Part 2: Porosity and Wear Mechanisms in Graphitized Cathode Material", *Light Metals 2005, 757-762*
7. F. Hiltmann, P.M Patel and M.M Hyland, "Influence of Internal Cathode Structure on Behavior during Electrolysis Part 1: Properties of Graphitic and Graphitized Materials", *Light Metals 2005, 751-756*
8. S. Wilkening and P. Reny, "Erosion Rate Testing of Graphite Cathode Materials", *Light Metals 2004* , 597-602

Light Metals 2006 *Edited by Travis J. Galloway* ***TMS (The Minerals, Metals & Materials Society), 2006***

STUDY ON CATHODE PROPERTY CHANGES DURING ALUMINUM SMELTING

Fengqin Liu[1,2], Yexiang Liu[1]

[1]Central South University; Changsha, Hunan 410083, China

[2]Zhengzhou Research Institute of Chalco; No.82 Jiyuan Rd.; Shangjie, Zhengzhou, Henan 450041, China

Keywords: spent cathode, reduction cell, physical property, microstructure, porosity, penetration, graphitization

Abstract

The properties of the spent cathode samples from the reduction cells with different working life have been studied in this paper. The chemical and mineralogical compositions of spent cathodes and their physical properties have been analysed by relevant test instruments and XRD, and their microstructures have been observed under SEM. The study results show that the apparant density, compressive strength of the cathode samples are enhanced, and their electrical conductivity is even increased by 30-50% after undergoing aluminum reduction process. It is found from SEM and XRD analysis that there appears a large amount of scaly graphite formed in the spent cathodes and some fluorides penetrating into the gaps and pores in the cathodes. It is suggested from the discussion on the testing results that the cathode blocks with higher strength, lower porosity and poorer permeation performance and containing a lower amount of large pores and cracks should be used in order to prolong the cathode working life.

Introduction

About 80% of cathode area in the reduction cells is the surface of prebaked cathode blocks made from calcined anthracite and graphite scrap, while 20% of cathode surface area is covered by ramming paste.

The cathode materials directly contact molten salt bath of high molar ratio at high temperatures and undergo strong corrosion and penetration of the molten cryolite and even sodium permeation and reactions during baking and early starting stages of the cells[1]. In the aluminum reduction process the following chemical reactions will take place between the cathode materials and penetrated chemical substances:

$$Al+6NaF=Na_3AlF_6+3Na$$

$$4Al+3C=Al_4C_3$$

Thus, the cathode blocks expand due to the chemical reactions. Reaction products are deposited and gaps and cracks appear inside cathodes, which leads to easier erosion of the cathodes. The mechanical erosion happens on the cathode surface by flowing of aluminum and bath melts and aluminum tapping operation. Both chemical corrosion and mechanical erosion undertaken by the cathode materials are the main factors to influence the cathode life.

The penetration of bath and metal into cathodes and the chemical reactions taking place in the cathodes during aluminum smelting process have been studied by many researchers[2],[3],[4]. The properties of cathodes have been changed greatly in the smelting process[5].

It is found by a series of research work that carbon cathode materials are catalytically graphitized during aluminum smelting. The graphitization has been characterized by X-ray diffraction. The graphitization and molten salts penetration bring about the great change in electrical and thermal conductivity of the cathodes. The performance changes of different kinds of industrial cathode materials with different operation duration and temperature variation have been studied as well. Alkali metal expansion measurement has been applied to study the attack on the cathode materials. Substitution of less than 5 mol% KF for NaF did not affect the alkali metal expansion. When 20 mol% of NaF is replaced by KF the expansion will be increased. Substitution of LiF for Na F resulted in slightly reduced the expansion.

The penetration of bath into cathode materials is closely related to the properties and porosity of the cathodes. The porosity distributions are quite different for the cathode blocks produced by different plants and processes. Some mathematical models are developed for evaluation of the performance of the cathodes with different porosity structures and cell life time.

Experimental

The cathode samples are drilled and taken from the different parts in the spent cathodes of stopped reduction cells, which are of different types, from different smelters and with different cell life. The cathode cylinder samples drilled in smelters are 100-120 mm high and their diameter is 60 mm. The samples from smelters are further processed to the standard sizes for tests: φ 50×100mm, φ 50×40mm and φ 50×20mm, and then they are ready for all the tests.

The ash content, apparent density, real density, compressive strength and electrical conductivity of all the samples are analyzed. Some representative samples are chosen for the tests that the microstructures of the cathode samples and the formation and micromorphology of the substances penetrating inside cathodes can be observed by scanning electronic microscopy, while their mineralogical compositions can be tested by X-ray diffraction.

Results and Discussion

The routine property changes of the spent cathodes

The test results of this work for the properties of 5 spent cathode samples is shown in Table 1. The apparent density is increased from 1.58-1.60 g/cm^3 before smelting to 2.10-2.30 g/cm^3 after smelting. The ash content is increased from 4-6% to 35-40%, which is caused by the penetration of a large amounts of cryolite, NaF and CaF_2 and formation of some amount of AlFe alloys. Some substances, such as cryolite, penetrated into the cathodes bring about density increase and less porosity, which leads to about 100% increase of compressive strength.

It is found by X-ray diffraction that a large amount of amorphous carbon inside cathodes is transformed to graphite carbon after smelting for some time, which causes great increase of electrical conductivity of the cathodes. The special electrical resistance of the cathodes after smelting at normal temperature is reduced greatly from 40-45μ Ω m to 20-25μ Ω m.

The chemical and mineralogical compositions and microstructure changes of the spent cathodes

The chemical composition analysis and X-ray diffraction tests for carbon cathodes both before and after smelting process have been carried out. The results show that there are about 90-95% of carbon, 2-4% of oxygen and small amount of Al, Si and S in the cathodes before smelting. In the spent cathodes after smelting the average carbon content is reduced to about 60%, while others are NaF, cryolite, CaF_2 and some newly crystallized Na(K) containing alumina (approximately represented as $Na(K)Al_{11}O_{17}$), which have penetrated into the cracks, gaps and caves in the cathodes during smelting.

Fig.1 shows that the yellow powder taken from spent cathode surface and expansion cracks in the cathodes contains a great amount of fluorides including cryolite, NaF and CaF_2, some Na and K containing alumina and Al_4C_3.

Fig.2 and Fig.3 are the X-ray diffraction analysis of upper layer of the spent cathode samples taken from the different Chinese smelters.

Table 1 The Property Changes of the Cathodes after Smelting Process.

Routine Properties	Before Smelting	After Smelting				
		1	2	3	4	5
Apparent Density, g/cm^3	1.56-1.58	2.13	2.10	2.20	2.17	2.20
Compressive Strength, MPa	36-38	62.4	71.0	82.0	58.3	54.4
Ele. Resistance, μ Ω m	40-42	20.4	19.7	18.8	20.1	22.0
Ash Content, %	4-6	35.2	37.0	37.8	33.9	41.0
Mineralogical Composition	C	Graphite, Cryolite, NaF, CaF_2, Al_4C_3, $Na(K)Al_{11}O_{17}$ etc.				

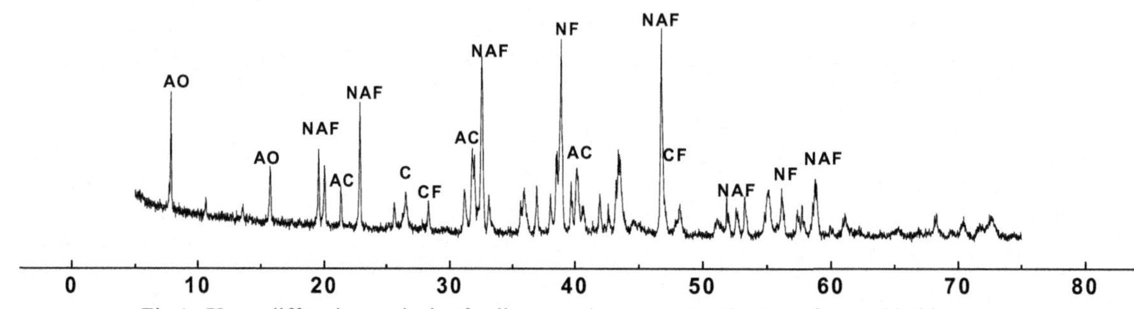

Fig.1 X-ray diffraction analysis of yellow powder on spent cathode surface and in big cracks
NAF—cryolite; NF—NaF; CF—CaF_2; C—carbon; AO—$Na(K)Al_{11}O_{17}$; AC—Al_4C_3.

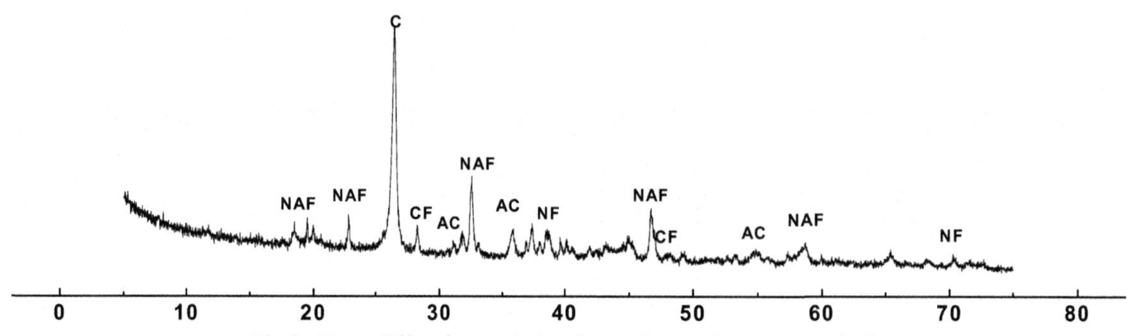

Fig.2 X-ray diffraction analysis of upper layer of the spent cathode
NAF—cryolite; NF—NaF; CF—CaF_2; C—carbon; AC—Al_4C_3.

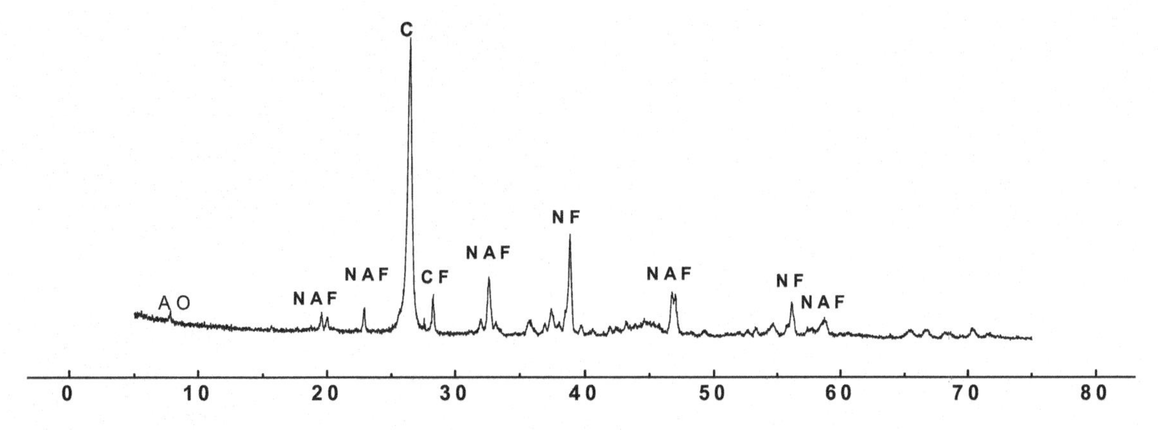

Fig.3 X-ray diffraction analysis of upper layer of spent cathode
NAF—cryolite; NF—NaF; CF—CaF_2; C—carbon; AO—$Na(K)Al_{11}O_{17}$.

It can be found from Fig.2 and 3 that there usually exist various fluorides, such as cryolite, NaF and CaF_2, and sometimes small amount of Al_4C_3 or Na and K containing alumina in the spent cathodes.

Fig.4 and 5 taken by Scanning Electronic Microscopy(SEM) show that there are a lot of substances penetrating into the cracks, gaps and caves in the microstructures inside of spent cathode, some of which take the smooth and irregular shapes without edges and corners and are recognized as cryolite (Na_3AlF_6) with general size ranges of 20-200μm by energy spectrum analysis (Fig.7).

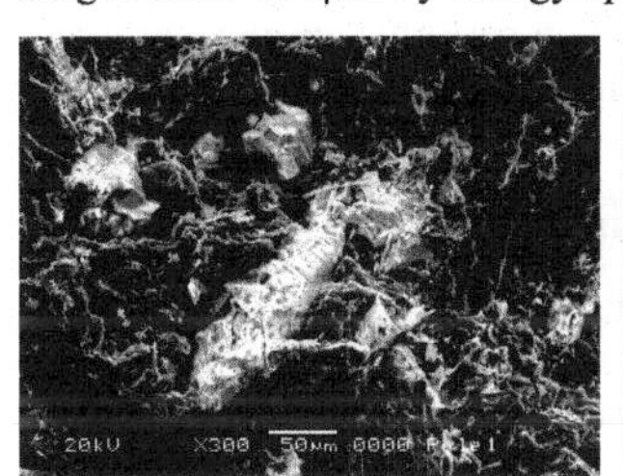

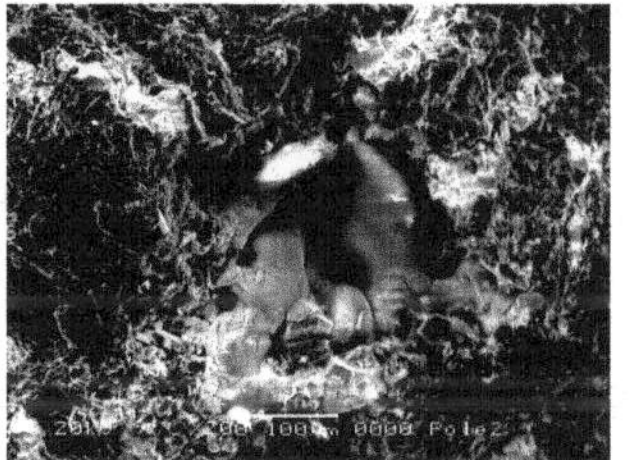

Fig.4 Cryolite in cathode gaps. Fig.5 Cryolite in cathode pores.

Fig.6 shows there is some gap between the layers of carbon material in the layer on layer cathode structure and some cryolite is filling the gap.

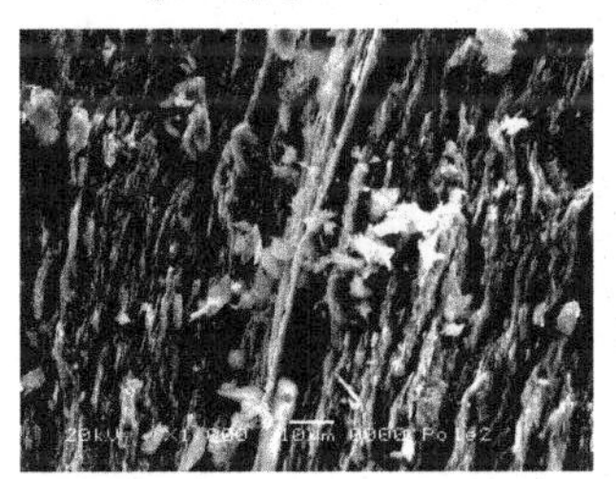

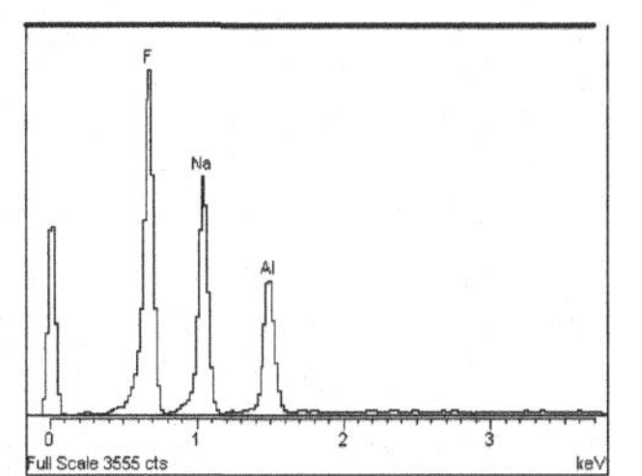

Fig.6 Layer on layer structure and cryolite in the gap. Fig.7 Energy spectrum of cryolite in spent cathodes.

A great number of NaF grains are scattered about inside of spent cathode materials as strip shape, irregular shape or pillar shape, which can be seen in Fig.8 and 9 respectively.

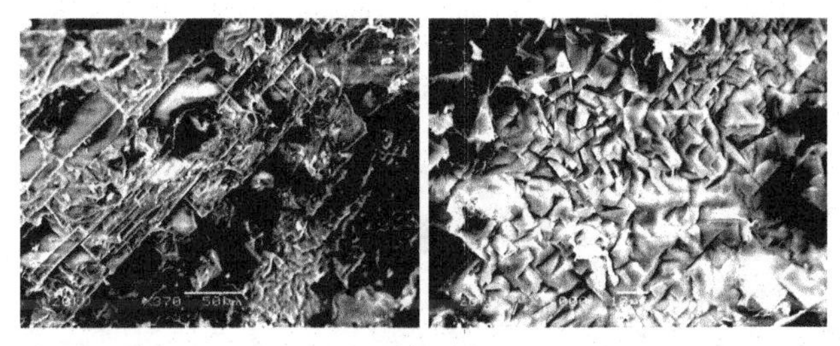

Fig.8 NaF grain of strip shape. Fig.9 NaF of irregular shape.

CaF_2 can be found existing in spent cathode material as the grains with irregular shape, which can be seen in Fig.10. Fig.11 shows its chemical composition by energy spectrum.

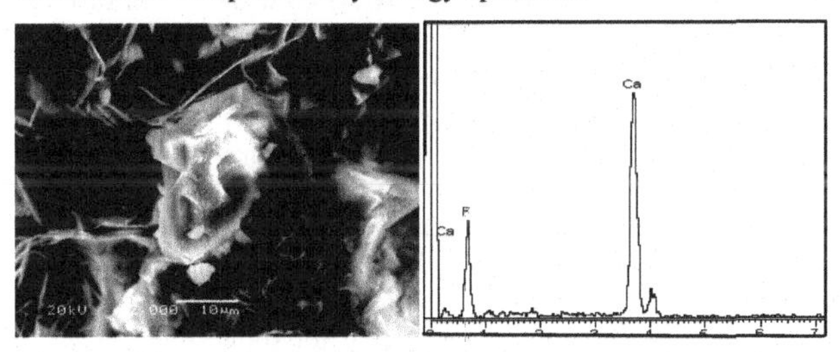

Fig.10 CaF_2 of irregular shape. Fig.11 Energy spectrum of CaF_2.

Fig.12 and 13 show that there are a lot of newly well crystallized Na and K containing beta-alumina with typical prism and sheet shapes and with size range of 10-100μm forming in some pores and gaps in the spent cathodes with the approximately chemical composition of $Na(K)Al_{11}O_{17}$.

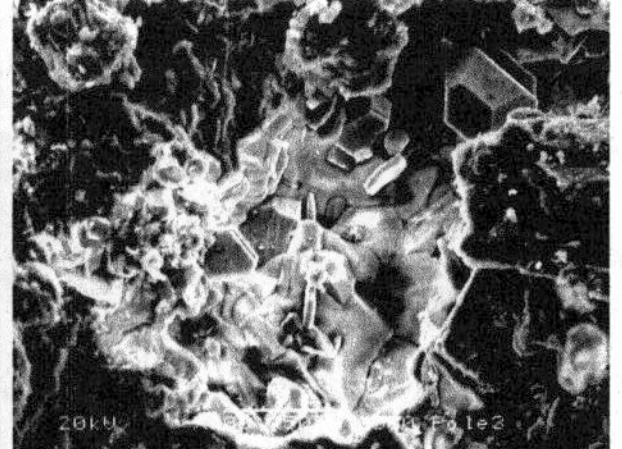

Fig.12 Na containing alumina with the prism shapes. Fig.13 Na containing alumina with the sheet shapes.

There can be found both Na and K in its crystal structure, which came from the cell bath penetration. And it is obvious that the well crystallized Na (K) containing alumina is not the same as the alumina added from alumina feeding system of smelters and it is assumed that some Na in the newly crystallized alumina in the cathode caves is substituted by K. Fig.14 is the energy spectrum analysis of the newly formed Na (K) containing alumina.

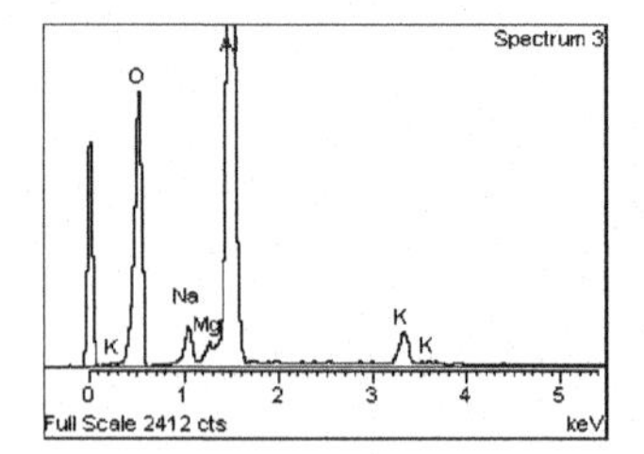

Fig.14 Energy spectrum of Na (K) containing alumina.

Aluminum carbide (Al_4C_3) usually appears in the yellowish surface of spent cathodes and expansion cracks between and inside of cathodes. Under SEM it takes an irregular shape of crystallized grains and sometimes has a smooth surface, as shown in Fig.15 and 16.

Fig.15 Irregular shape of Al_4C_3.Fig.16 Al_4C_3 with smooth surface.

There are some papers on the properties and wear mechanism of graphitized cathodes[6],[7]. Some kind of graphitization for the amorphous carbon in the smelting operations has been observed. It is clearly shown in Fig.17 and 18 that the carbon in some areas of the spent cathodes has been graphitized and lots of well crystallized flake graphite have been formed, which can be the main reason why the electrical conductivity of the spent cathodes can be improved greatly after smelting for some time. Fig.21 is the energy spectrum analysis result of the flake graphite. It can also be found out that there always appear cryolite(Fig.17 and 18), NaF(Fig.19) and CaF_2(Fig.20) around the formed flake graphite, which probably implies there might be some relation between carbon graphitization and the fluoride environment. The graphitization of the carbon in the cathodes during aluminum smelting process is quite interesting and the further study on its graphitization mechanism will provide some theoretical basis for improvement of graphitized cathode production.

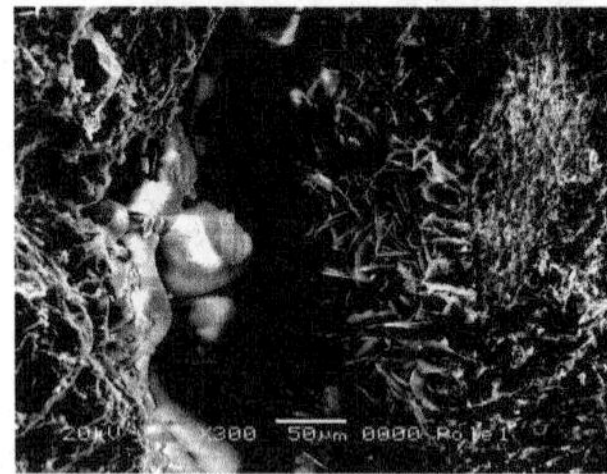

Fig.17 Newly formed flake graphite and cryolite nearby(1). Fig.18 Newly formed flake graphite and cryolite nearby(2).

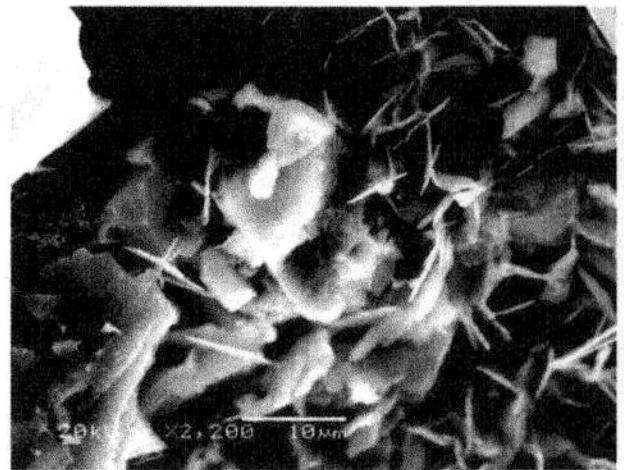

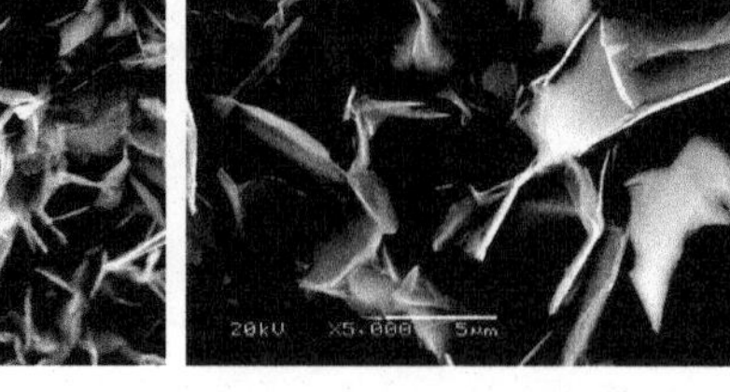

Fig.19 Newly formed flake graphite and NaF (on the left). Fig.20 Newly formed flake graphite and CaF_2 (on the right).

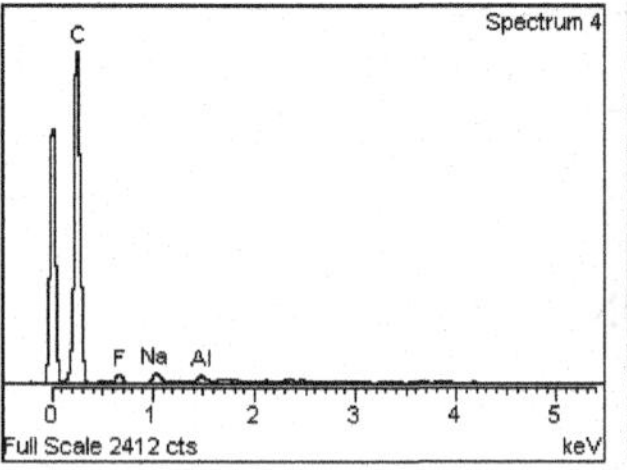

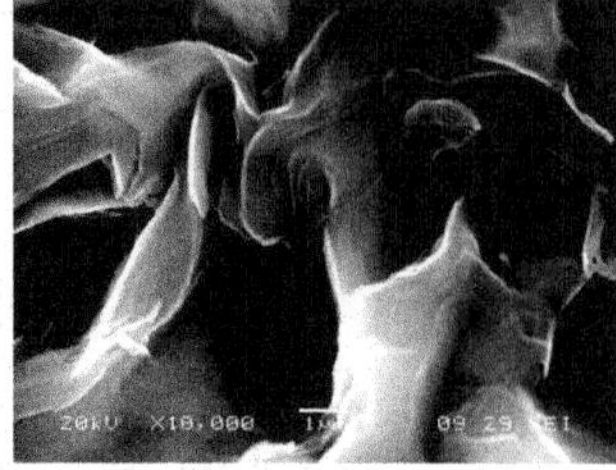

Fig.21 Energy spectrum of flake graphite in the spent cathodes. Fig.22 Morphology of flake graphite in the graphitized cokes.

The morphology of flake graphite in the spent cathode is compared with that in the graphitized cokes in Fig.22. The morphology and the energy spectrum of flake graphite in graphitized cokes is almost the same to that in the spent cathodes.

It is suggested from above discussions on the substances formed and penetrating into the cathode materials during aluminum smelting process that the cathode blocks with higher strength, lower porosity and poorer permeation performance and less larger internal gaps and caves should be used in order to prolong the cathode working life.

Conclusions

(1) The great changes of various physical properties, chemical compositions, internal microstructures and carbon morphology and crystalline in the cathodes made from calcined anthracite and graphite scraps take place during the aluminum smelting process.

(2) The apparent density and compressive strength of the cathodes are greatly increased during smelting. And their electrical conductivity is increased by 50%.

(3) As the main content of bath, Na_3AlF_6, NaF and CaF_2 are found to be penetrating into cathode blocks. Some newly formed Al_2O_3 containing small amount of Na and K precipitates in the caves of the cathodes, which can be presented as $NaAl_{11}O_{17}$ approximately.

(4) A great number of flake graphite pieces have been observed in the cathodes after smelting under SEM. And Na_3AlF_6, NaF and CaF_2 always exist around the newly formed flake graphite pieces. It can be supposed that the graphitization process of carbon in the cathodes can be accelerated by existing fluorides.

(5) Reducing the porosity in the cathode blocks is an important way to reduce bath penetration, improve cathode quality and prolong cell life.

References

1. Tyke Naas and Harald A. Øye, "Interactions of Alkali Metal with Cathode Carbon", *Light Metals 1999*, 193-198.

2. R. Shamsili and H.A. Øye, "Melt Penetration and Chemical Reactions in Carbon Cathodes during Aluminium Electrolysis", *Light Metals 1994*, 731-738.

3. Pierre-Yves Brisson et al., "Revisiting Sodium and Bath Penetration in the Carbon Lining of Aluminum Electrolysis Cell", *Light Metals 2005*, 727-732.

4. Sergey A. Khramenko et al., "Effect of Porosity Structure on Penetration and Performance of Lining Materials", *Light Metals 2005*, 795-799.

5. Morten Sørlie, Hermann Gran and Harald A. Øye, "Property Changes of Cathode Lining Materials during Cell Operation", *Light Metals 1995*, 497-506.

6. Frank Hiltmann, Pretesh Patel and Margaret Hyland, "Influence of Internal Cathode Structure on Behavior during Electrolysis , Part 1: Properties of Graphitic and Graphitized Material", *Light Metals 2005*, 751-755.

7. Pretesh Patel, Margaret Hyland and Frank Hiltmann, "Influence of Internal Cathode Structure on Behavior during Electrolysis, Part 2 : Porosity and Wear Mechanisms in Graphitized Cathode Material", *Light Metals 2005*, 757-762.

Light Metals 2006 *Edited by Travis J. Galloway* ***TMS (The Minerals, Metals & Materials Society), 2006***

WETTING AND CRYOLITE BATH PENETRATION IN GRAPHITIZED CATHODE MATERIALS

Anton V. Frolov[1], Alexander O. Gusev[1], Nikolai I. Shurov[2], Nina P. Kulik[2], Leonid V. Sitnikov[2],
Larissa M. Babushkina[2], Victor P. Stepanov[2], Yurii P. Zaikov[2], Andrei P. Khramov[2], Vyacheslav B. Malkov[2]
[1]Engineering-Technological Center Ltd, RUSAL, Pogranichnikov st. 37/1, Krasnoyarsk, 660111, Russia
[2] Institute of High Temperature Electrochemistry, S.Kovalevskaya st. 20, Ekaterinburg, 620219, Russia

Keywords: graphitized cathode materials, wetting, cryolite bath penetration.

Abstract

A special electrochemical cell was used in order to measure angles of wetting on different materials depending on cathode current density. Wetting of all samples was found to improve with cathode current rise. The dynamics of decrease in contact angles depends on the material type and open porosity. The investigation of bath penetration was carried out at 960°C under argon atmosphere in a special closed container made of the same cathode materials. Penetration depth was found to depend on open porosity. Under the same conditions, the impregnated materials have a smaller penetration depth. The distribution of bath components depending on the distance from the interface boundary has been investigated by the method of X-ray electron-probe microanalysis. The conclusions on the absorption mechanism of different elements were made on the basis of bath components concentration in pores.

Introduction

The service life of cathodes in the electrolyzers for the aluminum production is determined by the intensity of the interaction of the used materials with molten bath and liquid aluminum. This interaction includes both the chemical reactions and the solidification of substances in pores [1-5]. The action of many factors can be superimposed on each other, accelerating or decelerating the wear. For example, the formed carbides fill the pores and, thus, cause the material densification, but subsequently they are dissolved in the bath and lead to the carbide erosion. The intercalation of sodium into graphite not only deteriorates the cathode by swelling and deformation, but also promotes the graphitization, which acts in the opposite direction.

By laboratory tests and analysis of the cathodes, which served for different terms under industrial conditions, the corrosion resistance of cathodes was shown to be substantially determined by the macrostructure and microstructure of cathode material – by its density, porosity, and degree of graphitization [6-7]. However, the effect of these factors on the service life of the materials is not always unambiguous. For example, an increase in the degree of graphitization decreases the sodium expansion of cathode, but decreases the resistance to carbide erosion. This field still needs a more detailed investigated for the determination of the regularities permitting the selection of optimum cathode material.

In the present work, we studied the interaction of three cathode-block samples with liquid media of electrolytic bath at the initial stage, which in many respects determines the rate of physical and chemical wear. This stage includes the propagation of liquid phase through the solid surface (wetting) and the capillary suction of the bath into the pores and cracks of the carbon material (impregnation). The better the material is wetted, the more rapidly and deeper it is impregnated with bath, the sooner the cathode made from this material comes out of action.

The most important criterion, which determines the presence or the absence of wetting and impregnation, is the contact wetting angle Θ, which is sensitive to the structure and composition of the solid electrode surface. The cryolite-alumina melts with predominantly ionic chemical bonds poorly wet pure graphite with covalent bonds, i.e. $\Theta > 90$. The maximum contact angle is typical of the perfect graphite single crystal with the minimum surface energy. However, the industrial graphitized carbon materials consist of nonuniform and irregularly aggregated imperfect crystallites. The surface energy of such solid bodies is determined by a complex of factors, which include the sizes of crystallites and their spatial orientation, the presence of uncompensated bonds, and the structure of intergranular and intragranular pores. The presence of large and extended open pores in the cathode material is one of the causes for its impregnation with bath, sometimes at a depth of several centimeters already at the electrolyzer start-up phase.

Upon current passage, aluminum and sodium deposited on the cathode interact with carbon. This increases the surface energy of solid surface, improves wetting, and accelerates impregnation.

Upon the investigation, the real aluminum production conditions were tried to simulate as far as possible. For the study of wetting, current was passed through the cathode and weighed the meniscus of the liquid leaking onto the sample. In this case, unlike the method of a lying drop, no visual observations are required, and the cell construction is substantially simplified. Furthermore, the experimental difficulties caused by a small drop volume, which complicates the introduction of additional electrodes for the polarization of three-phase boundary and the measurement of its potential, were avoided.

At low current densities, the formation of the meniscus of liquid was governed only by the phenomenon of wetting. An increase in the current density decreased the Θ value and led to an intense impregnation of the material with bath. In this case, the change in the electrode weight was caused by the leakage of meniscus onto the surface and by the penetration of liquid phase into the open pores.

To study the impregnation of cathode materials with the bath, the crucible method was used: any current was not passed, but, after filling aluminum powder into the pot in addition to bath, were attained the conditions providing the deposition of sodium. In the presence of the alkali metal, which can form interlayer compounds (intercalates) with graphite [8], the penetration of bath into the graphitized carbon materials was facilitated.

Experimental

Initial materials

The samples of three graphitized cathode blocks A, B, and C, close in density and porosity, were taken for the study. The apparent density of materials B and C impregnated with organic resins was 1.73 and 1.74 g/cm^3, respectively, and the apparent porosity was 16 and 14%, respectively. The porosity of sample A was the same as that of sample B, but the density was substantially smaller, 1.67 g/cm^3. Figure 1 shows the surface of the materials at a magnification of ×15. Sample A is characterized by large round pores caused by an incompact stacking of the filler grains in the carbon paste [9]. The same pores, but in a substantially smaller number, are observed in material C. At the surface of sample B, the circular voids are connected with each other by long wavy shrinkage cracks.

According to the data of X-ray electron-probe microanalysis, the materials contain (wt %) up to 5.3 Fe; ~0.2 S; ~0.2 Si.

The composition of the used bath was close to the industrial one, $[93,5\%(NaF\text{-}AlF_3)_{CR=2,3}+6\%CaF_2+0,5\%MgF_2]+5\%Al_2O_3$. It was prepared by air melting of artificial commercial cryolite with a cryolite ratio of 1.8 and chemically pure ("kh. ch." grade) fluorides.

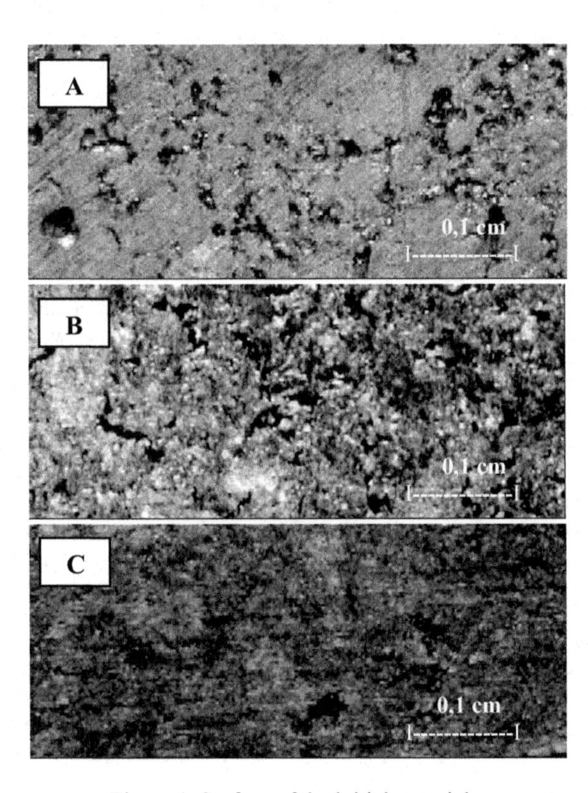

Figure 1. Surface of the initial materials: A, B and C

Wetting

The design of the electrochemical cell for measuring the contact angles at the three-phase boundary is given in Figure 2. The bath was put into a glass-graphite crucible, which was used as an auxiliary electrode upon polarization. The platinum wire immersed into the same melt was used as a reference electrode.

The measurements were performed in an atmosphere of dry purified argon, which was blown through the cell at a constant rate. As the cell was heated to a temperature of 960°C, the electrode to be studied was lowered down until the contact with the melt. After the establishment of equilibrium, we recorded the weight of the liquid meniscus on the sample and the stationary potential. The measurements were repeated by passing DC of cathode direction through the sample in the galvanostatic regime. The experiment was stopped when the balance readings became unstable. For the calculation of the contact angle, we used the relationship $\Theta = \arccos(mg/L\sigma_{lg})$, where m is the weight of the meniscus of liquid at the sample, σ_{lg} is the surface tension of the liquid at the boundary with the gas, and L is the perimeter of wetting.

The surface tension of melt was calculated from the equation obtained by the statistical treatment of a large experimental data array with a standard deviation of ±1,76 mJ m^{-2} [10]. This equation does not allow for the contribution of magnesium fluoride to the σ_{lg} value; therefore the error in our calculations was somewhat higher, about 4 mJ m^{-2}. The current density was taken as the value of I/s, where I is the current passing through the cell and s is the electrode end area. The weight of meniscus was determined with a "Sartorius" electronic microbalance with an accuracy of 10^{-4} g.

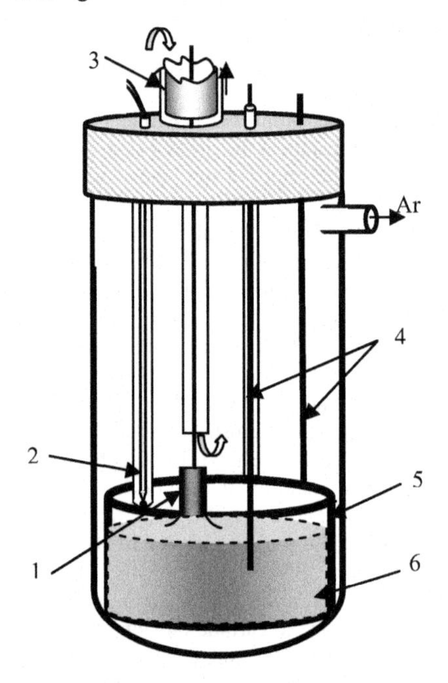

Figure 2. Scheme of electrochemical cell for measuring the wetting contact angles: (1) working electrode, (2) thermocouples, (3) mobile vertical section, (4) platinum wire, (5) glass-carbon crucible, (6) molten bath.

Bath penetration

The materials under study were used for the preparation of cylindrical samples with a blind hole of 40 mm in diameter and 35 mm in depth. They were densely filled with the mixture of the powders of bath and aluminum in a weight ratio of 5:3. In the absence of Al powder, we failed to reveal the penetration of bath into the pores.

The crucibles were closed by ground covers from dense reactor graphite and hermetically sealed by carbonaceous glue. Then they were installed in a stainless steel container and held in an atmosphere of dry purified argon at a temperature of 960°C for 24 hours.

At the end of the experiment, the samples were cut along the longitudinal axis, and the sections of walls and bottom were ground by emery paper. The impregnation depth was determined from the maximum distance of the bath spots from the edge of cavity.

The fragments of the surfaces of side-wall and bottom crucible parts were examined with a JSM-5900 LV scanning electron microscope. The X-ray electron probe microanalysis of elements in surface sections of about 0.0122 mm^2 in area was performed by the selection of the pores partially or completely filled with bath and present at different distances from the phase boundary We obtained the element distribution maps and the profiles of the intensities of characteristic X-ray radiation along the scan line.

Results and discussion

Wetting

In the absence of polarization, the contact wetting angles of the investigated materials were 83-90°. According to literature data, the Θ value obtained by the sessile drop method for graphite in the contact with the cryolite with CR = 3 is between 127° and 144° [11-15]. The wetting is improved as cryolite ratio decreases, alumina is introduced into the bath, the degree of crystallinity decreases, and graphite purity decreases. The allowance for all these factors give the Θ values in reasonable agreement with those obtained experimentally.

The passage of cathode current improved the wetting of all samples (Figure 3a). The changes in the cathode potential in this case are shown in Figure 3b. At a low current density, the measured angle is close to the equilibrium contact wetting angle. With growth of a current the surface of an electrode is polarized the more strongly, than more densely a material. The point is that the recorded apparent current density is substantially smaller than the true current density for the porous sample A and is close to it for dense samples B and C. Therefore, the shift of the electrode potential to the negative side and a decrease in the contact angle occur more rapidly for denser materials B and C. At a current density of 0.05-0,1 A/cm^2, the potential at these electrodes reaches the value sufficient for the deposition of aluminum, which interacts with carbon. Compared with graphite, the formed Al_4C_3 carbide, which is the compound with ionic bonds, is better wetted by the melt. The wetting curves of samples B and C exhibit a decrease in Θ. Similar processes, but with a certain retardation, occur also on more porous cathode A.

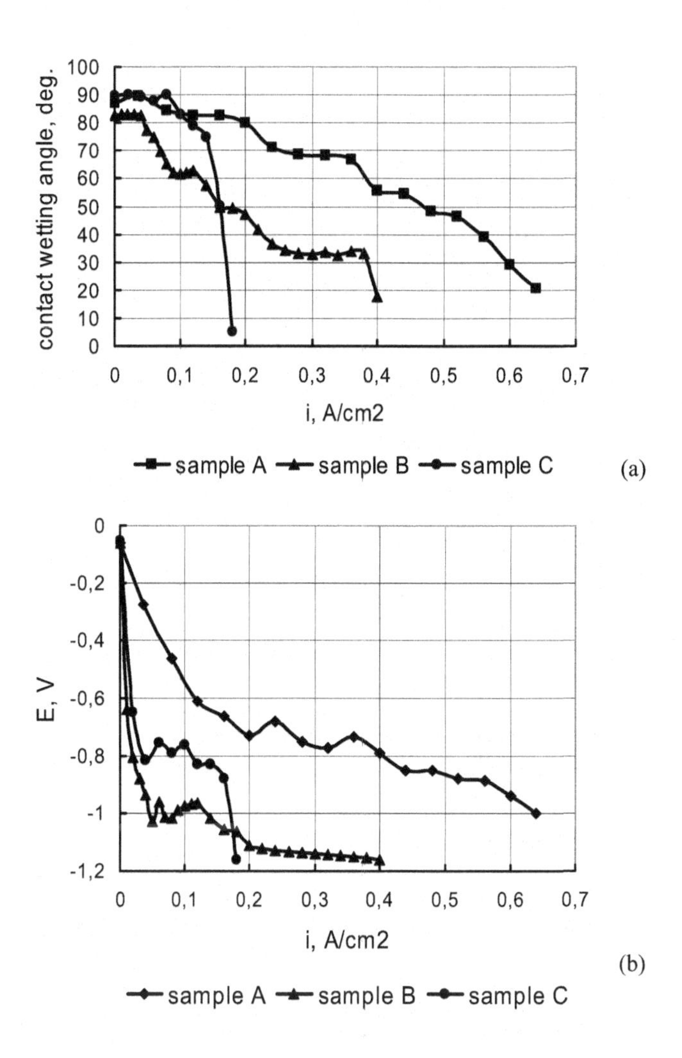

Figure 3. Contact wetting angle (a) and potential (b) as a function of the cathode current density in the samples at a temperature of 960°C

A further increase in the current density leads to the cathode deposition of sodium and calcium along with aluminum. These metals also can react with carbon: sodium forms intercalates and calcium forms carbides CaC_2. In addition to carbides, other compounds with a high fraction of ionic bonds (sulfides or silicides) can be present on the electrode surface. They are formed as a result of the reduction of impurities existing in the block materials. The quantity of sulfur-containing impurities increases upon the impregnation of blocks; therefore, the wetting of samples B and C upon the current passage is improved more substantially than that of samples A. However, the experimental determination of the contact angle becomes virtually impossible because the sample weight varies. These variations can be caused by cathode surface exfoliation resulting from the intercalation of the alkali metal.

Bath penetration

No visible damages were observed at the outside surface of the crucibles from the materials under study. Figure 4 shows the sections of samples A and C. About a half of the cavity was occupied by the aluminum regulus, whose surface most frequently consisted of well distinguishable solidified drops.

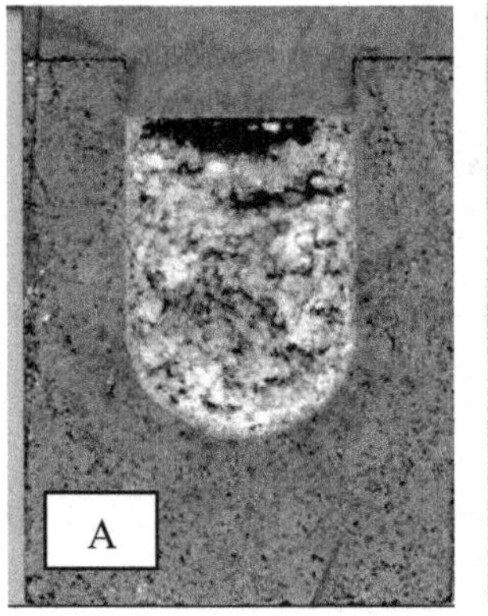

Figure 4. Cross section of crucibles from materials A and C after contact with the bath for 24 hours

The entire internal surface of the crucible was covered by a layer of yellow and, in places, brown colors. Similar layers were investigated in many works [e.g., 1]. Their basic component is aluminum carbide Al_4C_3. Upon storage in air, this layer eventually becomes white and crumbles as a result of hydrolysis under the action of moisture:

$$Al_4C_3+12H_2O = 4Al(OH)_3\downarrow+3CH_4\uparrow \quad (1)$$

The introduction of a water drop into the layer of yellow substance on the internal surface of the crucibles leads to an immediate violent reaction accompanied by bubbling. In addition to the aluminum carbide, calcium carbide CaC_2, which also can be hydrolized in air by the reaction $CaC_2 +2H_2O = Ca(OH)_2\downarrow+ C_2H_2\uparrow$, can be present at the surface of the carbon material. Saltlike type carbides (metanid and acetylenide) can form mutual solid solutions; this facilitates their joint precipitation at the phase boundary. The specific hydrogen sulfide smell accompanying cutting of the crucibles indicates the presence of refractory sulfides of aluminum, calcium, and sodium. The hydrolysis of Na_2S, CaS, and CaC_2 leads to the formation of alkali. They cause the raspberry color of the phenolphthalein solution moistening the filter paper applied to the cut surface of the cathode material impregnated with bath in the region of the cavity edge.

At a certain distance from the carbon block interface, the phenolphthalein test does not exhibit any coloring. This means that the content of metallic sodium in the cathode material at this distance is negligibly small. However, the bath remaining in the internal cavity of crucible or in the pore acquired raspberry color upon wetting with the phenolphthalein solution. The alkaline reaction can be caused by sodium or its sulfide dissolved in the bath.

The impregnation depth was about 7.4 mm for sample A. The increase in the material density (from 1.67 to 1.74 g/cm^3) in going to sample B with the same porosity (16%) decreases the impregnation depth to 6.0 mm, and the decrease in the porosity to 14% (material C) at the virtually same density (1.74 g/cm^3) also decreases the impregnation depth (to 5 mm).

Figure 5a shows the X-ray radiation maps of the element distribution near a large pore on the surface of sample C. After contact with the bath for twenty-four hours, the melt partially filled the large pore, penetrated into the smaller and microscopic cavities.

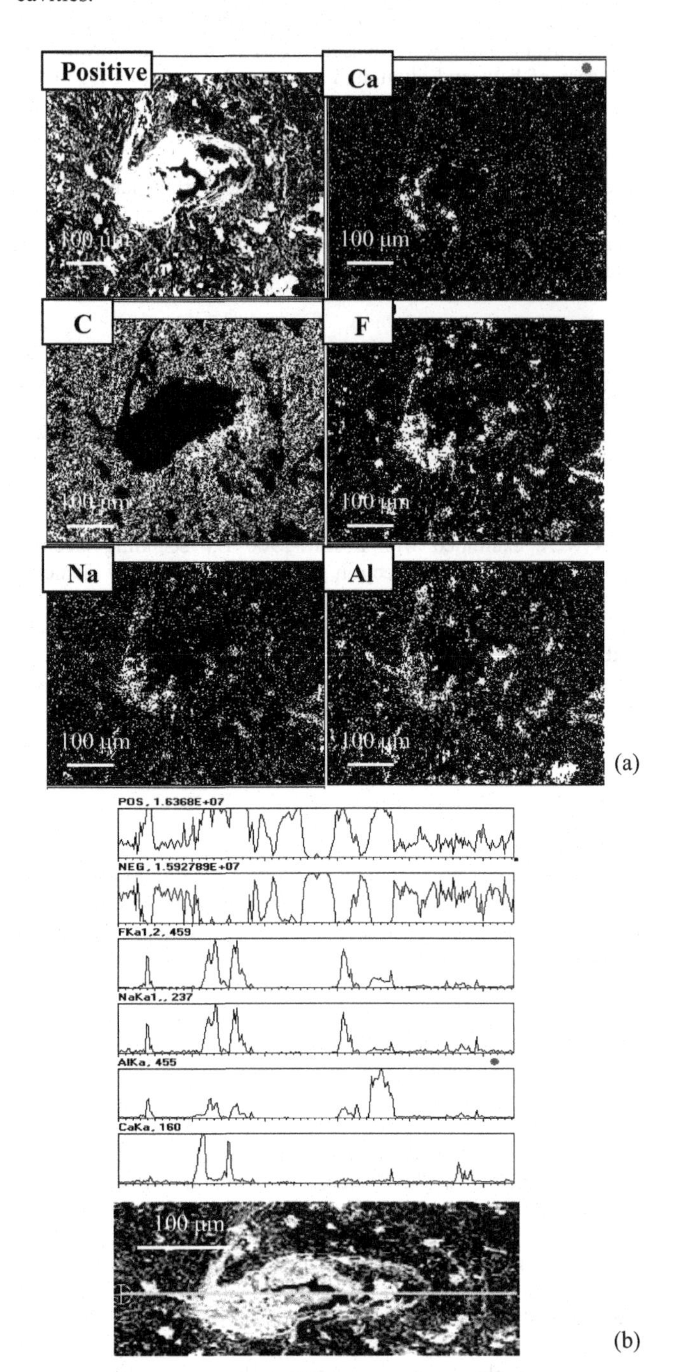

Figure 5. (a) Maps of the distribution of elements and (b) the profiles of the intensities of characteristic X-ray radiation along the scan line, which intersects the large pore on the surface of sample C after the contact with the electrolyte.

The bath, which reacts with liquid aluminum, is the sole source of sodium entering into the material under the conditions of our experiments:

$$Al_{liq.}+3Na^+ \rightleftarrows 3Na\,(Al)_{liq.}+ Al^{3+} \quad (8)$$

$$Al_{liq.} + 2Na^+ \rightleftarrows 3Na_2{}^+_{solution} + Al^+ \quad (9)$$

In this case, the active particles appear, which can react with graphite with the formation of the interlayer compounds-intercalates:

$$Na_2{}^+ + xC_s \rightarrow NaC_{x\,s} + Na^+ \quad (10)$$

$$Na_{solution} + xC_s \rightarrow NaC_{x\,s} \quad (11)$$

Intercalated sodium is characterized by a high diffusion mobility in the graphite interplanar spaces; therefore, it easily migrates into the depth of the material both through the bodies of graphite grains and through more porous binder. Even the Na content that is as small as in our experiments substantially improves the wetting of the graphitized carbon and its impregnation with bath.

The concentrations of Al, Ca, Na, and F in the pores, which are present near the block surface (at a distance below 1 mm), do not exceed their concentrations in the initial melt. These pores are characterized by a noticeable mass exchange with the bath volume in the crucible. This provides a partial removal of carbides (and sulfides) from the graphitized carbon as a result of their dissolution in the cryolite.

Conclusions

Graphitized cathode materials close in density and porosity in the absence of polarization are approximately equally wettable by the bath of industrial composition (Θ=83-90°). However, they differ in the character of the change in the contact angle upon the passage of current: the denser is the material, the more rapid is the shift of the potential of its surface to the negative side and the sharper is the cathode portion of the wetting curve. The decrease in the contact angle with increasing current density is related to the formation of aluminum and calcium carbides (and sulfides, especially in the case of the impregnated cathode blocks) on the surface of graphitized carbon and by the penetration of sodium into its depth.

The depth of the impregnation of the investigated blocks decreases with increasing density and decreasing apparent porosity (A>B>C).

The X-ray electron-probe microanalysis of the contents of the pores filled with bath showed that their composition depends on the distance of pore from the bath/cathode interface of the sample. If the microcavity is present at a distance smaller than 1 mm from the block surface and has an active bath exchange with the bulk electrolyte, then the composition of the bath solidified in it does not differ from the initial one. In deeper pores, the concentration of sodium and fluorine is also close to the concentrations of these elements in the initial bath, but the contents of aluminum and calcium are higher by a factor of 1.5-3. This means that sodium is accumulated in the pores generally in the form of fluorides, while aluminum and calcium are contained as carbides, which can cooperatively precipitate on the interphase surface. The distribution of elements over the pore profile confirms the existence of CaC_2 and Al_4C_3 near the walls. In addition, sulfides Al_2S_3, CaS, and Na_2S can be formed at the phase boundary in the impregnated materials.

No noticeable quantity of sodium intercalated into graphite was found under the conditions of our experiments on the impregnation without current passage; however, a small sodium radiation was recorded in the element distribution maps for the sample sections characterized by a total absence of fluorine radiation.

References

1. Xian-an Liao, H.A. Øye, "Phisical and wear of carbon cathode materials", *Light Metals*, 1998, p.667-674.
2. P.-Y. Brisson, G. Soucu, M. Farard, H. Darmstadt, G. Servant, "Revisiting sodium and bath penetration in carbon lining of aluminum electrolysis cell", *Light Metals*, 2005, p.727-732.
3. S.A. Khramenko, P.V. Polyakov, A.V. Rozin, A.P. Skibin, "Effect of porosity structure on penetration and performance of lining materials", *Light Metals*, 2005, p.795-799.
4. A. Zolochevsky, J.G. Hop, T. Foosnæs, H.A. Øye, "Surface exchange of sodium, anisotropy of diffusion and diffusion creep in carbon cathode materials", *Light Metals*, 2005, p.745-750.
5. Y. Sun, Q. Wang, K.A. Rye, M. Sørlie, H.A. Øye, "Modelling of thermal and sodium expansion in prebaked aluminum reduction cells", *Light Metals*, 2003, p.603-610.
6. M. Sørlie and H.A. Øye, *Cathodes in Aluminium Electrolysis* (Aluminium Verlag, Düsseldorf, Germany, 1994), 408.
7. L.P. Lossius, H.A. Øye, "Melt penetration and chemical reactions in 16 industrial aluminium carbon cathodes", *Metallurgical and materials transactions*, 2000, 31B, p.1213-1224.
8. M.B. Rapoport, *Graphitized carbon interlayer compounds and their value for metallurgy of aluminum* (Moscow, M.:Tsvetmetinformatsiya, 1967), 67.
9. F. Hiltman, P. Patel, M. Hyland, Influence of internal cathodes structure on behavior during electrolysis. Part I: Properties of grapfitic and grapfitized material, *Light Metals*, 2005, p.751-756.
10. V. Danec, O. Patarak, T. Ostvold, Surface tension of cryolite-based melts, *Canadian Metallurgical Qarterly*, 1995, v.34, №2, p.129-133.
11. J. Thonstad., P. Fellner, G.M. Haaberg, J. Hives, H. Kvande, A. Sterten, *Aluminium. Fundamentals of the Hall-Heroult process.* (3rd edition. Aluminium-Verlag, Dusseidorf, 2002).
12. K. Grjotheim, C. Krohn, M. Malinovsky, K. Matiasovsky, J. Thonstad, *Aluminium electrolysis. Fundamentals of the Hall-Heroult process* (2rd edition. Aluminium-Verlag, Dusseidorf, 1982).
13. E.A. Zhemchuzhina, A.I. Belyayev, Influence of the imposition of direct current on the wetting of graphite with cryolite and cryolite-alumina melts, *Izv. Vyssh. Uchebn. Zaved.*, Tsvetnaya Metallurgiya, 1961, no. 5, pp. 123-132.
14. J.B. Metson, R.G. Haverkamp, M.M. Hyland, Jingxi Chen, The anode effect revisited, *Light Metals*, 2002, p.239-244.
15. L.M. Babushkina, L.B. Sitnikov, N.P. Kulik, B.P. Stepanov, Yu.P. Zaykov and A.O. Gusev, Wetting of carbon and oxide materials with cryolite-based melts as a function of polarization, *Rasplavy*, 2004, no. 6, pp. 63-76.
16. N.G. Ilyshenko, A.I. Anfinogenov, N.I. Shurov, *Interaction of metals in ion melts* (Moscow: Nauka, 1991), 176.

Light Metals 2006 *Edited by Travis J. Galloway* ***TMS (The Minerals, Metals & Materials Society), 2006***

SODIUM PENENTRATION INTO CARBON-BASED CATHODES DURING ALUMINUM ELECTROLYSIS

Jilai Xue, Qingsheng Liu, Jun Zhu, Wenli Ou
School of Metallurgical and Ecological Engineering, University of Science and Technology Beijing
Xueyuan Road 30, 100083 Beijing, China

Keywords: Sodium Penetration, Carbon-Based Cathodes, Aluminum Electrolysis

Abstract

Experimental investigation of sodium penetration into carbon-based cathodes during aluminum electrolysis was carried out in a laboratory cell. Preliminary results show that the rate of sodium penetration is reduced on the cathodes with addition of B_2O_3 or TiB_2+B_2O_3. Fluoride salts originated from cryolitic melt penetration were presented in the cathode bodies after electrolysis. B_2O_3 or TiB_2+B_2O_3 were found within the areas of the binder phase of the cathode, which may play a role in reducing the rate of sodium penetration.

Introduction

Carbon is still the dominant material for cathodes in aluminum reduction process for today. For the carbon cathodes, sodium penetration is widely considered as the main cause for cathode deterioration [1, 2]. The poor wettability by aluminum also limits the possibility to reduce the cell voltage drop by narrowing the anode-cathode distance. Therefore, a stable, wettable cathode would provide opportunity for improving the aluminum reduction technology [3].

In the early work, carbon/TiB_2 composites were investigated for possible use as cathode materials [4, 5], and better wettability and wear resistance were obtained [6]. However, sodium penetration into the cathodes remains as a problem. The relative high cost of TiB_2 cathode is another reason that blocks its application in the aluminum industry.

The intention of this work is to find out a medium-term, economically acceptable solution for advancing the wettable cathode technology. The main attracting point of the carbon/TiB_2 composites as wettable cathode material is that it can be made in a process similar to the ones used to fabricate industrial cathodes for today, so that the so-made cathodes could retain most of the advantages of the pure carbon cathodes but have a better aluminum-wettability and resistance to sodium attack. Using this technology strategy, the initial investment for cathode fabrication facilities and implementation of the cathode of this kind in smelter would be minimized.

This paper represents a continued effort to make better reduction cathodes with appropriate additives into the carbons. Experimental investigation of sodium penetration into carbon-based cathodes during aluminum electrolysis was carried out in a laboratory cell. Materials characterizations of the cathode samples were carried out before and after the aluminum electrolysis. The objective of this investigation is to look at the effects of the selected additives on the process of sodium penetration into the carbon-based cathodes.

Experimental

Materials and preparation of the carbon-based cathodes

Electrocalcined anthracites (industrial products), TiB_2 (chemical grade) or / and B_2O_3 (chemical grade) were mixed with a constant amount of carbonaceous binder (industrial product). The green samples were formed with a pressure up to 22 MPa and were prebaked at 150°C. Then they were heat treated up to 1200°C in a vessel packed with fresh carbonaceous powders. The cylindrical samples were 60 mm in length and 15 mm in diameter.

Set-up for sodium penetration during aluminum electrolysis

Figure 1 shows the experimental set-up for sodium penetration during aluminum electrolysis. The electrolysis cell was placed in a vertical tube furnace at a temperature of 960°C and an atmosphere of argon gas. A constant current was provided by a MPS302 DC power supply and the current density at cathode was 0.5 A/cm^2. The cylindrical cathodes were immersed 2 mm in the cryolitic melts and rotated at a speed of 180 r.p.m. All penetration tests were performed in the melts with a cryolite ratio of 4, 5% Al_2O_3 and 5%CaF_2.

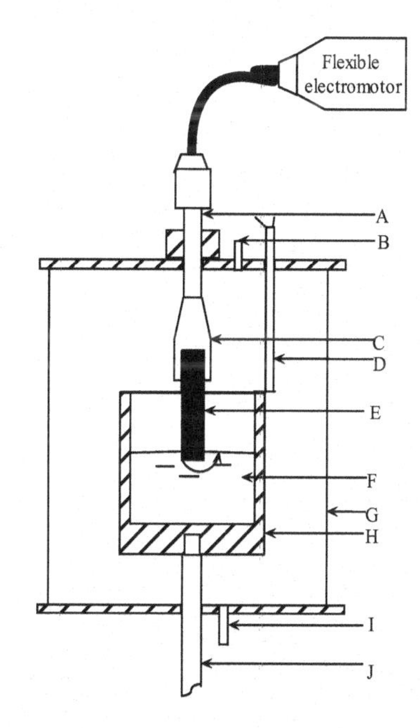

Figure 1. Experimental set-up for aluminum electrolysis: A) cathode steel rod, B) gas outlet, C) graphite support, D) thermocouple, E) carbon-based cathode , F) cryolite melt, G) furnace, H) graphite crucible (as anode), I) gas inlet, J) anode rod

Characterization of the cathode samples

The apparent density of the cathode samples was determined by measuring their geometric dimensions. After aluminum electrolysis, the cathode samples were cut open along their axis and the sodium penetration depth was measured by a phenolphthalein test against their cross-section.

The cross-section of the cathode samples were inspected by SEM-EDS, and the backscatter or secondary electronic image, X-ray mappings of titanium, aluminum, sodium, oxygen, carbon and fluorine were made simultaneously. XRD analysis was also applied to identify the penetrated chemicals in the specimens from the slices of the cathodes.

Results and Discussion

Sodium penetration into the cathode samples

Table I shows that the apparent density of the carbon-based cathode samples varies with the selected additives. The apparent density was found to increase with complex addition of TiB_2 $+B_2O_3$, while decrease with addition of B_2O_3. It was noted that the TiB_2 addition was a contributing factor to the increased density.

Table I. Compositions and Apparent Density of the Carbon-Based Cathode Samples

Sample No.	Composition	Apparent Density
C-01	Anthracite + binder	1.39 g/cm^3
CB-01	Anthracite + binder + B_2O_3	1.18 g/cm^3
CTB-01	Anthracite + binder + TiB_2 +B_2O_3	1.53 g/cm^3

Table II. Penetration Depth of Sodium into the Carbon-Based Cathode Samples after Aluminum Electrolysis

Sample No.	Penetration Depth,* cm	
	At 1 hour	At 2 hour
C-01	2.0	2.7
CB-01	1.6	2.0
CTB-01	1.1	1.5

* The measured penetrated depth varied in the range of 7% for three parallel runs of sample CTB-01.

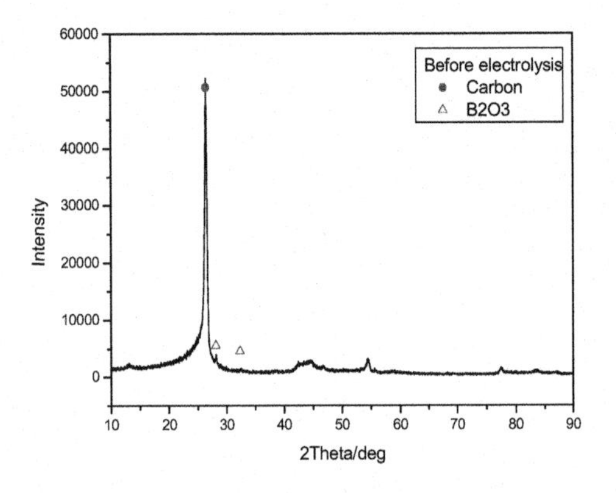

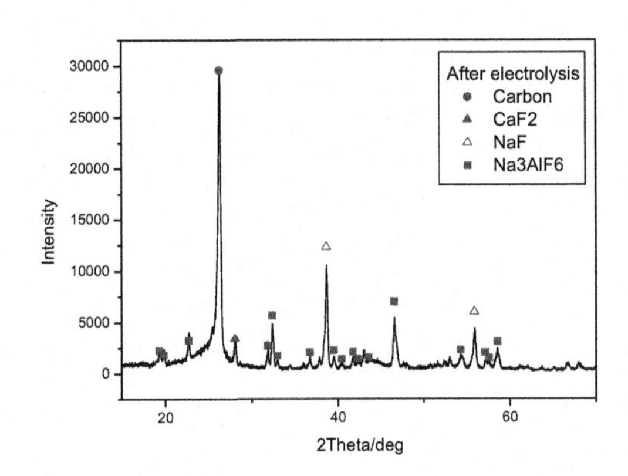

Figure 2. XRD analysis of the carbon + B_2O_3 cathode before and after aluminum electrolysis (3 hours at 0.5 A/)

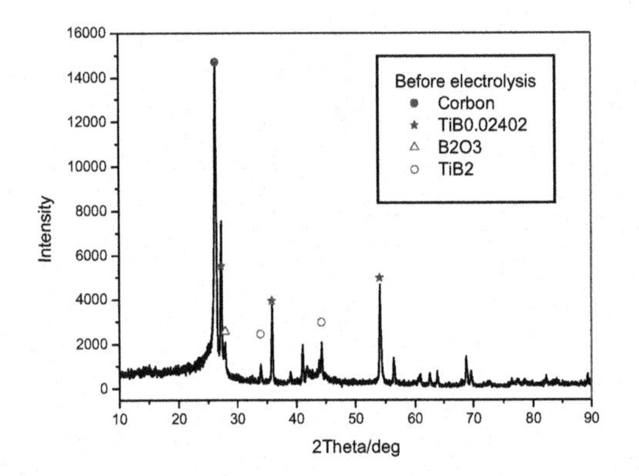

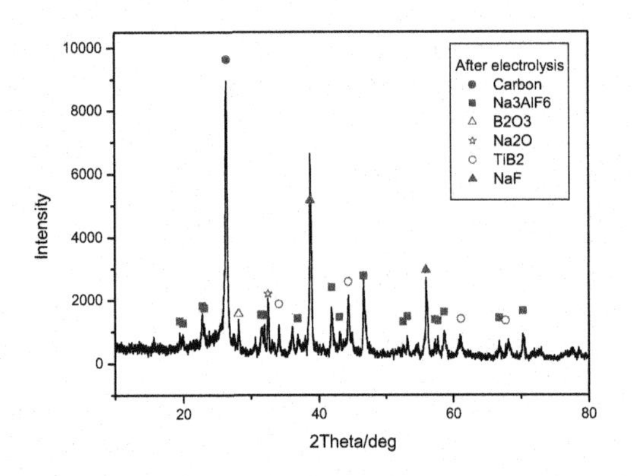

Figure 3. XRD analysis of the carbon + B_2O_3 cathode before and after aluminum electrolysis (3 hours at 0.5 A/)

During aluminum electrolysis metallic sodium is generated by a reaction

$$3\ NaF(l) + Al\ (l) = AlF_3(l) + 3\ Na(in\ C) \qquad (1)$$

It is well known that the sodium can penetrate into the inner part of the carbon cathode [1, 7, 8].

Table II presents the penetration depth of the sodium against the time of aluminum electrolysis. The penetration rate decreased significantly with B_2O_3 addition alone. The combination of B_2O_3 addition with TiB_2 also made the penetration rate lower than that with the pure carbon.

TiB_2 is usually applied as an additive to improve the wettability of the carbon cathode by liquid aluminum. The preliminary results above show a potential of B_2O_3 additive that can be used to inhibit sodium penetration. The experimental investigation suggests the complex addition of TiB_2 +B_2O_3 may offer possibility to make a better cathode in both wettability and resistance to sodium penetration.

XRD analysis of the electrolyzed cathode samples

Figure 2 shows the XRD analysis of the carbon cathodes with addition of B_2O_3, which indicates which indicates the presence of fluoride salts, NaF and Na_3AlF_6.in the cathodes after aluminum electrolysis. Figure 3 also shows strong presence of these salts within the cathode bodies after electrolysis. These suggest that cryolitic melts can penetrate into the carbon+B_2O_3 or the carbon+TiB_2+B_2O_3 cathodes.

In the pure carbon cathodes, such a melt penetration usually occurs as a process following the penetrating front of sodium [1]. Even though B_2O_3 and TiB_2+B_2O_3 have demonstrated an effect on reducing the sodium penetration, a similar effect on the melt penetration might not follow.

Again, the effects of the additives on the melt penetration and its association with the sodium penetration could differ from the pattern known for the pure carbon cathodes. This is still under investigation.

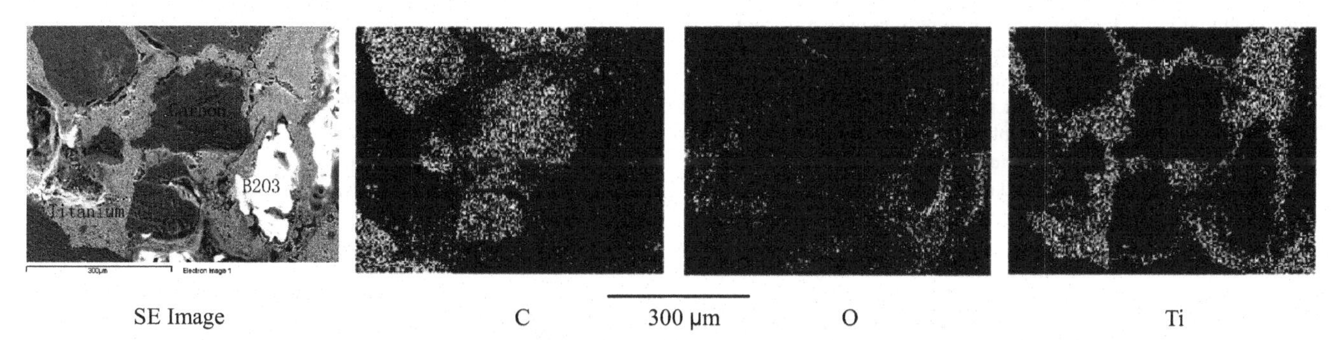

Figure 4. SEM-EDS Image and mappings of the carbon+TiB_2 +B_2O_3 cathode

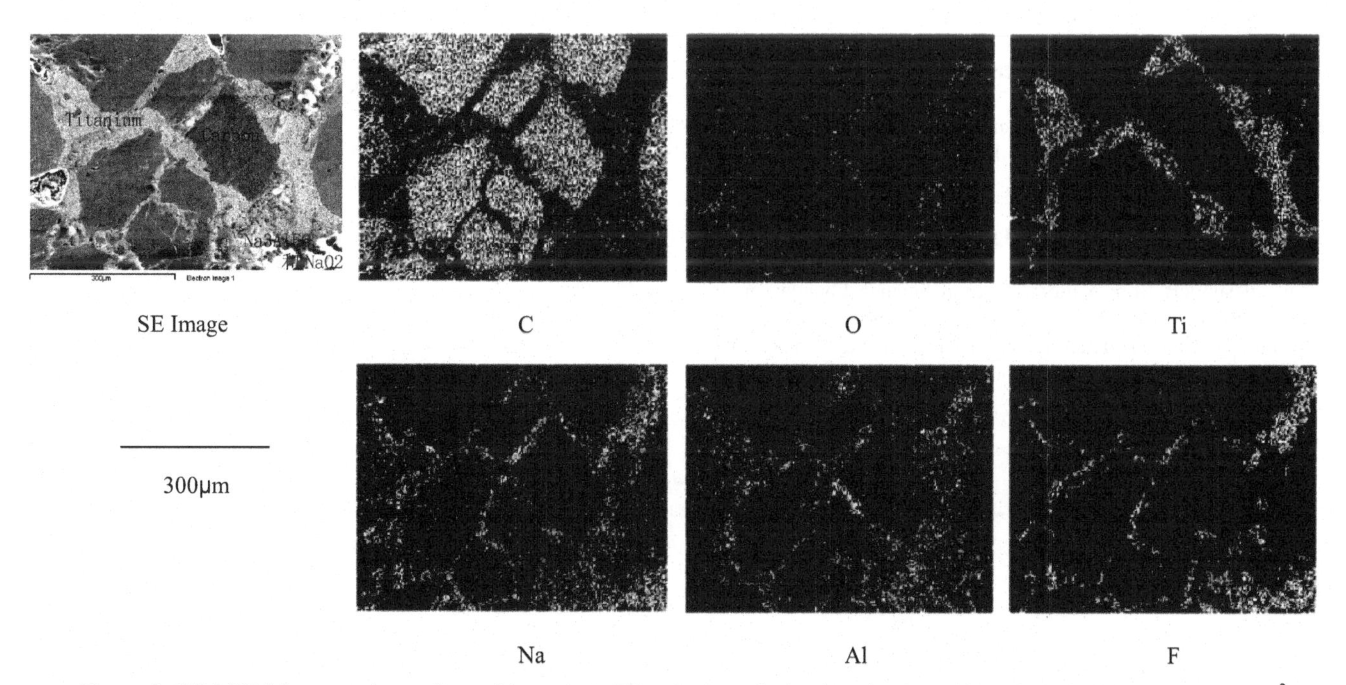

Figure 5. SEM-EDS Image and mappings of the carbon+TiB_2 +B_2O_3 cathode after aluminum electrolysis for 3 hours at 0.5 A/cm^2

SEM-EDS Observation on the cathode samples

Figure 4 shows the microstructure of the carbon+TiB_2+B_2O_3 cathode, where the carbon aggregates are the major phase. The fine particles of TiB_2 are well distributed in the continuous phase of carbonaceous binder. B_2O_3 is also located within the binder phase.

In Figure 5, the cathode sample was taken from the same cathode product as shown in Figure 4, after aluminum electrolysis for 3 hours. The specimen was taken from a location 5 mm above the electrolyte. SEM-EDS analysis shows the microstructure remained unchanged after the electrolysis, but new foreign elements, such as Na, Al and F appear within the cathode body. Most of them are located together in the areas of the binder phase, suggesting that the sodium, aluminum and fluorine exist in the form of NaF, AlF_3 and Na_3AlF_6, as seen in the XRD spectra in Figure 2 and Figure 3.

It is evident that the amount of Na in the binder areas is much higher than that in the carbon grains, Figure 5. However, it should be pointed out that there is relative less Na in the Ti-rich area of the binder phase. The porous boundaries are usually considered as a diffusion channel for Na or melt penetration [9, 10]. By adding B_2O_3 or TiB_2 +B_2O_3, the binder phase seems to become denser and harder than before. The chemical stable additives may fill in the diffusion channel at the boundaries and make a resistance to sodium penetration.

Conclusions

- Preliminary experimental results show that rate of sodium penetration into the carbon based cathode is reduced with addition of B_2O_3 or TiB_2+B_2O_3.
- Fluoride salts from melt penetration are presented within the bodies of carbon based cathodes with B_2O_3 or TiB_2+B_2O_3 additives after aluminum electrolysis.
- B_2O_3 or TiB_2+B_2O_3 are found within the areas of binder phase where they may play a role in reducing the rate of sodium penetration.

Acknowledgement

Financial support from National Natural Science Foundation of China (NSFC), funding for Doctor Degree from Ministry of Education of China and funding from University of Science and Technology Beijing are gratefully acknowledged.

References

1. M. Sørlie and H. A. Øye, *Cathodes in Aluminium Electrolysis*, 2nd ed., Aluminium-Veriag, Dusseldorf,
2. 1994.
3. H. A. Øye, "Materials Used in Aluminum Smelting," Light Metals 2000, Warrendale PA, USA: TMS, 2000, 3-15.
4. R. P. Pawlek, "Aluminum Wettable Cathode: An Update," Light Metals 2000, Warrendale PA, USA: TMS, 2000, 449-454.
5. J. Xue and H A. Øye, "Investigating Carbon/ TiB_2 Materials for Aluminum Reduction Cathodes," JOM, 44 (11) (1992), 28-34.
6. J. Xue and H A. Øye,"Wetting of Graphite and Carbon/TiB_2 Composites by Liquid Aluminium," Light Metals 1993. Warrendale AP, USA: TMS, 1993, 631-637.
7. J. Xue and H A. Øye, "Al_4C_3 Formation at the Interface of Al–Graphite and Al-Carbon/TiB_2 Composites," Light Metals 1993, Warrendale PA, USA: TMS, 1993. 631-637.
8. P.-Y. Brisson, G. Soucy and M. Fafard, "Revisiting Sodium and Bath Penetration in the Carbon Lining of Aluminum Electrolysis Cell," Light Metals 2005, Warrendale PA, USA: TMS, 2005, 727-732.
9. P. Brilloit, L P. Lossius and H. A. Oye, "Melt Penetration and Chemical Reaction in Carbon Cathodes during Aluminium Electrolysis," Light Metals 1993, Warrendale PA, USA: TMS, 1993, 321-330.
10. C. Krohn, M. Sorlie and H. A. Oye, Light Metals 1982, Warrendale PA, USA, 1982, 311.

SiC IN ELECTROLYSIS POTS: AN UPDATE

Rudolf P. Pawlek
TS+C, 14 Avenue du Rothorn, CH – 3960 Sierre, Switzerland

Keywords: SiC, chemical stability, wear mechanism

Abstract

The physical and chemical properties, the bond types and the resistance to bath, aluminium, carbon dioxide and air attack of SiC are reviewed, outlining especially the use of non-oxide bonded SiC refractories as sidewall materials for aluminium electrolysis cells. State-of-the-art is nitride-bonded SiC. This type of SiC sidewall lining material is particularly suitable for the operation of modern electrolysis cells equipped with point feeders for automatic alumina feeding to the bath. It is also a key material in upgrading older cells to higher current and anode size.

Introduction

This update is a continuation of [1, 2] and covers the period 1995 to 2005. The literature explains the required physical and chemical properties: the sidewall needs to

- Resist molten aluminium [3]
- Resist molten bath [3]
- Resist CO_2/CO and HF vapours [3]
- Have low porosity [4-7]
- Have very high strength [8]
- Have reasonably high thermal conductivity [8, 9]
- Have high bulk density [6]
- Have high electrical resistivity [6, 10]

Binders

Silicon carbide based refractories can be divided into four kinds of materials according to their bonding phases [11]. These phases include silicon carbide, silicon nitride, silicon oxynitride and SiAlON.

SiC

According to Skybakmoen et al [3], self-bonded SiC was superior to all other materials tested, but some local and uneven corrosion occurred where the more reactive binding phase was preferentially attacked [10]. The weakest part of the SiC bricks seems to be the binder. SiC-based materials show high oxidation resistance in general by forming of a protective layer of SiO_2. In the gas/melt interface and in the vapour zone one may expect faster removal of SiO_2 by dissolution; corrosion resistance then depends on the on the base materials.

SiAlON

Cao et al [12] investigated β-SiAlON/SiC composites in a molten cryolite melt to see how SiC content, corrosion kinetics and mechanism affected corrosion resistance. They found that the corrosive reactions between β-SiAlON/SiC matrix and cryolite are retarded due to the formation of $NaAlSiO_4$ with a high viscosity on the surface of β-SiAlON/SiC matrix.

Si_2N_2O

According to Welch [10] some oxynitride species show better short-term resistance, and perform satisfactorily when the anode effect frequency is reduced. However, electrolyte penetration results in partial or total dissolution of the protective oxide formed, resulting in long-term structural degradation. Cheng et al [11] found that Si_2N_2O and SiAlON-bonded SiC have less corrosion resistance against cryolite melts than Si_3N_4 bonded SiC.

Si_3N_4

However, compared with the other binders for silicon carbide refractories, nitrides are the least corroded [13], because they have an extremely high electrical resistance and therefore do not suffer from the electrochemically driven processes. Therefore silicon nitride bonded silicon carbide is the material most used in the primary aluminium industry and counts as state-of-the-art sidelining material. According to Skybakmoen [7] the Si_3N_4 phase is thermodynamically more unstable than the SiC-phase and it has been claimed that the Si_3N_4 phase is oxidised first in a Si_3N_4 bonded SiC material.

Manufacture

The manufacture of Si_3N_4 bonded SiC has been reviewed several times [7-9, 14, 15]. Si_3N_4 bonded SiC materials are produced by blending SiC particles (mm size) with fine-grained silicon metal powder and a binder. SiC may have some content of SiO_2, Si, Fe, and C coming from the production process. The bricks are shaped by pressing/vibration, then dried to remove moisture. The bricks are sintered in furnaces in a flowing nitrogen atmosphere at temperatures of more than 1400°C. During the nitriding process silicin reacts with nitrogen, forming Si_3N_4 as a binder between the SiC grains. The final product will then consist of of large SiC grains (mm size) with a microporous binder phase of Si_3N_4. Typical composition is 72-80 wt.% SiC and 20-28 wt.% Si_3N_4. High purity (better than 97%) SiC grains must be used. For shaping the bricks, vibratory presses are recommended so as to avoid stratification.
These materials have to develop the right amount of nitride with the least residual silicon metal and other contaminants like oxygen and aluminium phases. Good control of the nitriding process is therefore crucial. The physical properties are also important, as excessive porosity is detrimental to thermal conductivity and to oxidation resistance.
Wu [15] proposed a reverse reaction sintering process. In this process the Si_3N_4 is firstly converted into reactive oxides, then react sintered. The new sintering process is accomplished by the reactive sintering of newly formed Si_2N_2O or SiO_2 phases. The

products have excellent physical and chemical properties due to their dense structure and to in-situ formed oxides or sub-oxides which fully fill the boundary between Si_3N_4 and SiC particles.

Oxidation resistance

As demonstrated by Jorge et al [16], oxidation resistance is the critical parameter for the performance of silicon nitride bonded SiC in aluminium reduction cells. A non-oxide Si_3N_4 can be oxidized in a relatively reductive atmosphere (such as CO) [17] to form silica that will easily be corroded by cryolite. According to Ivey [8] unprotected SiC therefore oxidizes rapidly in the temperature range of 800-1150°C. As Welch [10] reported, the ignition temperature for the binder face can be as low as 200°C. For low oxygen supply rates, this results in preferential oxidation and powdering of the material, creating both cavities and markedly changed thermal resistance. The oxidation resistance of SiC is a function not only of the bond system, but also of porosity and pore size.
However, in materials based on SiC and exposed to the gas phase, Hagen et al [18] observed a protective layer of sodium aluminosilicate fluoride glass. Enhanced corrosion resistance results from the tendency of silicon to form viscous oxyfluoride glasses, which kinetically retard both the oxidation and dissolution of the oxidised product.

Wear mechanism

According to Andersen et al [19] the wear mechanism of Si_3N_4 bonded SiC bricks seems to combine attack by Na, HF, oxidation and moisture. Below bath level, oxidation and sodium play a major role, but bricks there deteriorate least. At bath level the attack is severe. Above bath level the bricks disintegrate, but reaction products are almost absent. Here HF vapour seems to transform the oxidation products into volatile fluorides and other gases.
The main attack agents of cell materials in general are cryolite, liquid aluminium, $NaAlF_4$, CO and CO_2. Instead sodium hydroxide was found in SiC and therefore it appears that sodium metal or sodium oxide combined with moisture causes chemical attack. HF gas probably also causes chemical attack. Only at the bath level were Na-Al-silicates.
Bath acidity greatly influences chemical resistance. Corrosion increases with increasing cryolite ratio (increasing sodium content in the bath). Skybakmoen and co-workers [7] propose the following reaction sequence:

- Sodium penetration
- Oxidation to Na_2O and formation of $Na_2O.xSiO_2$
- Swelling/descaling/dissolution of the reaction products
- New reaction cycle

This reaction cycle is assumed to be especially damaging for materials with high oxygen content.
The older the SiC block, the faster it oxidates and loses its thermal conductivity. Jorge et al. [9] distinguish different steps:
The first step involves oxidation without any infiltration but with a fast drop in thermal conductivity of the Si_3N_4 bonded SiC material.
The second step involves continuous infiltration, wear and a slight decrease of thermal conductivity. Thus at ultimate failure, the average thermal conductivity had dropped by approximately 38%.
When combined with joint openings and air gaps, the drop of thermal conductivity leads to a higher hot face temperature and to a less stable frozen ledge. Thus, the sidewall material is more often directly exposed to molten and gaseous bath components. These gaseous components and even the molten bath may penetrate to directly attack the steel shell. The subsequent metal exfoliation creates an insulating layer that dramatically increases the risk of a tap-out of the pots.

Inert anodes

Nitride-bonded silicon carbide sidelining was also tested with oxygen evolving anodes [7]. The test result shows clearly that a significantly higher degree of corrosion occurs under O_2 atmosphere than under CO_2 atmosphere. The Gibbs energy of reaction is higher in the case of O_2 because of its higher oxidizing potential. According to Wang et al. [20] the oxidation resistance shows little difference at 850°C and 750°C, but is much worse at 950°C. According to the testing of dynamic corrosion by the bath at 950°C, the corrosion of silicon carbide bricks speeds up obviously.

Chemical/oxidation test cells

A laboratory test cell for measuring the chemical/oxidation resistance for SiC-based materials was developed by Skybakmoen et al. [7] Their test cell exposed the materials to all corrosive environments: liquid aluminium, cryolite melt, its vapours (mainly $NaAlF_4$) and CO_2/CO from the anode. The tested materials normally undergo corrosion in the gas phase (zone above bath) much like that observed in industrial cells. Strongly corroded samples even showed visible corrosion in the upper bath level zone. In some cases the test samples were completely corroded in the gas zone.
However, a more realistic test to study the chemical attack on Si_3N_4-bonded SiC sidewall bricks in an electrolysis cell system should involve Na-metal together with oxygen, moisture, and NaF in a temperature range between 300 and 1000°C [19]. The Si_3N_4-bonded SiC brick is not resistant to the combination of moisture, Na-metal, HF and oxidation. The combination of these agents is one of the most aggressive environments for oxides, especially for SiO_2.

Solutions for electrolysis pots

Nitride-bonded SiC is now used in cell construction in more than 30 plants worldwide and in thousands of electrolysis cells ranging in amperage from 30 – 350 kA [8, 21]. The major reasons for the extensive use of SiC are to

- prevent air burn
- increase productivity
- extend pot life.

The excellent corrosion resistance of nitride bonded SiC, combined with its extremely high strength and reasonably high thermal conductivity allow an increase in amperage and anode size while in most cases also extending cell life. The ability to use thinner shapes and to provide extra protection in the event of loss of frozen ledge favours SiC in this application.
To maximise benefits from nitride-bonded SiC refractories in aluminium cell construction, it is necessary to examine the economics and heat balance [22]. Heat balance is important not only for the best operation of the cell but also to maximise the life of nitride-bonded SiC sidewalls as well.

Clearly, care must be taken in designing sidewalls in order to maximise the benefits of nitride-bonded SiC refractories.
Using nitride-bonded SiC slabs in electrolysis pots [23 - 25], Refrax® Arc was developed by [16, 26]. This silicon nitride-bonded silicon carbide offers improved oxidation and cryolite resistance. Improved cryolite resistance may have a real influence on the functional characteristics of the sidewall, i.e. the thermal conductivity at any time and the corrosion resistance when the frozen ledge disappears.
These authors developed another improved silicon nitride-bonded silicon carbide material, Sicfrax [9], which offers regular performance in the target range of index: oxidation gain below 2% and corrosion index below 35% according to the cryolite test on pre-oxidised samples of blocks up to 100 mm thick.
Curtis et al. [27] proposed the use of a composite carbon-silicon carbide sidewall block cathode. The composite block provides the opportunity to extend the potlife of cells operating with conventional carbon sidewalls, or it can offer excellent cost saving while maintaining the desired operational results in cells using full size silicon carbide bricks. The composite block consists of a calcined anthracite carbon block, of nitride-bonded silicon carbide and of a glue or carbonaceous cement. The advantage of the silicon carbide layer is to provide excellent resistance to erosion and oxidation of the sidewall lining, while the carbon provides the added flexibility in design and specification in case the blocks need machining.
Combined side wall protector blocks were proposed by Zienkowicz [28] to replace the ramming paste of the big joints in aluminium reduction cells. He perfected a new manufacturing technique to form the new combined sidewall protector blocks using a proprietary technique to bond silicon carbide material to a preformed carbon block containing variable graphite additions. The combined sidewall protector block can be manufactured in block sizes of up to 800 mm x 500 mm maximum, this limit corresponding to the allowable weight for manual handling and installation.

Conclusions

Without protection a sidelining made of silicon nitride bonded silicon carbide is – under severe test conditions – chemically not stable and will corrode, the binder matrix being the weak component.
When considering revamping aluminium electrolysis cells using nitride-bonded SiC it is important to profit from their excellent corrosion resistance, extremely high strength, and reasonably high thermal conductivity to increase amperage and anode size while also extending cell life.
Using silicon nitride-bonded silicon carbide as sidelining material when using inert anodes is problematic, as the sidelining will be corroded, because the Gibbs energy of reaction and the oxidizing potential are higher in the case of oxygen.

Acknowledgments

The author gratefully acknowledges with thanks the help of Mrs. G. Brichler and Mr. A. Bushnell during the preparation of this manuscript.

References

1 R. P. Pawlek, "SiC in aluminium electrolysis cells", Light Metals 1995, ed. J. W. Evans (TMS, Warrendale, Pa), 527-533
2 R. P. Pawlek, "SiC in aluminium electrolysis cells", VIII. Al Sympozium, 25. – 27. Sept. 1995, Slovakia, Ziar nad Hronom – Donovaly, 177-189
3 E. Skybakmoen et al., "A new corrosion test method for sidewall materials used in aluminium electrolysis", VIII. Al Sympozium, 25. – 27. Sept. 1995, Slovakia, Ziar nad Hronom – Donovaly, 195-200
4 E. A. Cortellini, "Ceramic carbide or nitride lining for sidewalls in aluminium electrolysis cells", US patent 5,560,809 (26 May 1995)
5 A. A. Sviridov, S. A. Podkopaev, and L. N. Ruzhevskaya, "Blocks for sidelining of electrolysers with increased resistance to sodium penetration and to oxidation", Proceedings VIII Int. Conf., Krasnoyarsk, Sept. 10-12, 2002, "Aluminium of Siberia 2002", 43-44
6 B. Gao, Z. Wang and Z. Qiu, "Corrosion tests and electrical resistivity measurement of SiC-Si_3N_4 refractory materials", Light Metals 2004, ed. A. T. Tabereaux (TMS, Warrendsle, Pa), 419-424
7 E. Skybakmoen et al., "Quality evaluation of nitride-bonded silicon carbide sidelining materials", Light Metals 2005, ed. H. Kvande (TMS, Warrendale, Pa), 773-778
8 J. L. Ivey and P. F. Boilly, "Use of silicon carbide refractories in the aluminium industry", Siberian Aluminium – 96, ed. P. Polyakov (1997), 87-99
9 E. Jorge, O. Marguin and P. Temme, "Si_3N_4 bonded SiC refractories for higher aluminium reduction cell performance", Proceedings IX Int. Conf. Sept. 9-11 2003, "Aluminium of Siberia 2003" Krasnoyarsk, 166-177
10 B. J. Welch, "Limitations of carbon-carbon composite electrodes and optional composite developments for aluminium smelting", 9th Cimec-World Ceramics Congress, Ceramics: Getting into the 2000's, Pt. D, ed. P. Vincenzini (Techna Srl, 1999), 647-659
11 Z. Cheng et al., „Comparison of cryolite resistance of silicon carbide materials", Proc. VIIth Australasian Aluminium Smelting technology Conf. and Workshop, Melbourne, Australia, 11-16 Nov. 2001, 7pp.
12 S. H. Cao et al., "The corrosion behaviour of β-SiAlON/SiC composites in cryolite molten solutions", *J. Cent.-South Inst. Min. Metall.* 25 (2) (1994), 202-207
13 B. J. Welch, M. M. Hyland and B. J. James, "Future materials requirements for the high-energy-intensity production of aluminium", *JOM* 53 (2) (2001), 13-18
14 N. Zienkovicz, "High performance ceramics and refractories", TMS-Industrial Aluminium Electrolysis, Quebec City, 8-12 Sept. 2003
15 W. P. Wu, "Reverse reaction sintering of Si_3N_4/SiC composite materials for aluminium electrobath", *Guisuanyan Xuebao* 32 (12) (2004), 1524-1529
16 E. Jorge et al., "Optimization of Si_3N_4 bonded SiC refractories for aluminium reduction cells", *CN Refractories* 5 (2001), 74-78
17 J. Zhao et al., "The properties of Ni_3N_4 bonded SiC material for aluminium electrolysis cell", Light Metals 2000, ed. R. D. Peterson (TMS, Warrendale, Pa), 443-447

18 E. Hagen, M.-A. Einarsrud and T. Grande, "Chemical stability of ceramic sidelinings in Hall-Heroult cell", Light Metals 2001, ed. J. L. Anjier (TMS, Warrendale, Pa), 257-263
19 F. B. Andersen et al., "Wear of silicon nitride bonded SiC bricks in aluminium electrolysis cells", Light Metals 2004, ed. A. T. Tabereaux (TMS, Warrendale, Pa), 413-418
20 Z. W. Wang et al., "Corrosion resistance of SiC insulation side wall for aluminium reduction cell", *Dongbei Daxue Xuebao, Ziran Kexueban* 23 (4) (2202), 345-347
21 J. Schoennahl et al., "Optimization of Si_3N_4 bonded SiC refractories for aluminium reduction cells", Light Metals 2001, ed. J. L. Anjier (TMS, Warrendale, Pa), 251-255
22 M. M. Ali and A. A. Mostafa, "Improvement of the aluminium cell lining behaviour by using SiC", Light Metals 1997, ed. R. Huglen (TMS, Warrendale, Pa), 273-278
23 G. F. Vedernikov, A. L. Yurkov, and L. V. Krylov, "Testing and mastering of new lining materials for design of high amperage prebaked anode pots", Proceedings VIII Int. Conf. Sept. 10-12, 2002, "Aluminium of Siberia 2002", 121-123
24 B. Ruan et al., "Silicon carbide tiles for sidewall lining in aluminium electrolysis cells", *China's Refract.* 8 (2) (1999), 6-8
25 H. Lu and Z. Qiu, "Clean production process for aluminium electrolysis in 21st century", *Kuangye* 9 (4) (2000), 62-65, 83
26 E. Jorge, P. Temme and B. Kokot, "The use of nitride bonded SiC in aluminium reduction cells and latest developments", Proc. VII Int. Conf. Sept. 10-12, 2002, in Krasnoyarsk of "Aluminium of Siberia 2001", 100-105
27 E. I. Curtis, P. D. Mascieri and A. Tabereaux, "The utilization of composite carbon-silicon carbide sidewall blocks in cathodes", Light Metals 1996, ed. W. R. Hale (TMS, Warrendale, Pa), 295-301
28 N. Zienkowicz, "Combined side wall protector blocks", *Aluminium* 81 (4) (2005), 323-324

A NEW TEST METHOD FOR EVALUATING Si_3N_4-SiC BRICK'S CORROSION RESISTANCE TO ALUMINUM ELECTROLYTE AND OXYGEN

Cao Xiaozhou, Gao Bingliang, Wang Zhaowen, Hu Xianwei, Qiu Zhuxian
(Northeastern University, College of Materials and Metallurgy, Mail box 117, Shenyang, Liaoning 110004 China)

Keywords: SiC- Si_3N_4, Chemical resistance, Sidelining material, Aluminum electrolysis

Abstract

In the paper a new test method was applied to evaluate the bath resistance and oxygen resistance of silicon nitride bonded silicon carbide bricks that were used for sidelining in aluminum electrolysis pots. In test, carbon dioxide gas was input to the molten electrolyte in which part of sample was immersed in order to simulating the factual operating conditions of cells. Based on the method the bath resistance and oxygen resistance of refractory bricks from several Chinese refractory factories were tested. The method was also used to evaluate the bath resistance of different parts of one brick. The test result showed that the inside of the brick had worse bath resistance than the outside, which may be caused by lower content of Si_3N_4 and higher content of impurities in the inner part of the brick.

Introduction

Today, the biggest prebaked-anode cell for primary aluminum production is 350kA in China. Compared to the small cells, the thermal field distribution is very different in the big cells. The side wall of the big cell requests the higher heat dissipating capacity, which enables a protective solid electrolyte layer to be formed on the surface of the side wall, thus protecting the side wall material from corrosion.

Side wall materials should have some properties: high heat conductivity, good mechanical robustness at high temperature, good oxidation resistance, good heat shock resistance, good corrosion resistance. It is more important for large-scale electrolysis cell. The carbon side wall has poor erosion and corrosion resistance, which leads to damage of side wall easily and short service life.

Si_3N_4-SiC materials have many excellent properties, such as high hardness, good resistance to wear, high thermal conductivity and good insulativity[1-4], they were applied widely in large-scale prebaked-anode type cell.

There are many methods to evaluate the performance of Si_3N_4-SiC materials used for aluminum cell and other metallurgical industry, including static corrosion[5-10] tests and dynamic corrosion tests. Skybakmoen[11-12] and Wang[13] investigated dynamic corrosion by different methods, separately.

The dynamic corrosion experimental set up adopted by Egil Skybakmoen is shown in Figure 1. During electrolysis, CO_2 bubbles generated on the bottom of carbon anode and released from the electrolyte, which cause the fluctuation of the electrolyte. At the condition, CO_2 and high temperature air and fluctuating electrolyte cause the samples to be corroded during the test. The anode was renewed after some time because of consumption [11-12].

Wang used"Rotating-CO_2-cryolite"test method[13] to investigate dynamic corrosion of SiC materials, the test device was shown in Figure 2. The samples were rotated in molten cryolite at 30rpm in CO_2 atmosphere, which lead to the dynamic corrosion of samples.

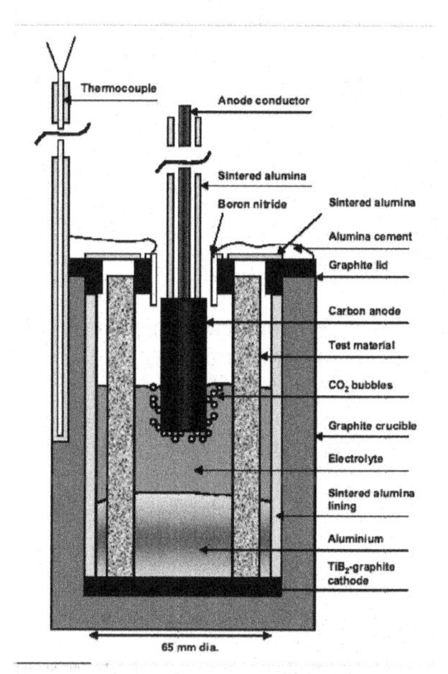

Figure 1. Schematic graph of the dynamic corrosion experimental set up adopted by Egil Skybakmoen[12]

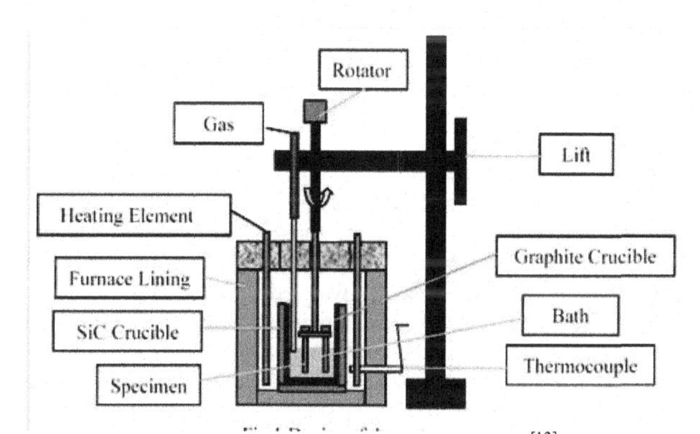

Figure 2. Device of Rotating-CO_2-cryolite method[13]

In this paper, a new dynamic corrosion test has been designed to evaluate the chemical resistance of Si_3N_4-SiC sidelining bricks. SiC bricks provided by several Chinese refractory factories were tested by the method. The corrosion resistance comparison between inner of brick and outer of brick was investigated.

Experimental

Sampling

A slice of thickness 20 mm was from the central part of the brick. As shown in Figure 3, we cut some samples from it for the corrosion resistance test, mineral phase analysis, chemical analysis and physical property test. Number 1-4 （10 mm×20 mm×100mm） for chemical corrosion test, number 5 （20 mm×20 mm×75mm） and number 6-7 （10 mm×30 mm×20mm） for mineral phase and chemical analysis.

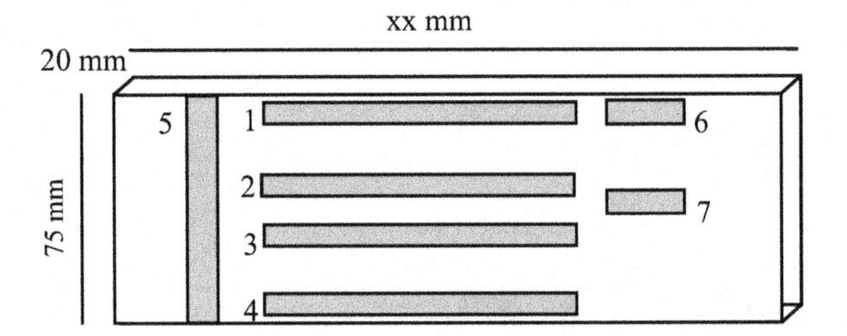

Figure 3. The schematic graph of sampling for chemical resistance test, chemical composition and phases measurements.

Figure 4 shows that the cross section color of SiC brick is not uniform. The color of central parts of brick is deeper than that of the peripheral. The chemical composition and phases of inner and outer were analyzed.

Figure 4. Cross section morphology of the SiC-Si_3N_4 brick.

Chemical resistance test

In industrial aluminum electrolysis cells, the CO_2 gas generates on the bottom of anodes and releases from the molten electrolyte. Sidelining materials undergoes the most severe corrosion at the gas/bath interface. In order to simulate the actual work condition of electrolysis cells, the CO_2 gas was led into the bath through a pipe. The gas pipe made of corundum was posited vertically in the middle of the cell, and the pipe end was posited 3cm below surface of the electrolyte. Samples were partly immersed in the molten electrolyte, and arranged in a circle. During tests, the CO_2 gas released from the electrolyte and caused the fluctuation of electrolyte. The samples were attacks dynamically by gases and electrolyte. The schematic graph of test cell was shown in figure 5.

The inner diameter of graphite crucible is 90mm and its height is 125mm. The graphite crucible was protected from oxidation by a steelless crucible and covered with a steelless lid with a hole. The crucible containing the melt was heated to 950℃ in the electric furnace, and the holding time was 48h. The CO_2 gas flow rate was controlled at 1-2 l/min. Due to dissolution of Al_2O_3, the corundum pipe was lowered 2 cm every 2h. The temperature was measured by $PtRh_{10}$-Pt thermocouple.

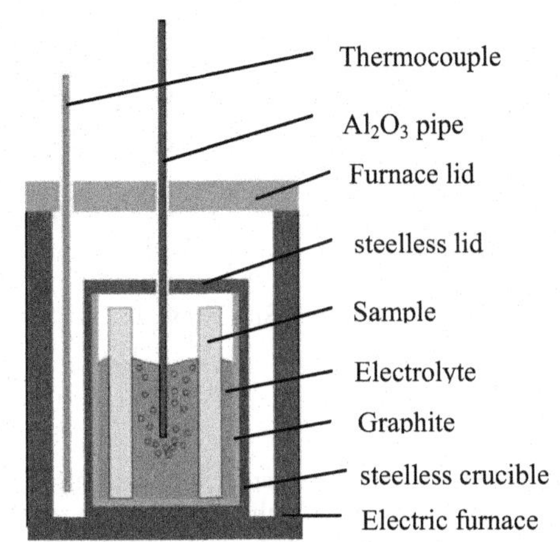

Fig.5 Experiment equipment for the corrosion of dynamic electrolyte

The composition of the melt was cryolite with NaF/AlF_3 molar ratio of 2.4 (90 wt%), CaF_2(5 wt%) and Al_2O_3(5 wt%). Because of electrolyte vaporization, it is necessary to add electrolyte to maintain composition stability and quantity balance of the electrolyte during the experiment.

After test, samples were taken out and rapidly cooled to room temperature and washed with 7%mass $AlCl_3$ aqueous solution to remove the adhering electrolyte.

The degree of corrosion was determined by the volume loss of the samples during the test. See equation 1. According to ISO 5017 standard, volumes of the samples were measured and calculated.

$$\text{Volume loss}[\%]=[(V_1-V_0)/V_0]\times 100\% \quad (1)$$

Where: V_0= volume before test

V_1= volume after test

Result and Discussion

Based on this method, sidelining materials come from five refractory factories were tested. The results are listed in Table 1.

Table 1. The volume loss and apparent porosity of the samples

Sample No.	Volume loss (%)	Open Porosity-1 (%)	Open Porosity-2 (%)	Flow rate (l/min)
A_1	4.99	14.49	3.91	2
A_2	7.30	14.73	5.28	2
B_1	6.81	11.36	5.73	2
B_2	6.75	10.52	4.75	2
C_1	3.86	15.58	6.25	2
C_2	2.12	15.93	6.55	2
D_1	1.75	10.36	5.85	2
D_2	8.55	12.26	6.55	2
D_3	8.86	12.29	6.02	2
D_4	14.22	12.99	5.91	2
E_1 (i)	4.65	13.91		1
E_2 (i)	4.39	13.84		1
E_3 (o)	2.07	10.10		1
E_4 (o)	2.62	10.65		1

Open porosity-1was determined before test.
Open porosity-2 was determined after test.

The electrolyte permeated in the samples in the corrosion process which led to reduced apparent porosity after test. C factory samples (C_1, C_2) have higher open porosity, but corrosion resistance is good. However, samples (D_2, D_3) from D factory have lower open porosity, but corrosion resistance was bad. There is no inevitable correlation between open porosity and the degree of corrosion.

Photos of sample before and after the dynamic test are shown in Figure 6. The samples were severely corroded at the gas/bath interface. Because CO_2 released from the liquid, which made liquid fluctuate greatly. The middle part of the sample was attacked by CO_2 gas and electrolyte altogether, so the degree of corrosion was serious in the zone. CO_2 gas flow rate also has influence on corrosion resistance of Si_3N_4-SiC materials and the following chemical reactions are proposed.

$$SiC + 3CO_2(g) = SiO_2(l,s) + 4CO(g) \quad (1)$$

$$Si_3N_4 + 3CO_2(g) = 3SiO_2(l,s) + 2N_2(g) + 6CO(g) \quad (2)$$

However the influence of electrolyte composition on corrosion resistance of Si_3N_4-SiC materials was not studied. Maybe it has great influence on corroding sidelining materials.

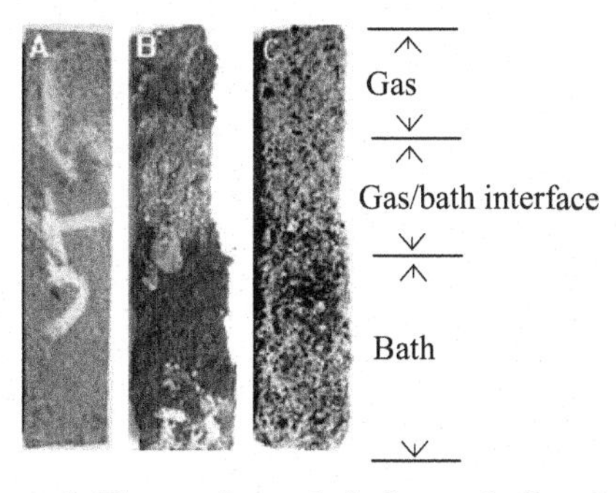

Figure 6. Photos of sample before and after chemical resistance test

A-original sample; B-corrosion sample; C-clean-up sample

Corrosion resistance difference between inner and outer of the brick

Samples taken from the central and peripheral part of brick (E_1,E_2,E_3,E_4) were subjected to phase determination and chemical composition analysis. Chemical concentrations of SiC and Si_3N_4 were determined by analyzing contents of carbon element and nitrogen element. The phases were determined by XRD. The results are given in Table 2.

Table 2. Chemical compositions and XRD analysis results of inner and outer of the brick

	Central	Peripheral
SiC(wt%)	77.67	71.33
Si_3N_4(wt%)	16.75	21.75
XRD		
α-Si_3N_4(comparision)	Less	More
β-Si_3N_4(comparision)	More	Less

As seen from Table 2, the inner part of the brick has higher content of SiC than outer part, which may be the reason why the inner part has the deeper color.

According to Table 1, samples E_1 and E_2 (inner part) showed worse chemical resistance than samples E_3 and E_4 (outer part), which is contrary to the results obtained by Skybakmoen[11]. But we think it is not a paradox. Because in the experiments carried by Skybakmoen et al, the samples with Si_3N_4 21%mass showed better chemical resistance than that of samples with Si_3N_4 25%mass. In our case, the samples showed better chemical resistance if the concentration of Si_3N_4 is around 21%. We think that the optimum concentration of Si_3N_4 may be in the range of 20~22%mass. It is necessary to do further research.

Conclusion

A new test method was applied to evaluate the chemical corrosion resistance of SiC-based material in electrolyte. The test result is close to the other two kinds of test methods.

The chemical content difference of SiC leads to colour variation between central part and peripheral part of brick.

The relationship between Si_3N_4-SiC materials composition and corrosion resistance was investigated. Bricks with Si_3N_4 20~22% mass will have good corrosion resistance.

Acknowledgement

The authors appreciate the financial support from Liaoning Natural Science Foundation (Project No.20031014). We also sincerely acknowledge Mr. Yu Yaluo for his valuable contribution to this study.

Reference

[1] Zhuxian Qiu. Produce Aluminium in Prebake Cell. Metallurgical Industry Press, 2005.

[2] Andreas Sonntag. New R-SiC extends service life in kiln furniture[J] . *Am Ceram Soc Bull* , 1997 , 76(11) : 51-54.

[3] Pawlek.R.P, "SiC in aluminium electrolysis cells," Light Metals 1995, 527-533.

[4] Karl A S. Yuanhua Wang. SiC and High Technology Ceramics. Foreign Refractories, 1990, 15 (4): 1-8.

[5] Quanxiang Chen. "SiC Composite Material for Aluminum Electrolysis Cell side wall". Light Metals 1997 (11)

[6] Zhao.JG, Zhang.ZP , Wang.WW , et al. "The properties of Si_3N_4-boned SiC materials for aluminium electrolysis cell," Light Metals 2000, 443-447.

[7] Fickel.A.F, Kramss.J.S, Temme.P.W, "Silicon Carbide refractories for aluminium reduction cell linings," Light Metals 1987, 183-188.

[8] Gao Bingliang, Wang Qiquan, Qiu Zhuxian, et al. "Possibility of application of SiC refractory in aluminum reduction cells," *Light Metals*, 2001, 270(4): 40-43. (in Chinese)

[9] Wang Zhaowen, Gao Bingliang, Zhao Bingyang, et al. "Corrosion resistance of SiC insulation sidewall for aluminum reduction cell," *Journal of Northeastern University (Natural Science)*, 2002, 23(4): 347-350. (in Chinese)

[10] Eirik Hagen, Mari-Ann Einarsrud and Tor Grande, "Chemical stability of ceramic sidelinings in Hall-Heroult cells," Light Metals 2001, 257-263.

[11] E.Skybakmoen, Lisbet I.Stoen,Jannicke H. Kvello and Ove Darell. "Qulity Evaluation Of Nitride Bonded Silicon Carbide Sidelining Materials", Light metals 2005,773.

[12] Skybakmoen.E, Guldbrandsen.H, Stoen.L.I, "Chemical resistance of sidelining materials based on SiC and carbon in cryolitic melts-a laboratory study," Light Metals 1999, 215-222.

[13] Wenwu Wang, Junguo Zhao, Jiancun Deng, Guohua Liu. "Test Method For Resistance Of SiC Material To Cryolite"8th Australasian Aluminium Smelter Technology Conference And Workshop Yeppoon, Australia, 3-8 October 2004.

Light Metals 2006 *Edited by Travis J. Galloway* ***TMS (The Minerals, Metals & Materials Society), 2006***

TEST METHOD FOR RESISTENCE OF SiC MATERIAL TO CRYOLITE

Junguo Zhao , Zhiping Zhang , Wenwu Wang , Guohua Liu
Luoyang Institute of Refractories Research (LIRR), Luoyang, 471039, P.R.China

Key words: Test method, SiC materials, cryolite erosion

Abstract

The 'Rotating-CO_2-Cryolite' test method (rotating specimens in molten cryolite in CO_2 atmosphere) has been designed to simulate the actual working conditions in reduction cells when no side-ledge is present. The influences of temperature, holding time, CO_2 flow rate, atmosphere inside furnace (CO_2, air and argon gas) and rotation speed upon the cryolite resistance of Si_3N_4-SiC materials have been studied. The study shows that the optimized test parameters are 1000°C, holding time 24h, CO_2 flow rate 1.0L/min and rotating speed 45rpm.

Introduction

In recent years, in order to save energy and raise productivity, Si_3N_4 bonded SiC material has been used as major sidelining material on large pre-baked reduction cells and has become a good substitute for traditional carbon sidelining. Nevertheless, its quality directly affects the service life of reduction cells. Nowadays, there have been three major test methods to evaluate the cryolite resistance of Si_3N_4 bonded SiC material, as follows:

(1) Polarized Test [1]

Simulating the actual working condition of industrial reduction cells, but the variations of technical parameters, especially fluctuation of carbon consumption, have a certain impact on the test results.

(2) CO_2-Cryolite Test [2]

CO_2 was flowed through the furnace at 60ml/min. The test specimens were in static state. Corrosion of the specimens was retarded due to adherence of reaction products to the surfaces of the specimens. In addition to that, the corrosion time was only 2h. Therefore, the corroded specimens did not change remarkably.

(3) Stirred Finger Test [3]

Compressed air was flowed through the furnace during the test, and the specimens were stirred in molten cryolite. However, the holding time was as long as168h.

A test method for evaluating corrosion resistance of SiC materials, i.e. 'Rotating-CO_2-Cryolite' test (rotating specimens in molten cryolite in CO_2 atmosphere), has been devised, by which quantitative test results can be obtained within a shorter period. The influences of test parameters, such as temperature, holding time, CO_2 flow rate, furnace atmosphere (CO_2, air, Ar), and rotation speed on corrosion resistance have been investigated and the parameters of the test have been optimized.

Experimental

Test specimens

T-shaped Si_3N_4 bonded SiC specimens with dimensions of 20/30×20×120mm were prepared by molding. The phase composition was analyzed by XRD , indicating that the major phase was α-SiC and α-Si_3N_4, with only about 1%β- Si_3N_4. No Si_2N_2O was detected. This shows the oxygen content of the specimen is quite low. By testing the apparent porosity, specimens with close apparent porosity (14~14.5%) were chosen to conduct the corrosion test in order to minimize the influence of the fluctuation of apparent porosity.

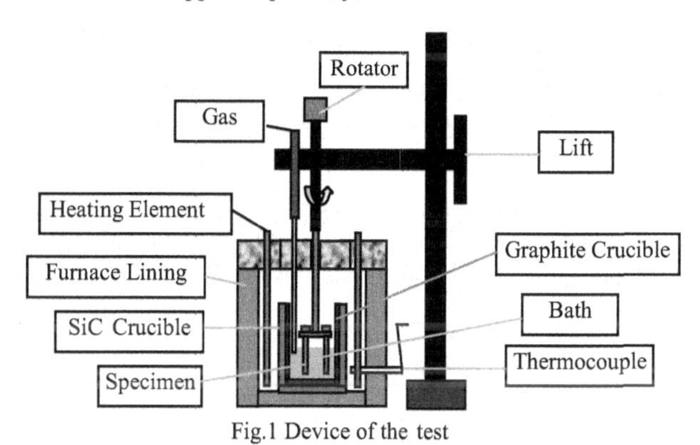

Fig.1 Device of the test

Test device and procedure

The test device is shown in Fig.1. The inner diameter of graphite crucible is 160mm and its height is 245mm. The test temperature is measured by PtRh10-Pt thermocouple and controlled within ±2 °C, automatically. The device allows for simultaneous testing of four specimens that are rotated at certain speed and exposed to molten cryolite, its vapour and CO_2. This test method is the so-called 'Rotating-CO_2-Cryolite'. The characteristics of the test method are as follows:

(1) Dynamic rotation

Specimens rotate at certain speed. For one thing, affected by the shearing stress, the adhering products on the surface of the specimens enter into cryolite melt quickly, which results in continuous corrosion on the fresh surface. For another, rotating the specimens promotes the evenness of cryolite melt and Al_2O_3, CO_2 dissolved in it, making the specimens immerged into the melt eroded uniformly.

(2) Large CO_2 flow

CO_2 flowing through cryolite melt is accurately controlled by micro-flowmeter. The minimum flow rate of CO_2 in this test is 0.5L/min, 8 times the CO_2 flow rate in 'CO_2-Cryolite Test'. Stable CO_2 flow rate can be maintained during the test.
The composition of cryolite melt (bath) is shown in table 1. Put the graphite crucible holding 5kg electrolyte (with SiC crucible outside) into an electric furnace. Raise the temperature at a rate of 8□/min to the test temperature and hold it for 1.5h to melt cryolite entirely, then immerge the specimens partly into the bath and rotate them at certain speed. Meanwhile, dip the Al_2O_3 tube vertically into the bath 3cm below the surface and begin to blow CO_2. Due to the dissolution of Al_2O_3, lower the tube 1cm every 1.5h. Ten minutes before finishing the test, immerges the whole specimen into the bath to wipe off the sediment on the surface of the specimens in gas phase area. Take out the specimens rapidly after the test, cool them down to room temperature, then dip them into solution with 12% $AlCl_3$ for 48h to wipe off the specimen surface sediment.

Table1. Composition of the bath

Composition	Cryolite (molecular ratio 2.3)	CaF_2	Al_2O_3
Weight %	88%	5%	7%

The influences of temperature, holding time, CO_2 flow, furnace atmosphere and rotation speed on cryolite resistance of the specimens were analyzed.

The degree of corrosion is determined by the volume loss of the specimens. Measure the volume in accordance with ISO 5017. The volume loss is calculated by the following formula:

Volume loss [%] = $[(V_0-V_1)/V_0]\times 100\%$
Where: V_0 — volume before
V_1 — volume after

Results and discussion

The appearance of the corroded specimens shows that the sections in bath and gas phase areas were slightly eroded, while the section at the gas/bath interface was severely worn out (see Fig.2), this may result from the dissolution and washout of oxide product by cryolite melt combined with oxidation of CO_2 and air. The degree of corrosion depends on temperature, corrosion duration, CO_2 flow rate, furnace atmosphere and rotation speed.

Influence of temperature on corrosion resistance

Fig.3 shows the relationship between corrosion resistance of the specimens and temperature. It is obvious that temperature change exerts much influence on corrosion resistance. As the temperature rises, the volume loss increases proportionally. Every 50°C raised, volume loss will increase to 1.7times the previous one.
When test temperature is too low, although it approaches actual electrolysis temperature (940°C–960°C), the corrosion is not noticeable, volume loss is only 2.5%. As temperature is up to 1000°C, volume loss increases rapidly to 4.4%. Actually, when reduction cells run well, Si_3N_4-SiC side lining is hardly corroded due to the protective side ledge covering its surface. However, when anode effect occurs, electrolysis temperature rises abruptly up to 990–1000°C, resulting in ledge melting and becoming thinner. The SiC side lining materials are eroded even more severely at such a high temperature.

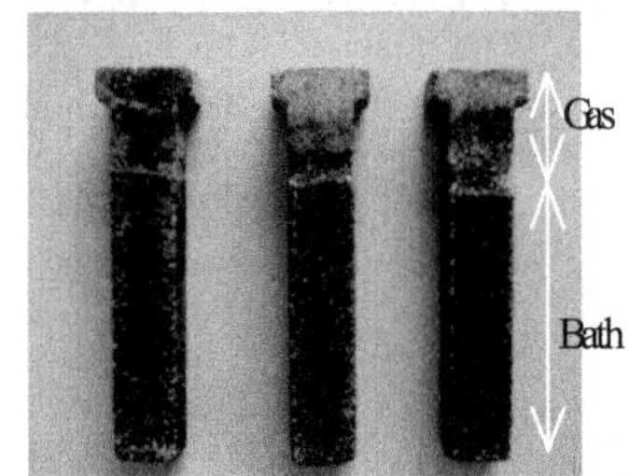

Fig.2. The appearance of the typical specimens after test

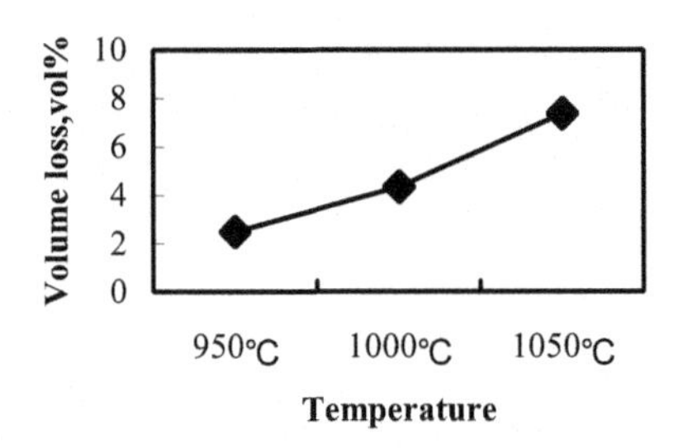

Fig.3. Corrosion resistance vs Temperature
(CO_2 flow 1.0L/min, holding time 24h, 30rpm)

When test temperature is too high (1050°C), specimen surface is uneven after corrosion test. SiC grains are exposed and isolated, volume loss up to 8%. This is probably due to the intensified oxidation resulting from the rising temperature. Because the cryolite resistance of Si_3N_4 bonded SiC materials depends on oxidation degree, the more oxidized, the poorer the cyrolite resistance is [1].

Influence of holding time on corrosion resistance

Fig.4 shows the relationship between corrosion resistance and holding time. It is obvious that holding time has a linear proportion relation with corrosion resistance. That is, with the holding time prolonged from 16h to 32h, volume loss increases from 3% to 5.8%. On average, volume loss within one hour does not differ remarkably, around 0.2%.

Influence of CO_2 flow on corrosion resistance

Fig.5 shows the relationship between corrosion resistance of the specimens and CO_2 flow. With CO_2 flow increased from 0.5L/min to 1.0L/min, volume loss goes up from 3.6% to 4.4%, while with CO_2 flow from 1.0L/min to 1.5L/min, volume loss only increases to 4.7%. It shows that, in this flow range, the slight fluctuation of CO_2 flow has little influence on corrosion.

Influence of furnace atmosphere on corrosion resistance

Fig.6 shows the relationship between corrosion resistance of the specimens and furnace atmosphere. It is obvious that the influences of CO_2 and air on corrosion resistance are equivalent. Volume loss is 4.4% and 4.6% respectively. Argon has the least influence on corrosion, with volume loss only 1.1%, as is confirmed by the appearance of the eroded specimen (see fig.7). The specimen in flowing argon has been eroded very slightly in gas/bath interface. The part immerged in the bath has smooth surface, without sign of being eroded.

Influence of rotation speed on corrosion resistance

Fig.8 shows the relationship between corrosion resistance of the specimens and the rotation speed. It is obvious that as the rotation speed increases from 30rpm to 60rpm, the volume loss gradually increases from 4.4% to 5.8%, basically showing a linear scaling up trend. But when the rotation speed adds continuously from 60rpm to 90rpm, the volume loss stops growing but slightly decreases, from 5.8% to 5.6%. Viewing from the 3-dimensional appearance of the eroded specimens, it shows that the faster the rotation speed is, the wider the eroded part is and the lower its erosion depth is. The high rotation speed may create gas insulation between the specimens and the molten cryolite, thus retarding further corrosion to the specimens. When the rotation speed comes to 90rpm, decrease of the volume loss might relate to that.

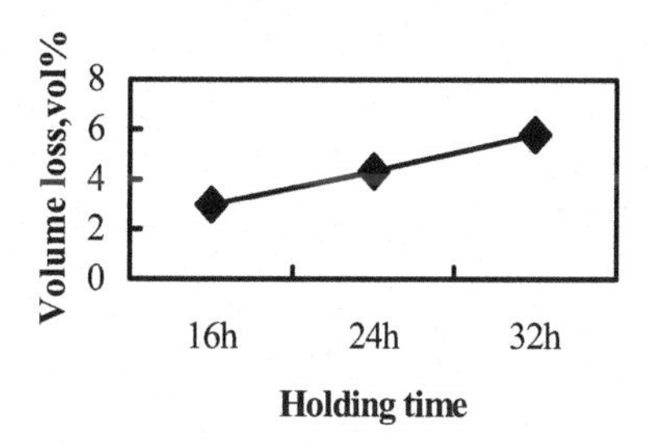

Fig.4. Corrosion resistance vs Holding time
(CO_2 flow 1.0L/min, 1000°C, 30rpm)

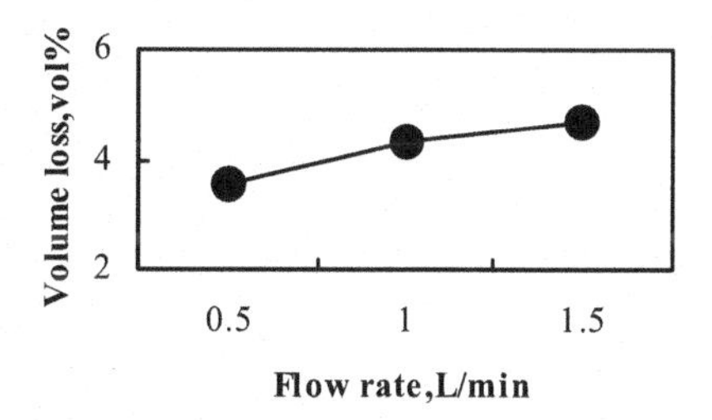

Fig.5. Corrosion resistance vs CO_2 flow
(Holding time 24h at 1000°C, 30rpm)

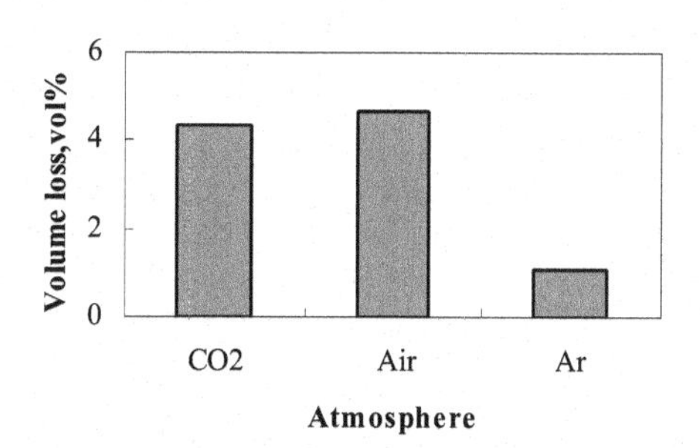

Fig.6. Corrosion resistance vs Furnace atmosphere
(Flow 1.0L/min, holding time 24h at 1000°C, 30rpm)

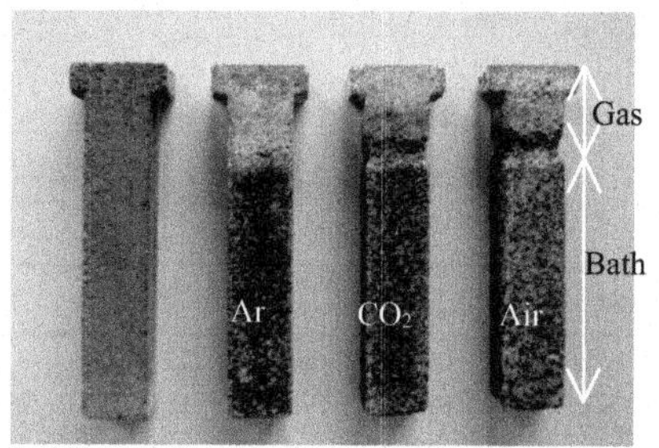

Fig.7. Appearances of the specimens in different atmospheres.

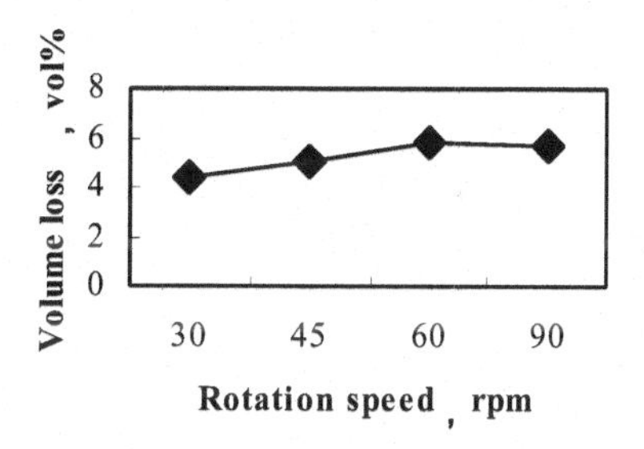

Fig.8. Rotation speed vs Corrosion resistance
(CO_2 flow rate 1.0L/min, holding time 24h at 1000°C)

Determination of the test parameters

According to above research results, the parameters of 'Rotating-CO_2- Cryolite' test have been determined initially, shown in Table 2.

Table 2 Parameters of corrosion test

Temp.	Hold time	Furnace atm.	Flow rate	Rotation speed
1000°C	24h	CO_2	1.0 L/min	45rpm.

When test temperature is too low and holding time too short, corrosion is inconspicuous. In the whole test process, no fresh bath is supplied once again. The higher the temperature is and the longer the holding time is, the more the fluoride volatilizes, and the more remarkably the composition of bath changes. Hold the specimens for 24h at 1000°C which is corresponding to the actual cell temperature (990°C~1000°C), after the anode effect. Then, the corrosion is noticeable, while the composition of the melt does not change much.

Electrolyte in industrial reduction cells is not in static state. Under the effect of electric field, magnetic field and heat field, the melt flows in whirlpool movement and washes the sidelining. Meanwhile, the sidelining is oxidized by CO_2 from anode. The influences of CO_2 and air on corrosion resistance are equivalent. In order to simulate the actual work condition of reduction cells, rotate the specimens at 45rpm, and flow CO_2 at the rate 1.0L/min that is sufficient for the CO_2 consumption in the whole test process.

Repetitive Test

A repetitive test has been conducted under the Table 2 parameter conditions, and the result is as follows:

Table 3 Results of the two tests (Volume Loss, %)

Specimen	1	2	3	4	Average value
First Test	4.9	4.8	5.2	4.7	4.9
Second Test	5.2	5.3	4.9	5.0	5.1

Table 3 shows that volume loss of the 4 specimens fluctuates within a range of 4.9±0.3 in the first test, while of 5.1±0.3 in the second test. Not only the corrosion test data of the specimens coming from the same furnace fluctuate slightly, but also the ones coming from different furnaces only have an error of 2% in volume loss in average. The data of the two tests put up fine repetition and recurrence.

Conclusion

(1) The 'Rotating-CO_2-Cryolite' test method has been well designed to simulate the actual working conditions in reduction cells when no side-ledge is present, and the corrosion of specimens in molten cryolite, its vapor and CO_2 atmosphere as well as under dynamic rotation conditions.

(2) It has determined the optimal test parameters as temperature 1000°C, holding time 24h, CO_2 flow rate 1.0L/min and rotating speed 45rpm.

(3) Repetitive test result has indicated excellent repetition and recurrence of the two corrosion tests data; as specimens from different furnaces only show an error of 2% in volume loss in average.

References

[1] Egil Skybakmoen, Henrik Gudbrandsen and Lisbet I.Stoen, *Chemical resistance of sidelining materials based on SiC and carbon in cryolitic melts- A laboratory study*, Light Metals (1999) 215-222.
[2] Eirik Hagen, *AlN and AlN/SiC ceramic sidelining materials in Aluminium electrolysis cells*, Norges teknisk-naturvitenskapelige universitet, 2000.
[3] *Stirred Finger Testing in Molten Bath - Running the Test*, Comalco Research & Technical Support.

CARBON TECHNOLOGY

Cathode Preheating/ Wettable Cathodes

SESSION CHAIR
Ketil Rye
Elkem Aluminium ANS
Mosjøen, Norway

THERMO-CHEMO-MECHANICAL MODELING OF A HALL-HÉROULT CELL THERMAL BAKE-OUT

Daniel Richard[1], Patrice Goulet[2], Marc Dupuis[3] and Mario Fafard[2]

[1] Hatch, 5 Place Ville Marie, Bureau 200, Montréal (Québec), Canada, H3B 2G2
[2] Laval University, Science and Engineering Faculty, Adrien-Pouliot Building, Sainte-Foy (Québec), Canada, G1K 7P4
[3] GéniSim Inc., 3111 rue Alger, Jonquière (Québec), Canada, G7S 2M9

Keywords: Numerical Analysis, Thermo-Mechanical Modeling, Transient Simulation, Electrolytic Cells

Abstract

Start-up of a Hall-Héroult cell is a delicate task. Modern practices for high amperage cells involve preheating the lining before the molten electrolyte is poured in. The optimum preheating method for a rapid production of metal and a long pot life is elusive.

Numerical modeling is an invaluable tool to gain insights into the complex phenomena taking place during start-up. The adequate modeling of the mechanical response of the lining is critical to detect risks of cathode block cracking or the development of gaps where liquids could leak. Taking into account the ramming paste baking, the quasi-brittle nature of carbon and the contact interfaces are examples of key elements to consider.

A finite element demonstration cell slice model was built and simulations of different thermal bake-out scenarios were performed using the in-house code ***FESh++***. Potential industrial application of the model is discussed.

Introduction

It is well accepted that start-up and early operation have a strong influence on the performance and life of a Hall-Héroult cell [1]. Generally, a preheating phase is necessary during start-up to ensure a smooth transition to normal operation.

The requirements for preheating, summarized in [2], are the following:

- The cathode block temperature must be high enough to:
  - Minimize bath freezing when bath is poured in. Freezing leads to an uneven current distribution and a potentially harmful unstable early operation.
  - Avoid large thermal gradients in the cathode blocks before bath is poured in. Large gradients may induce cracks.
- If the preheating rate is too fast or not uniform enough, large thermal gradients within the cathode blocks will occur and may also induce cracks.
- The paste temperature must be sufficiently high to avoid flash pyrolysis when bath is poured in.
- The lining must be maintained in compression at all times to ensure no gap is present in the lining as this would allow bath or metal penetration.

Although preheating methods in the industry vary, thermal bake-out using gas-fired or oil-fired burners has been shown to provide the most uniform temperature distribution in the lining [1,3]. The desired heat-up curve also varies, using one or more surface temperature ramps, and sometimes ending with a soaking time at constant temperature [1,2,4].

An experimental study was carried out on VS Søderberg cells to determine the optimal thermal preheating cycle [2]. Surface and sub-cathodic temperature measurements were performed for different ramps and final temperature. It was concluded that the best results were obtained with a low heat-up rate and a high final block temperature.

Unfortunately in practice, the pressure to produce metal as fast as possible calls for the shortest possible preheating time. This results in less than optimal heat-up curves for the cathode blocks and lining, and the mechanical response of the cell then becomes a critical limiting factor.

Important Aspects of Cell Mechanical Behaviour

Complex phenomena are taking place during the start-up and early operation of a cell. Predicting its mechanical response is not a simple task.

There are several difficulties in this task. For example, the interaction of the lining with the pot shell, the intrinsic mechanical behaviour of the lining materials, the transformations within the materials, and the cell construction are important and are changing during the bake-out period.

Numerical modeling is therefore an interesting tool to help provide insights into these complex phenomena and help in designing the optimal procedure for a given cell.

Lining and Pot Shell Interaction

The pot shell does not only serves as a structural container for the lining, but is also an integral part of the cell design. In addition to its thermal purpose, it must also maintain the lining in firm compression without inducing cracks or excessive deformation.

It is observed that in operation the pot shell deforms. The problem is in reality *strain-driven*. The shell and the lining deformation and stresses result from the complex interaction of their expansion (thermal and chemical) and their stress-strain response. Although during preheating the shell is not likely to deform very much, this does not change the fundamental nature of the problem. The dilatometric response of the materials and their stress-strain behaviour must therefore be known accurately.

Lining Mechanical Behaviour

Several dense refractory materials are characterized by a *quasi-brittle* behaviour. That is they can still bear some load after their peak stress has been reached, their strength and ductility increases with confinement, they are permanently deformed at only a fraction of their peak stress, and they are significantly stronger in compression than in tension. Concrete, dense bricks and carbon [5] are all quasi-brittle materials.

On the contrary, steel has the same response in tension and compression, regardless of the confinement. Therefore, it is unrealistic to assume a steel-like behaviour for the cell lining, as it will not provide an adequate tool to predict cracking.

Also, although the temperature is normally always increasing during preheating, it is still possible to undergo a local unloading. For example the load would decrease in the region surrounding a growing crack, or when a material experiences a contraction. For instance, this is the case with ramming paste as it is baked.

Therefore, it is important to account for thermally-, chemically- or mechanically-induced irreversible deformations in a material in order to capture the potential opening of gaps in the lining and to predict correctly the stresses.

Material Transformation and Time Response

During preheating, it is assumed that liquid bath is not present, so the effect of sodium and the associated chemical reactions within the refractory lining can be ignored.

However, castables will cure and ramming paste will start to bake and will undergo irreversible transformations of their microstructure. This will affect their thermal and mechanical behaviour. In general, these reactions also cause an irreversible volume change.

Ramming paste is undoubtedly a critical part of the cell, as it should seal the lining while accommodating some of the cathode blocks expansion. It is also difficult to characterize and to model. It was reported that during baking, most pastes first expand and then shrink. A plausible explanation is that the initial swelling is caused by a build-up of reaction gases while the subsequent shrinkage is due to the cokefaction of the binder phase [6,7].

The strength and stiffness of the paste increases by more than an order of magnitude during baking while its ductility decreases in the same proportion. Its behaviour evolves from being almost incompressible and plastic to that typical of quasi-brittle materials.

Ramming paste has also been shown to continue to deform under a constant load, a phenomenon known as creep [6]. Some of this additional deformation is recovered over time when the load is removed, but the rest of it is permanent. Creep relaxes the stresses in the material but also increases the risks of opening gaps, since it increases the deformation for a given load.

Cell Construction

Some of the lining materials are laid dry while mortar is used to join others. However, most interfaces cannot be assumed to be completely cohesive. Joint behaviour has a profound effect on the stiffness of the structure, and it is of paramount importance to characterize this correctly for the accurate prediction of a possible gap opening in the lining.

For most interfaces, the most conservative assumption is to neglect cohesion altogether. This means that this interface cannot sustain a tensile stress and that a gap will open under a tensile loading.

Finite Element Demonstration Model

For demonstration purposes, a realistic prebaked point-fed 300 kA cell design inspired from a VAW publication [8] was used. The thermo-electrical results, using ANSYS, are presented in [9] for normal steady-state operation.

Geometry, Assumptions and Simplifications

As a first step, a fully coupled cathode slice model was built using the finite element toolbox ***FESh++*** [10] to illustrate the effect of the heating rate at the center of the cell. The slice mesh represents a quarter cathode and its corresponding lining, shell and cradle, as shown in Figure 1. The shell and cradle are discretized using large rotation shell elements while the lining is discretized using 3D brick elements. The semi-graphitic cathode blocks (~30% graphite) are glued together, as can be seen from the absence of a small joint.

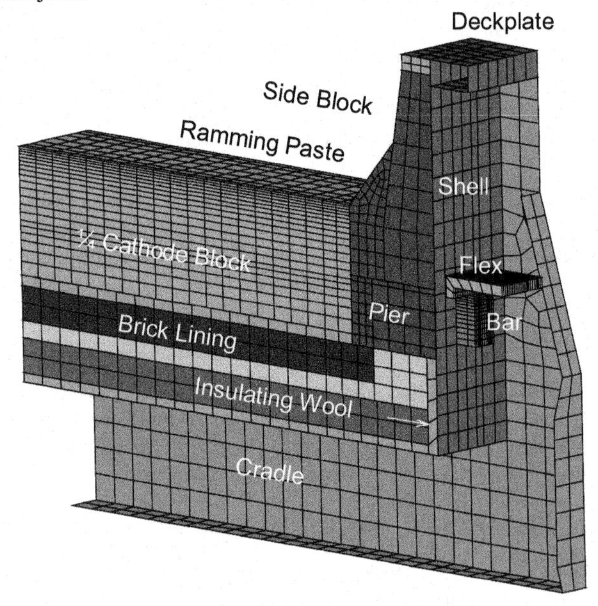

Figure 1 -Thermal slice mesh.

Contact mechanics is used between different parts of the lining, as can be seen from the non-concordant mesh at the interfaces between different parts (Figure 2).

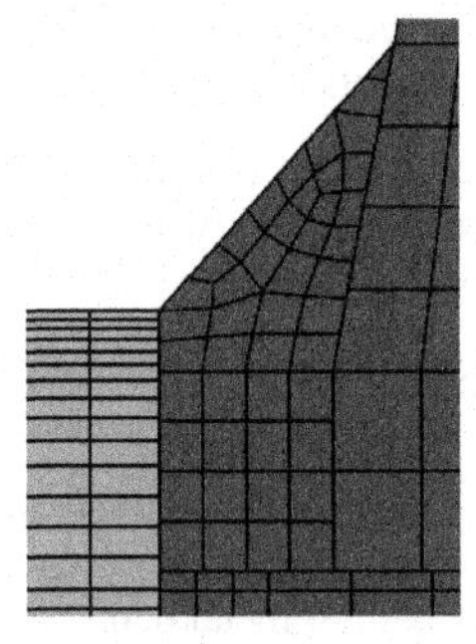

Figure 2 – Slice mesh detail: non-concordant interface mesh.

An additional thermal contact resistance (constant or contact pressure-dependent) can be used at an interface, for example to emulate the effect of a mortar joint. Thermal conductance values were estimated from [11]. The interfaces are assumed to be non-cohesive, *i.e.* they cannot sustain a tensile stress.

The cradles are welded to the shell, and the steel plate thicknesses were estimated from experience [12]. The collector bar is rodded with cast iron, and a simple geometry was assumed, as shown in Figure 3. At ambient temperature, an air gap is present between cast iron and carbon, and as the assembly heats up, thermal expansion of the parts eliminates this air gap.

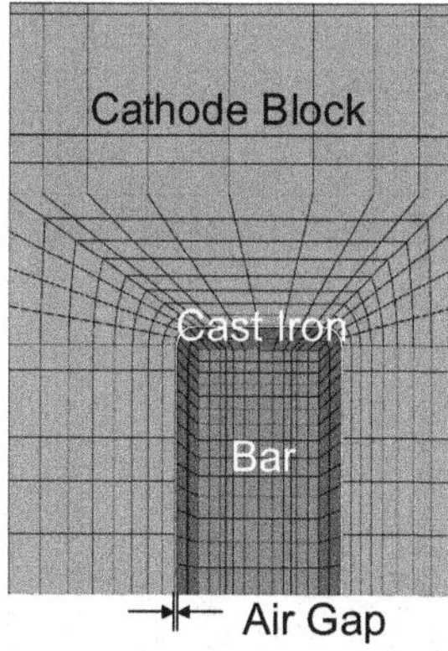

Figure 3 - Collector bar, cast iron and half-cathode block detail.

The brick lining under the cathode block is assumed to have no bending stiffness because in this design there is a layer of insulating refractory fibre wool that will absorb its thermal expansion in the horizontal plane. Therefore, the brick lining under the cathode does not contribute significantly to the mechanical response of the cell during preheating and is accordingly not solved in the mechanical problem. Conceptually, the cathode block and the pier are assumed to rest on springs of equivalent stiffness to the underlying brick lining. This is implemented in the finite element model by using contact mechanics to connect the shell floor to the bottom of the collector bar and pier. The mechanical mesh is shown in Figure 4. Note that the whole slice is solved for temperature (see Figure 1).

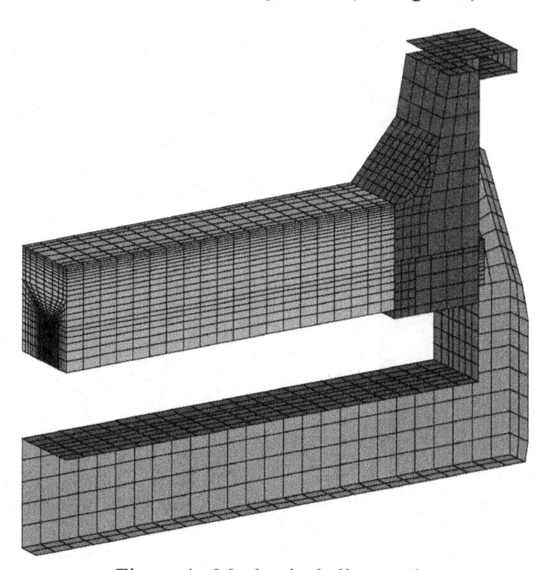

Figure 4 - Mechanical slice mesh.

Material Properties

The thermal properties were obtained from [9]. The mechanical constitutive laws are summarized in Table 1.

Table 1 - Assumed mechanical material models.

Material	Material Model	Reference
Cathode Block	Quasi-Brittle	[5]
Collector Bar	Elastic	[13]
Cast Iron	Elastic	[13]
Pier	Elastic	[13]
Ramming Paste	Reactive Quasi-Brittle	[6]
Side Block	Quasi-Brittle	[5]
Castable	Quasi-Brittle	[5]

Boundary Conditions and Loads

The only external mechanical load considered is gravity. It must be included to stabilize the problem, since the lining is mostly free to move in the upward vertical direction.

The dashed lines in Figure 5 represent planes S1, S2 and P3 on which symmetry conditions could be applied.

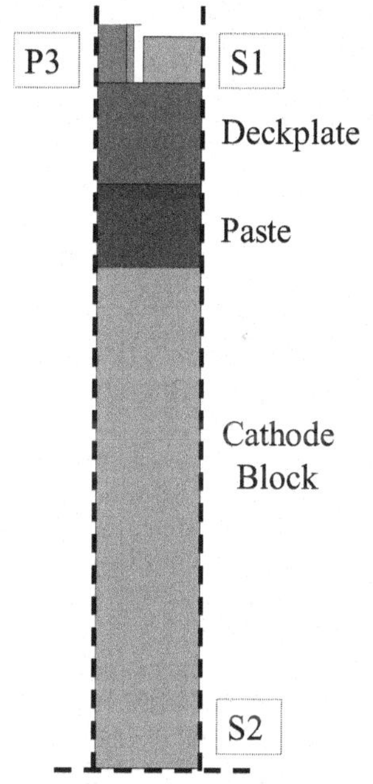

Figure 5 - Slice symmetry planes, top view.

Planes S1 and S2 are true symmetry planes. For the shell and cradle, P3 is obviously a true symmetry plane as well. However, the conditions for the lining on plane P3 are difficult to evaluate. It reality, the confinement on this plane is the result of the interaction between the lining and the shell along the length of the pot. For this study, the two extreme cases were considered for the lining on plane P3: symmetry conditions, and free to move.

The thermal boundary conditions for all external surfaces take into account natural convection and grey body radiation, using well-known semi-empirical correlations, and were taken from [9]. The surface of the ramming paste and the sidewall are insulated by crushed bath. A convection coefficient of 1 W/m^2K was used.

Given that the objective is not to design a gas-preheating burner system, but rather to study the transient mechanical response of the cell during preheating, the boundary conditions on the top surface of the cathode block were kept very simple. A heat transfer coefficient of 650 W/m^2K was assumed, with a gas temperature following a linear ramp-up to 955°C in 12, 24, 36 or 72 hours.

Results and Discussion

Thermal

Final temperatures at the top surface of the block in the center of the cell are within 20°C for all simulations. As expected, the thermal gradients within the cathode blocks are decreasing with the increasing preheating time. For instance, Figure 6 represents the difference between the surface and sub-cathodic temperature at the end of the cathode block (adjacent to the ramming paste, between the collector bars).

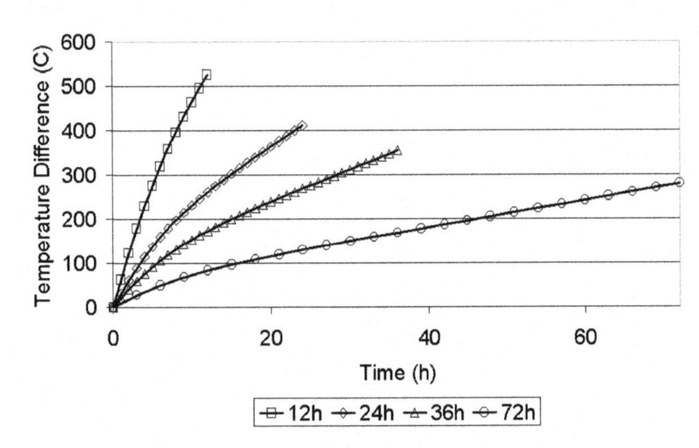

Figure 6 – Difference between surface and sub-cathodic temperature for different preheating times to 955°C.

Figure 7 shows the isotherms at the end of preheating to 955°C in 24 and 72 hours. The large thermal inertia of the lining is such that even after 72 hours of preheating, the shell floor and most of the shell sidewall are still cold (Figure 7b). Note that the discontinuity in the isotherms is a result of the thermal resistance of the interface between the parts.

An interesting effect of the transient diffusion of heat into the lining is that the extent of baking of the ramming paste changes dramatically with the preheating time, as shown in Figure 8. Note that the local baking index is defined as the paste local compressive strength normalized to its strength at full baking (refer to [6] for more details).

After 24 hours of preheating, some of the paste in the line of action of the cathode block expansion is still completely green (Figure 8a), *i.e.* it is still plastic. This is not the case for a preheating of 72 hours (Figure 8b). The baking of the paste must be taken into account in the design of a start-up method. Green paste deforms to accommodate the expansion of the cathode blocks, but it also shrinks when it starts to bake. The timing of this process must be such that no gap will open where bath could leak into the lining when the bath is poured in the cell after the preheating is completed.

a) 24 hours preheating to 955°C

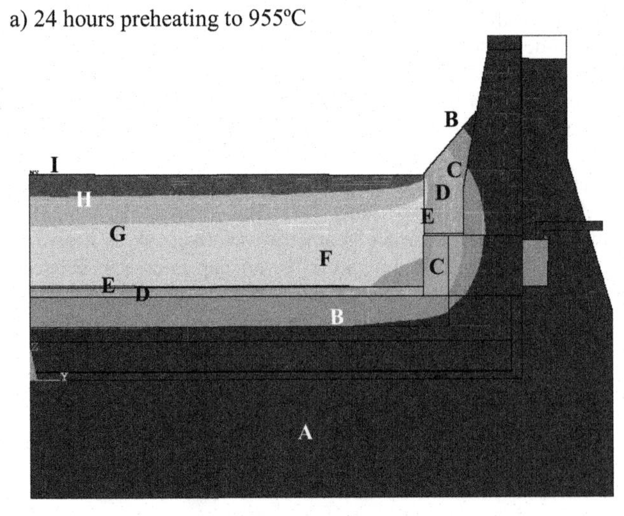

b) 72 hours preheating to 955°C

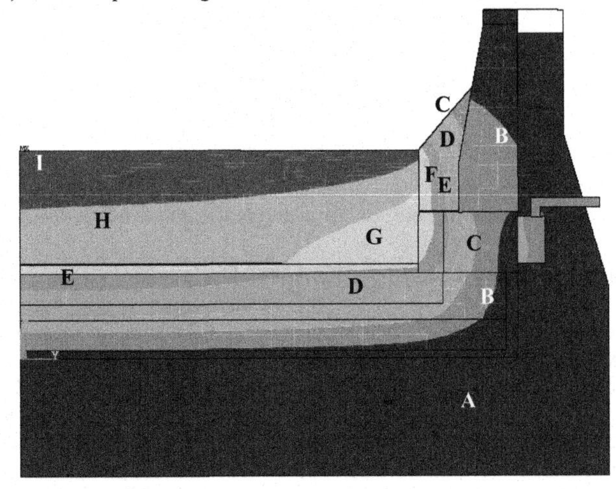

A:20 B:133 C:246 D:359 E:473 F:586 G:699 H:812 I:925 (°C)

Figure 7 – Isotherms at the end of preheating.

a) 24 hours preheating to 955°C b) 72 hours preheating to 955°C

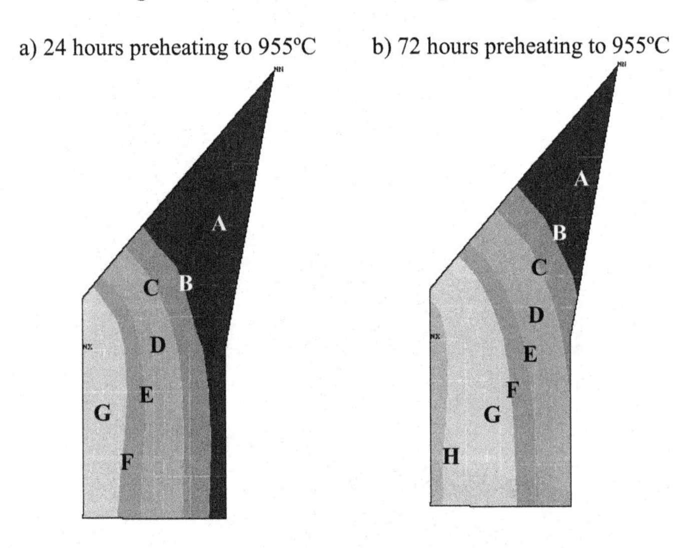

A: Green B:13 C:24 D:35 E:45 F:56 G:67 H:78 I:89 (%)

Figure 8 – Ramming paste baking index at the end of preheating.

Mehanical

The boundary conditions on the lining surface on plane P3 (Figure 5) have a small impact on the overall displacements along the width of the cell, but have a large impact on the stress-strain behaviour of the lining. This aspect will be covered later on.

The deformed shape of the lining at the end of the preheating can provide a strong indication on the formation of gaps where bath could potentially leak. For example, Figure 9 shows the final deformed shape for a 24 hours preheating to 955°C (corresponding to the temperature distribution in Figure 7a), with displacements amplified by a factor of 10. As expected, the cathode blocks are bending up and compress the ramming paste.

A gap opens between the ramming paste and the pier as the cathode block pushes up the paste (Figure 9, *A*). The thermally induced deformation of the shell also opens up a gap between the side block and the sidewall near the deckplate (Figure 9, *B*).

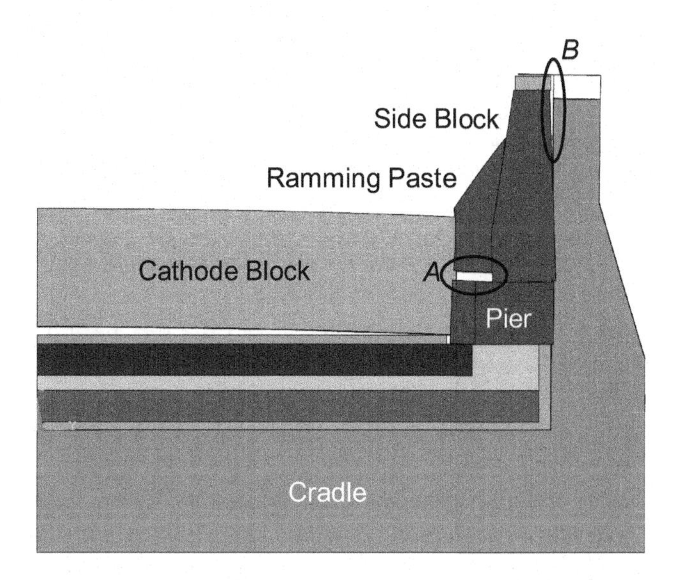

Figure 9 - Exaggerated deformed shape at the end of a 24h preheating to 955°C.

The impact of the boundary conditions on the lining on plane P3 (Figure 5) on the response of the cathode block was investigated. Two extreme cases were simulated: full symmetry and free to move. This corresponds respectively to an infinitely rigid shell, and to a perfect expansion joint along the length of the pot.

Faster preheating rates mean larger thermal gradients in the cathode blocks, which leads to tensile stresses as shown in Figure 10. For the studied cases where the lining surface on plane P3 is free to move, these stresses were not sufficient to crack the block.

Assuming a full symmetry boundary condition on plane P3, it was found that for all our studied cases, the cathode block is permanently deformed and cracks occur in the collector bar slot. A typical cracking pattern is shown in Figure 11. When the lining surface on plane P3 is free to move, no such crack develops.

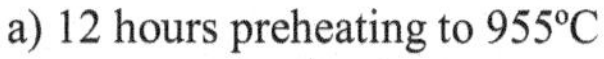
a) 12 hours preheating to 955°C

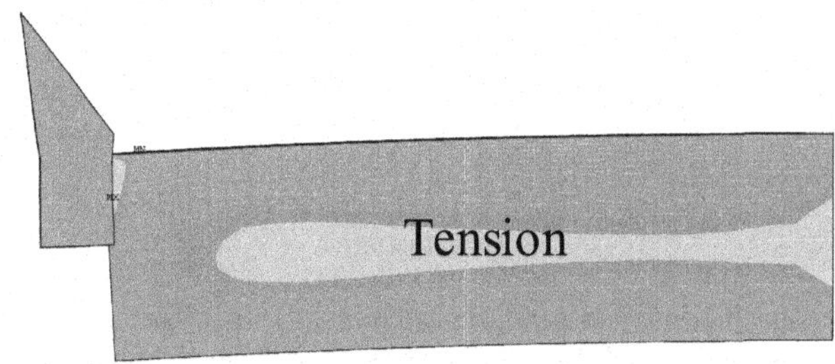

b) 24 hours preheating to 955°C

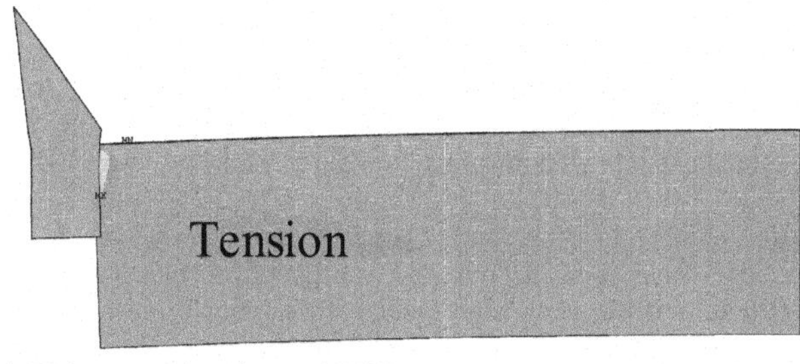

c) 72 hours preheating to 955°C

Figure 10 – First principal stress in cathode block and ramming paste at the end of preheating.

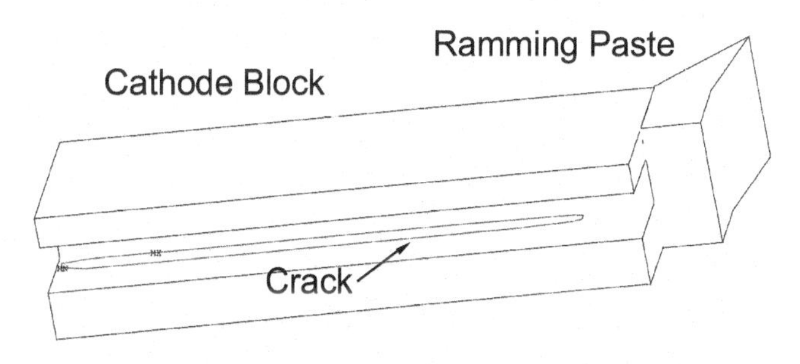

Figure 11 – Cathode block cracking for 24h preheating to 955°C, full symmetry conditions on plane P3.

It had been previously demonstrated that a slice model might be used for the thermo-electric design of a cell [14]. Given the significant difference of behaviour of the cathode block with the change in boundary conditions on plane P3, it is clear that a slice model cannot be trusted to accurately predict cathode cracking during preheating unless the shell provided no confinement along the length of the pot, or otherwise if it was infinitely rigid. As this is rarely the case in practice, at least a quarter cell should therefore be modeled. Obviously, this is also required to study the corner of the cell.

However, these results do show that a proper amount of expansion relief along the length of the cell must be built-in the design. Ramming paste small joints between cathode blocks will partially play that role. Predicting the resulting effect is not trivial. Further, to avoid leaks into the lining, the end walls must also provide an adequate confinement.

Conclusion

A finite element model of a Hall-Héroult cell slice was built using the in-house code ***FESh++***. Key features of the model include the baking of ramming paste, the quasi-brittle nature of carbon and contact interfaces.

The transient fully coupled thermo-mechanical solution on an Opteron 850 64-bit machine required approximately 2.6 GB of RAM and 2.9 to 3.3 hours of wall clock time per time step. The total wall clock time then varied between 35 and 120 hours.

Faster preheating rates mean larger thermal gradients in the cathode blocks. For the modeled configuration, this leads to a horizontal tensile stress zone in the cathode block. For the cases studied these stresses were not sufficient to crack the block but this highlights one of the possible cathode failure mechanisms.

From this work, it was shown that a slice model is not sufficient to detect cracking of the cathode blocks for practical cell designs. At least a quarter cell needs to be modeled to accurately represent the confinement of the lining along the length of the cell. However, it is possible to obtain valuable information with a slice model about the extent of ramming paste baking and the opening of gaps that would allow bath infiltration into the lining and possibly reduce pot life.

To help determine the optimal final cathode surface temperature and ramming paste baking profile, it is now necessary to study what happens when the bath is poured in the cell. The timing of the cathode block expansion and ramming paste shrinkage is critical to ensure no gap will form which would allow liquid to leak into the lining. Thermal shock of the cathode blocks and flash pyrolysis of the ramming paste must also be avoided. The desired conditions at the end of preheating could then be reverse engineered.

From the work presented here, it can be seen that a numerical model is an invaluable tool to gain insights into the complex response of a cell during start-up. Once validated with experimental measurements, it can be used in the optimisation of the heat-up practices. The results of the model could also be used to determine some control metrics in the operation, for example by relating the evolution of cathode bar temperature to the risks of cathode block cracking.

Acknowledgements

The authors wish to thank Jérôme Bédard from Laval University for helping with the literature review, the mesh, the input files and with running the simulations. We thank Lowy Gunnewiek from Hatch for his support and for the useful discussions and comments. The kindness of Professor Daniel Marceau from Université du Québec à Chicoutim is also acknowledged, for providing us with an access to his numerical computing facilities. Finally we thank the Natural Sciences and Engineering Research Council of Canada (NSERC) and REGAL for the granted partial financial assistance.

References

[1] M. Sørlie & H.A. Øye, "*Cathodes in Aluminum Electrolysis*", 2nd Edition, Aluminium-Verlag, 1994.

[2] C. Zangiacomi, V. Pandolfelli, L. Paulino, S.J. Lindsay & H. Kvande, "*Preheating Study of Smelting Cells*", in Proceedings of TMS Light Metals 2005, 333-336.

[3] H. Bentzen, J. Hvistendahl, M. Jensen, J. Melås, & M. Sørlie, "*Gas preheating and start of Søderberg cells*", in Proceedings of TMS Light Metals 1991, 741-747

[4] B. Rolofs, D. Eisma, G. Dickinson & D. Hunzinger, "*Thermal Bake-out of Aluminium Reduction Cells, a Technology for the Future*", in Proceedings of TMS Light Metals 2002, 343-346

[5] G. D'Amours, M. Fafard, A. Gakwaya & A. Mirchi. "*Multi-Axial Mechanical Behavior of the Carbon Cathode: Understanding, Modeling and Identification*", in Proceedings of TMS Light Metals, 2003, 633-639.

[6] D. Richard, G. D'Amours, M. Fafard, & M. Désilets. "*Development and Validation of a Thermo-Chemo-Mechanical Model of the Baking of Ramming Paste*", in Proceedings of TMS Light Metals, 2005, 733-738.

[7] F. B. Andersen, M. A. Stam & D. Eisma, "*A Laboratory Evaluation of Ramming Paste for Aluminium Electrolysis Cells*", in Proceedings of TMS Light Metals 2005, 739-744.

[8] V.A. Kryukovski, G.A. Sirasutdinov, J. Klein & G. Peychal - Heiling, "*International Cooperation and High-Performance Reduction inSiberia* ", JOM, 46(2), 1994, 23-25.

[9] M. Dupuis, "*Thermo-Electric Design of a 400 kA Cell: A Tutorial*", in Proceedings of TMS Light Metals, 2000, 297-302.

[10] M. Désilets, D. Marceau & M. Fafard, "*START-Cuve: Thermo-electro-mechanical Transient Simulation Applied to Electrical Preheating of a Hall-Héroult Cell*", in Proceedings of TMS Light Metals 2003, 247-254.

[11] P. Goulet, "*Modélisation du comportement thermo-électro-mécanique des interfaces de contact d'une cuve de Hall-Héroult*", Ph.D. Thesis, Laval University, November 2004, 289 pages.

[12] M. Dupuis & D. Richard, "*Study of the Thermally-induced Shell Deformation of High Amperage Hall-Héroult Cells*", in Proceedings of COM 2005, 35-47

[13] Y. Sun, K.G. Forslund, M. Sørlie & H.A. Øye, "*Modelling of Thermal and Sodium Expansion in Prebaked Aluminium Reduction Cells*", in Proceedings of TMS Light Metals 2004, 603-610.

[14] M. Dupuis, "*Thermo-Electric Analysis of the Grande-Baie Aluminum Reduction Cell*", in Proceedings of TMS Light Metals 1994, p. 339-342.

Light Metals 2006 *Edited by Travis J. Galloway* ***TMS (The Minerals, Metals & Materials Society), 2006***

Development of Coke-bed Preheating Method for 200kA cells

Shaher A. Mohamed1, M.M.Megahed2,H. Hashem1, M. El-Ghonimy1,
1.Aluminium Company of Egypt,(Egyptalum), Nag-Hammadi, Egypt, 83642
E-mail: r-d@egyptalum.com.eg
2. Cairo University, Faculty of Engineering

Key Words: Preheating, Bake-out, Coke Bed, Modeling, Shunt Rheostat.

Abstract

Many factors affect the potlife of aluminium reduction cells. Preheating and start-up are of the most significant parameters affecting cathode life. Different methods are used to preheat the cells. Coke bed preheating with shunt rheostat is one of the most popular methods.
The Aluminium Company of Egypt (Egyptalum) utilizes coke bed for preheating of aluminium reduction cells since 1991. Several modifications on the preheating process had been carried out. Coke bed height was reduced from 50 mm to less than 20 mm at present. In order to reduce the preheating current, different designs of shunting units have been used. This paper describes the different preheating conditions, as well as analyses and discussions of preheating measurements.
A finite element model is used to evaluate the different preheating schemes. The results indicate that the best scheme is the one which possesses less coke bed height and smooth increase in current.

Introduction

To ensure long potlife of the cathode in the aluminium reduction cell, it is necessary to have a cathode design with strong shell, a thermal design suited for expected operation, superior materials, high-quality construction techniques, carefully controlled bake-out and start up procedure and uneventful potroom operation. The contribution made by each of these variables to the overall potlife is debatable depending on the school of thought to which one ascribes.

Many authors have discussed the preheating of aluminium reduction cells: process, problem areas, methods, advantages, disadvantages, and potlife. Sorlie and Øye [1] discussed the different preheating methods and explained the ideas, advantages and disadvantages of each method. The remedies proposed to overcome some of the disadvantages were also indicated.

The main function of preheating is; to bake-out and precondition the lining materials by raising the temperature of the lining [1]. There are several requirements to achieve good preheating:

- Even temperature distribution and low temperature gradients across and through the cathode without producing abnormal thermal gradients across the working face of the cathode and down through the thickness of the lining [2].
- The rate of heating should be low enough to avoid thermal stresses within the cathode materials. [3]
- Ensuring pitch/binder pyrolysis of green rammed paste occurring in all green paste areas at a rate that results in optimum baked paste properties. [3]
- Avoiding air burn damage to the anode and cathode carbon materials
- Short preheating time
- Low preheating cost.

Clelland et al [4] correlated the data obtained from failed cell autopsies with the observations of cell design, preheating procedure and operating parameters to the ultimate cause of failure.

Dupius et al [5, 6] assessed the impact of time variation of cell current during coke-bed preheating upon cathode block temperature. The influence of power curve and coke bed geometry upon cathode block temperature and thermal stresses generated during preheating was discussed.

Dunn and Galadari [7] indicated that the resistance of the coke bed decayed substantially over the preheating period. The coke bed effectively defines the cell resistance. Both the voltage and power curves show blips when the shunt is removed, forcing more current through the cell.

Mohamed et al [8] used mathematical modeling to evaluate different schemes of current curves and coke bed thickness by conducting a transient heat transfer analysis by the finite element method. It was found that the best scheme is the one, which has the least thermal gradients.

Mohamed [9] explained a method of shunt resistance design. Thermoelectric analysis had been used to accomplish the calculations. Numerical procedure is used to calculate the necessary dimensions of resistance elements. Results indicated good preheating with shunt rheostat. Finite element is used to evaluate the different preheating schemes.

Zangiacomi [10] evaluated four different trials of preheating; each has its final temperature and preheating time. It is found that longer preheating time brings the following advantages: minimizes the thermal gradient after pull-up, more uniform control in terms of pot voltage and smoother pot start-up.

It has been found that the cost of preheating can be reduced, the preheating process can be improved, and start-up after preheating can be easier with a good preheating process. It has also been found that investing in preheating can significantly improve the potlife and hence improve plant profitability.

Preheating of Soderberg Cells at Egyptalum

The aluminium Company of Egypt (Egyptalum) started production in 1975, with a 150 kA, end-to-end Soderberg cell technology. The current was raised to 155 kA to increase production.

An old technique was used to preheat the Soderberg cells. Preheating starts with two oil burners working from one side of the end-to-end pot for two hours. Two other hours from the opposite direction was sufficient to provide an average cathode surface temperature suitable for metal addition. Then, as the cathode surface temperature increases, molten metal is added and the current passes for 48 hours to complete the baking process. This yielded a total preheating time of 52 hr. Some other details are used as covering cathode surface, and heating of cathode transverse walls. The cell is ready for start up after that.

Preheating of Prebaked Cells at Egyptalum

Six prototype, 200 kA, prebaked anode cells started production in 1991. The cells were part of a working Soderberg line. A by-pass busbar system was used to feed the working cells with the necessary current (155 kA). It has been decided to preheat the six cells with coke bed. This was the first time at Egyptalum to preheat the cells by Coke-bed preheating. Bake-out was conducted using a coke bed of 30 – 50 mm thickness, in a group of 2 pots each. The current passing through cell during preheating was controlled by means of silicon rectifier substations and the by-pass busbar system design. Table (1) shows the time duration for preheating, average preheating current and bottom temperature after 10 working days for cells # 601, 602, 603, 504, 605 and 606. It is observed that the bottom of cell # 603 has a higher temperature than # 601 and # 602. This can be correlated directly to the average preheating current, which has the smallest value for cells # 601 and # 602.

It concluded at that time the bake-out and start-up of pots # 601 and # 602 corresponded to the required current rise. More rapid current rise and longer bake-out time at the pots # 603-# 606 resulted in acceleration of coking process of the ramming mix followed by breakaway of the blocks from the peripheral joint.

Table 1.: Bake-out parameters for six prototype pots.

Pot #	Average Current, KA	Time Duration, hr	Bottom Temp, °C
601	150.2	45.2	149
602	157.9	53.2	116
603	195.5	44.0	214
604	195.5	52.2	170
605	199.6	52.1	166
606	200	70.0	182

Finally, it was recommended to adjust the preheating current during bake out using shunt rheostat, which was not available at that time; 1991 [11].

Finite Element Analysis

A finite element (FE) model was built to simulate Egyptalum reduction cell during preheating. A Plane-2D element with quadrilateral shape is employed. The element has four corner nodes with temperature as the only degrees of freedom per node. The element thickness was chosen to simulate cathode construction. More than 1100 elements comprising about 1000 nodes were utilized in building up the FE model.

Transient thermal analysis implies that temperature at any given point in the medium varies with time. In order to successfully conduct the required non-linear transient thermal analysis, temperature variation of thermal conductivity and specific heat over the temperature range during preheating should be prescribed.

The non-linearity of the problem is due to:

- Variation of material properties such as thermal conductivity and specific heat with temperature.
- Variation of the value of convective film coefficient with temperature
- Variation of heat generation rate and boundary conditions with time

The analysis starts with prescription of time variation of cell current. The present work considers three schemes of preheating by changing the cell current by means of an existing resistance elements. Table (2) lists the starting percentage current and coke-bed thickness employed in each scheme. Figure (1) illustrates the time variation of cell current for the three schemes.

Thermal energies are used to prescribe heat generation rate in cathode, anode and coke-bed in the FE model. Convection boundary conditions were prescribed on all outside surfaces of the FE model.

Table 2.: Details of Preheating Schemes

Scheme	Starting Current, %	Coke-bed Thickness, mm
A	30	40
B	35	33
C	40	27

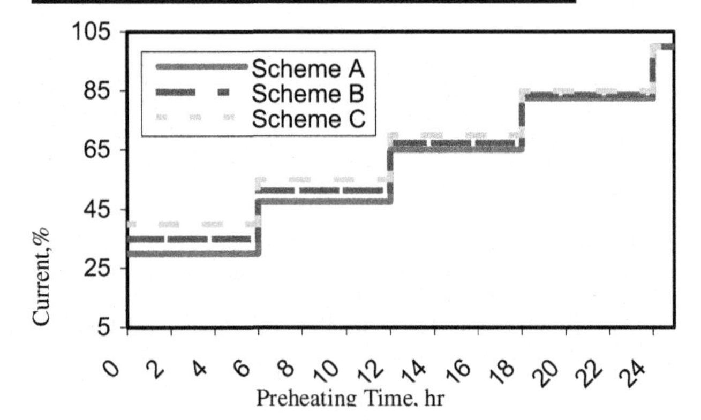

Fig. 1: Time variation of current %

A number of key points (1-6) were chosen at which temperature history is monitored for the three proposed schemes of preheating. Points 1 , 2 and 3 lie on cathode top surface central channel, mid of anode block and besides the ramming at the longitudinal wall. Points 4 , 5 and 6 lie on the bottom surface of carbon block and located exactly below points 1, 2 and 3 respectively.

Figures (2. a, b and c) show the time variation of temperatures at points 1, 2 and 3 for preheating schemes A, B, and C respectively. It is noted that during the first 18 hours of preheating, the three schemes yield comparable temperature history at cathode top surface. Subsequent to the first 18 hours, temperature differences are higher. Temperatures obtained for scheme A are higher than for scheme B and those for scheme B are higher than for scheme C. It is noted that the rates of temperature increase at points 1, 2 and 3 are closest to each other in scheme C. As a result of these differences in temperature rise rates, after 60 hours of preheating, values of temperature at point (2) are obtained as: 1253, 1081 and 925 °C in schemes A, B and C respectively. As a by product of the above results, it appears that in order to raise cathode temperature to the level of 900 °C, less than 50 hours of preheating are required. Preheating to 60 hours may increase temperature beyond the required level and therefore would increase temperature difference and resulting thermal stresses. The maximum temperature difference observed between points 1 and 2 in schemes A, B and C shows typical values of 509, 434 and 366 °C respectively. This gives the impression that scheme C is the best.

Figures (3. a, b and c) show the time variation of temperature difference between points (1-4), (2-5) and (3-6) for preheating schemes A, B and C respectively. It is worth re-emphasizing that: points (1-4), (2-5) and (3-6) denote vertical lines across thickness of carbon block. It is noted that the maximum temperature difference takes place at points (2-5) which are located below the middle of anode block. This maximum temperature difference attains its value after about 28 hours of preheating in the three schemes with temperature difference levels of 274, 225 and 182 oC for schemes A, B and C respectively. At times greater than 28 hours, the temperature difference starts to decrease reaching 190, 150 and 130 °C at 60 hours of preheating. It is also noticed that point 6 is hotter than point 3 at the end of preheating which may be attributed to the large amount of heat transferred to the outer parts of the cell and to the ambient. This value of heat transferred increases with temperature difference between internal lining and outer atmosphere.

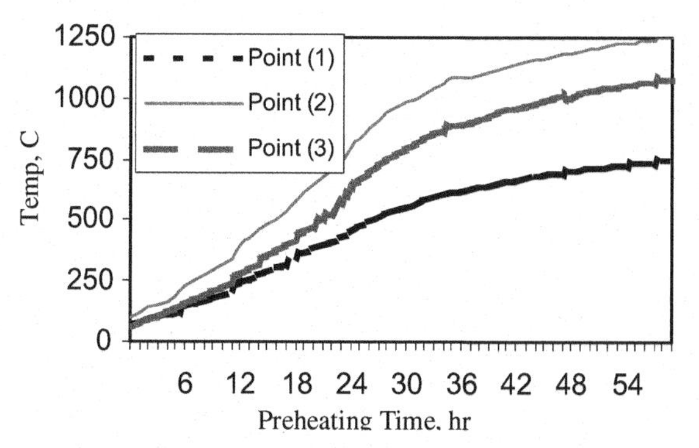

Fig. 2.a: Temperature-Time history along points 1,2 and 3 on cathode surface (scheme A)

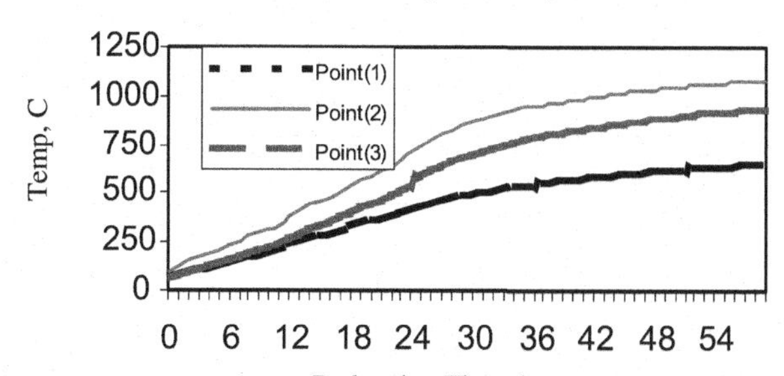

Fig. 2.b: Temperature-Time history along points 1,2 and 3 on cathode surface (scheme B)

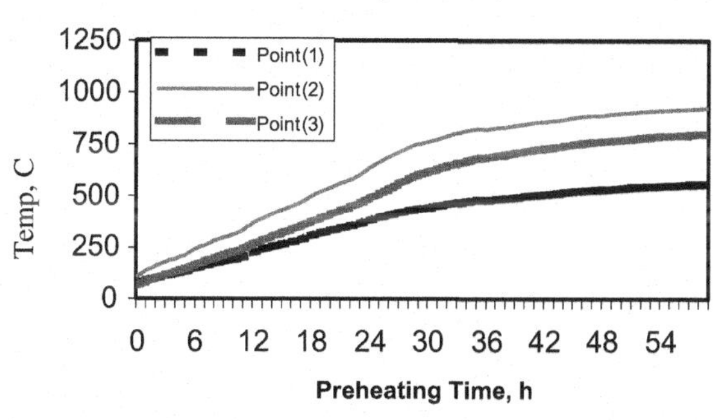

Fig. 2.c: Temperature-Time history along points 1, 2 and 3 on cathode surface (scheme C)

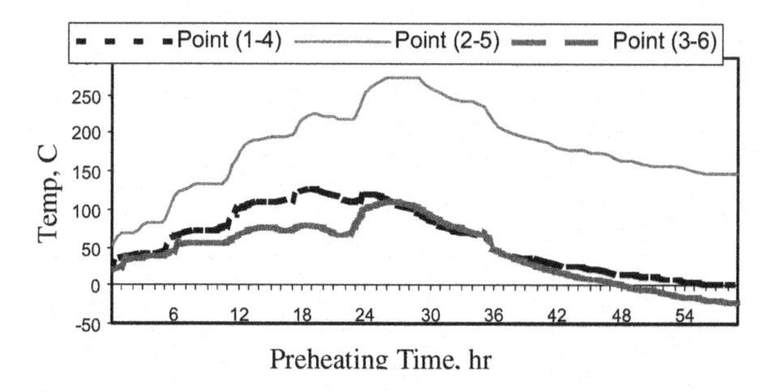

Fig. 3.a: Temperature difference - time history across upper and lower cathode surfaces (scheme A)

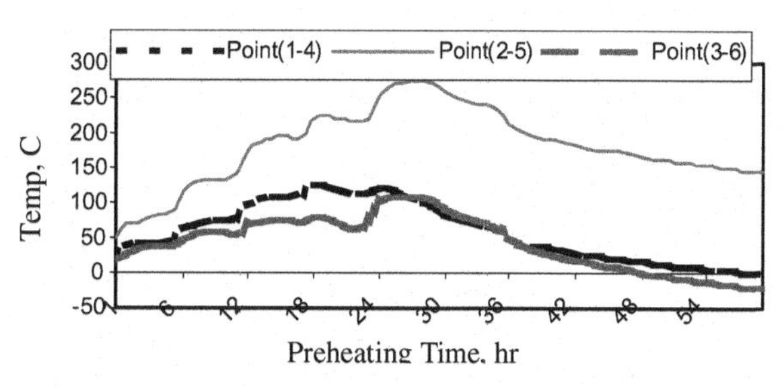

Fig. 3.b: Temperature difference - time history across upper and lower cathode surfaces (scheme B)

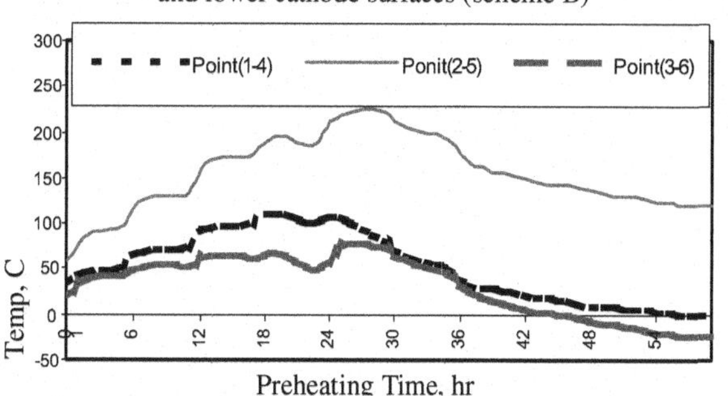

Fig. 3.c: Temperature difference - time history across upper and lower cathode surfaces (scheme C)

The difference that occurs in cathode surface temperature per unit time is defined as the heating rate, and such rate is preferred to be in the range of 5 to 50 °C/hr with optimum average value of 20 °C/hr [12]. Figure (4) shows the heating rate of cathode top surface for the three schemes. A jump in heating rate value is found at every change in heat added to the cell. The best scheme is the one which attains lower jumps in heating rate and an average value close to 20 °C/hr with lower standard deviation. Table (3) shows some important collected data for the three schemes. Excluding the first hour of preheating, scheme C is the best and scheme B is better than scheme A. The only disadvantage of scheme C is that the average heating rate is less than 20 °C/hr, for a preheating time of 60 hours. On the other hand, scheme C introduces an advantage over scheme A and B, since the standard deviation is the smallest.

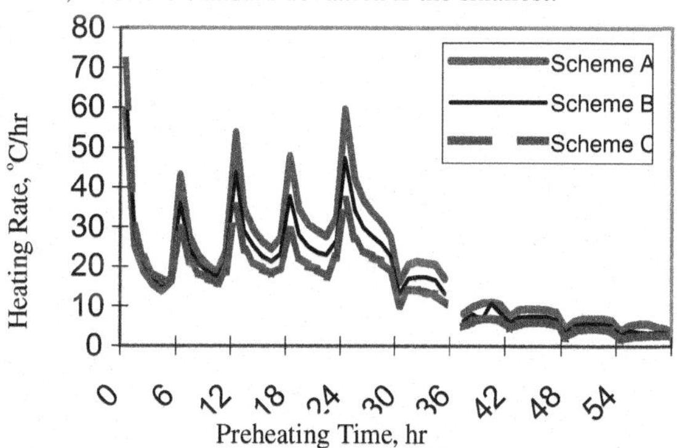

Fig. 4: Heating rate during preheating for the different preheating schemes.

Table 3: Heating rate data of the three schemes

Scheme	Average Value	Standard Deviation
A	20.6	14.03
B	17.7	12.6
C	15.1	11.77

Based on the above results, it appears that scheme # C provides the least temperature gradients in both the horizontal and vertical directions of cathode carbon block. This is likely to affect induced thermal stresses favorably and therefore, scheme # C is chosen for adoption.

Design of Shunt Rheostat

Table (4) gives the necessary data for cell and preheating process. These data are given by the teamwork in the production lines. Various metals are proposed for fabrication of resistance elements. Table (5) summarizes the advantages and disadvantages of the three alternative proposed materials. In addition to its heavy mass (87 tons), Cast Iron is brittle and hence may easily fracture during transportation.

Table 4: Available data for resistance design

Variable	Value
Line current, kA	200
Starting preheat current, kA	60-80
Number of risers	4
Total number of collector bars	32
Number of resistance units required	4
Number of collector bars joined to the upstream riser	11
Number of collector bars joined to the intermediate riser	5
Coke bed thickness, mm	30-50
Cell voltage, V	2.5-4

Aluminium and Ni-Cr are acceptable candidate materials for manufacturing of resistance elements. The required properties of Al and Ni-Cr are collected [13,14].

Aluminium has the advantage of being available at Egyptalum. Familiarity with its manufacturing methods is an additional advantage. It is important to notice that the use of Aluminium was found to be the most economical since it can be recycled in the worst case. For these reasons, aluminium had been chosen for the design of resistance elements.

Table 5: Advantages and Disadvantages of Proposed Materials

Material	Advantages	Disadvantages
cast-iron	Relatively low electric conductivity	High mass Brittle
Aluminium	Relatively low cost Easy to manufacture	High Volume High mass High electric conductivity
Nickel-Chrome	Low electric conductivity Low volume Low mass High melting point	High cost

Thermo-Electric Analysis

Design of shunt rheostat contains several calculation steps. Electric analysis considers the input data required for preheating and design of units, such as, coke-bed height, resistance element dimensions in addition to the starting preheating current. Thermal analysis provides the surface temperature of the resistance units, and verifies the working temperature.

1. Electric Analysis

A mathematical model [15] has been developed to solve the electric network for cells with different geometry. Sixteen cathode collector bars are connected in groups to the cathode busbars. Eleven collector bars passes current to the upstream riser and five to the intermediate risers through the cathode busbars. The current is then fed through the anode busbar of the next cell to the anode carbon blocks. It is important to notice that this network represents the cell during normal operation. The shunt resistances are connected between upstream, intermediate risers and the cathode busbar. A simplified method is used to calculate the current passing through each unit. The results of the simplified model were very acceptable as compared to the mathematical model results, Table (6)

Table 6: Calculated Current Values at Risers.

Type of Riser	Method of Calculation	
	Simplified	Mathematical Model [16]
Upstream Riser, kA	31.72	31.3
Intermediate Riser, kA	69. 78	68.8

2. Thermal Analysis

The above electric analysis yields dimensions, resistance and current values for the shunt units. Due to both resistance and current, heat is generated in the resistance elements. Thus, temperature of the resistance elements increases until the heat generated equals the heat lost to the outer boundary. The modes of heat transfer to the outer boundary media are mainly convection and radiation. For some materials, which have low specific electric resistance, length of the resistance was much higher than the available working space around the cell. It was then necessary, to reduce the space required by dividing the total length into several parts connected together in series and spaced with a small distance in between to allow for convection and some radiation to the atmosphere. Figure (5) shows an example of the practical proposed solution

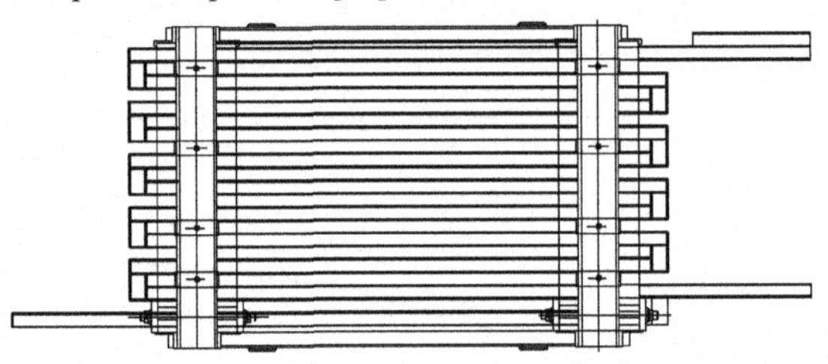

Figure 5: Schematic plan for shunt resistance compacted in short length.

Numerical Procedure

A computer algorithm was developed to determine the geometrical characteristics of the units, which comprises the rheostat. The input to the program consists of, electrical, thermal and physical properties of the candidate rheostat material, recommended values for cell current and voltages at the beginning of preheating, full line current and the available dimensions of rectangular cross sections usually employed for the construction of the rheostat units. In addition, an assumption is made to the rheostat starting temperature.

To determine the value of surface temperature, it is necessary to determine the value of convection heat transfer coefficient; which is a temperature and geometrical dependent. An iterative procedure is proposed to determine the surface temperature as follows:

1. Assume a suitable surface temperature
2. Extract necessary data from tables
3. Calculate the convection heat transfer values at the assumed temperature for the corresponding surface geometries
4. Calculate the surface temperature according to the calculated values of convection coefficients and including radiation to the outer atmosphere
5. Compare the calculated and assumed values of surface temperature; if convergence is achieved, otherwise; do the next iteration.
6. Find the exact value of specific electric resistance and then calculate new value of heat generation
7. Repeat steps 1 to 6 until convergence between assumed and calculated temperature within certain tolerance is achieved.

In the iterative problem in which the surface temperature is assumed, the corresponding film properties are extracted from tables, then calculation and comparison between assumed and calculated temperature is made. The problem is classified as a non-linear thermo-electric, with the following sources of non-linearity:

- Variation of material properties with temperature.

- Variation of convection boundary conditions with temp.
- Radiation from the outer surface is the last source.

Table (7) shows the data obtained to design a shunt unit, which carries 41250 ampere and 18750 ampere respectively to preheat a 200 kA cell at 80 kA starting current. Two of each unit is necessary to successfully conduct the required preheating. The data in the tables are to design aluminium shunt resistance units. These data were used to make the assembly and detailed drawing to enable preheating of Egyptalum pots with shunt rheostat for the first time.

Table (8) shows the design data for Nickel-Chrome resistance elements. The difference in mass, length, cross section is clear which makes the nickel chrome preferable. To conduct a successful preheating process it is required to use two of upstream and intermediate riser's shunt units. Therefore, a set of four units is necessary. It is important to notice the difference in number of resistances, which is related to the number of current steps. As the number of resistances increases the number of current steps increases and the amplitude of current jump decreases; so, smooth preheating process is achieved.

Table 7: Design data for Aluminium units for Shunt Resistance Elements.

Item	Up-stream Riser	nter-mediate Riser	Units
Potline Current	200,000	200,000	Ampere
Start baking current	80,000	80,000	Ampere
Shunt Current	41,250	18,750	Ampere
Baking Voltage (Start)	2.5-3.0	2.5-3.0	Volt
Coke Bed Thickness	27	27	Mm
Max. Current Density	1.0	1.0	Amp/mm^2
Calculated Current Density	0.9	0.5	Amp/mm^2
Material Density	2700	2700	Kg/m^3
Cross Section: Width Height Number of Sections	 70 650 1	 70 515 1	 mm mm Plate
Mass	9800	5700	Kg
Length	65	54	M
Calculated surface temp.	130	58	°C

Table 8: Design data for Nickel-Chrome units for Shunt Resistance Elements.

Item	Up-stream Riser	Intermediate Riser	Units
Potline Current	200,000	200,000	Ampere
Start baking current	80,000	80,000	Ampere
Shunt Current	41,250	18,750	Ampere
Baking Voltage (Start)	2.5-3.0	2.5-3.0	Volt
Coke Bed Thickness	27	27	mm
Max. Current Density	1.75	1.75	Amp/mm^2
Calculated Current Density	1.72	1.72	Amp/mm^2
Material Density	8300	8300	Kg/m^3
Cross Section: Width Height Number of Sections	 3 200 41	 2 200 19	 mm mm Plates
Mass	265	121	Kg
Length	1.35	1.35	m
Calculated surface temp.	450	450	°C

Results and Discussions

The aluminium shunt units had been extensively used for preheating of different pots from May 1993 to 1997. This indicates the design reliability and efficiency. It is necessary to bake-out some new Soderberg anodes from time to time. This process was carried out thermally and required the current to be disconnected at the end of preheating, thus wasting power and man-hours. The shunt rheostat was also employed successfully several times for anode-bake out, in addition to its original task.

Figures (6. a, and b) illustrate typical records of measured cell current and voltage during the first 24 hours of preheating. The preheating procedure consists of four phases each of 6 hours duration. The four rheostat units are disconnected consecutively at the end of each phase, i.e. every 6 hours. It is clear from the plots of current shown in Fig. (6 a and b) that when a unit is disconnected, instantaneous jumps in current and voltage are observed. Similar results are obtained by Dupuis [5] and Dunn [7]. A gradual increase is noted in jump amplitude as units are disconnected consecutively. The total equivalent resistance increases at every unit disconnection. The disconnected unit's current is distributed between the other remaining units and the cell. As the number of working units decreases the cell share, in this current, increases. This explains the reason for the increase in current jump amplitude. This type of jump is acceptable and is certainly better than cathode preheating by full line current.

It is also notable that a gradual increase in cell current occurs during each 6 hours phase. This is due to the decrease in cell resistance as the cell temperature increases with time. Cell voltage has the same behavior in the second, third and fourth phases of preheating (i.e. 6 - 24 hours). Similar results are obtained by Dunn [6].

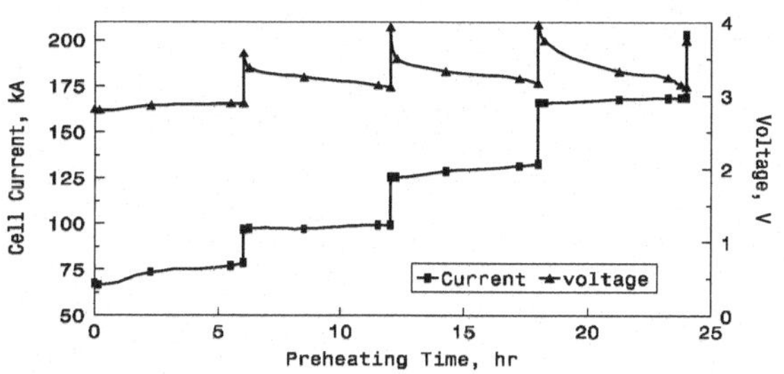

Figure 6.a: Measured Current & Voltage. vs. time during 24 hr of preheating.

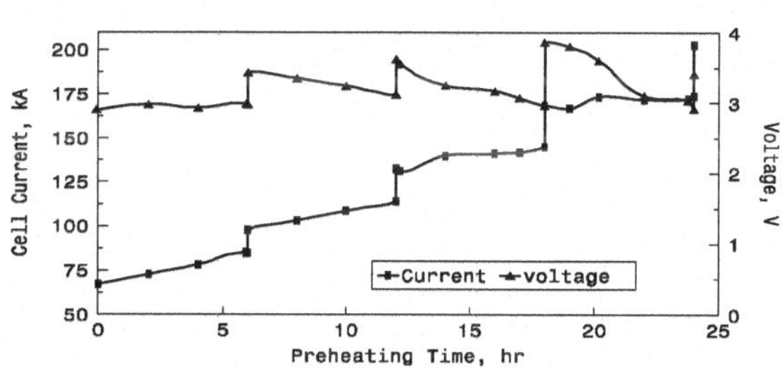

Figure 6.b: Measured Current & Voltage .vs. time during 24 hr of preheating for cell

Figures (7.a and b) show the time variation of cell current and voltage during the first 36 hours of preheating by the use of Ni-Cr shunt resistances. Measurements indicated that the cell current increases and the cell voltage decreases or remains nearly constant without such large jumps observed in preheating with aluminium shunts. Therefore, the results obtained by Ni-Cr shunts are better than those for aluminium shunts. However, both of the two shunts has its own disadvantages. Although, it is more reliable, the aluminum shunts causes current jumps at every unit disconnection in addition to its relatively high volume and mass, the Ni-Cr shunts are better in current loading, but requires more effort during 16-resistance disconnection in addition to the noise, which comes from its cooling fans.

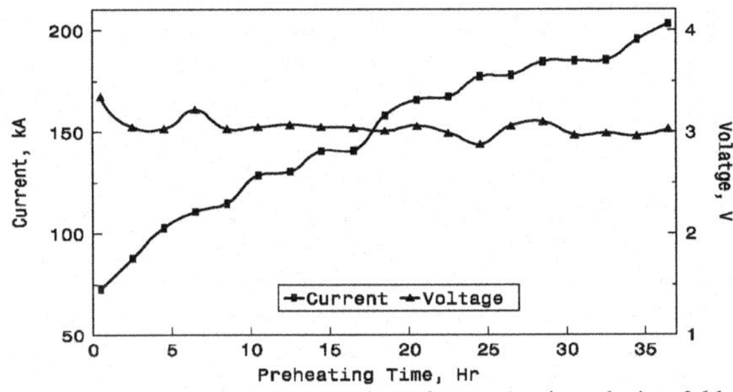

Figure 7.a: Measured Current & Voltage .vs. time during 36 hr of preheating

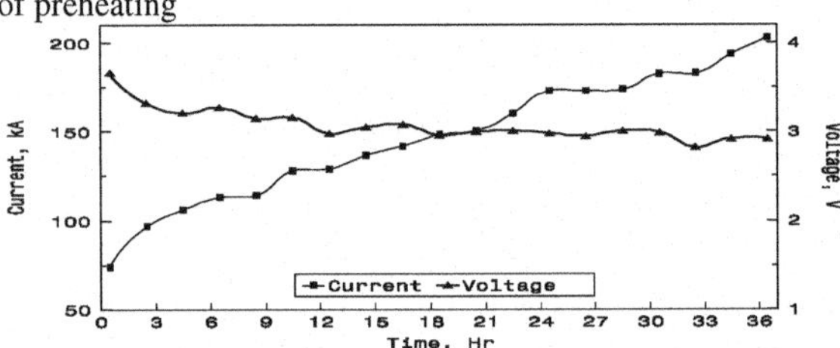

Figure 7.b: Measured Current & Voltage .vs. time during 36 hr of preheating for different cell

New Shunts

Although the Ni-Cr shunts overcome the disadvantages of aluminum shunts (heavy weight, huge volume, handling problem), the noise level and maintenance were a reason for Egyptalum to import mild steel shunts. The new shunts have the advantages of ease of connection, low mass and small volume. The units are cooled by natural convection.

A new cathode lining was introduced at Egyptalum in 1999. This lining makes it is possible to reduce the coke bed height to less than 20 mm. As a result of such development, the cell resistance was reduced and thus the current at start of preheating increased; Fig(8). To eliminate this problem a new design modification had been adopted to reduce the start preheating current from 135 kA to less than 100 kA.

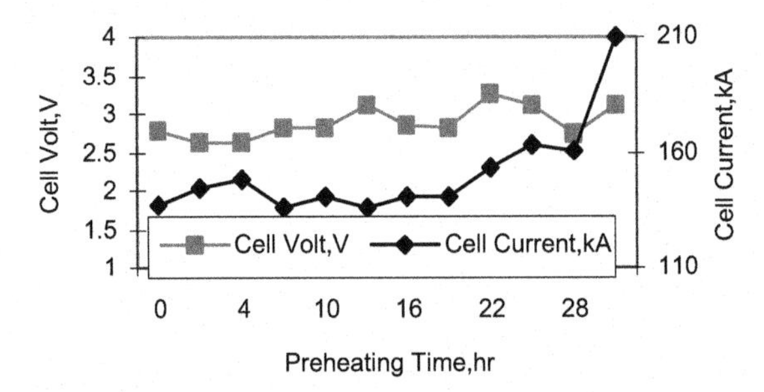

Figure 8: Measured Current & Voltage .vs. time of preheating for imported shunts

Conclusions

- Three schemes of preheating are evaluated by means of conducting transient heat transfer analysis by the finite element method. The scheme associated with the least thermal gradients is identified as the one possessing the highest starting current with least coke-bed thickness.
- One set of shunt resistances made of aluminum was designed, manufactured and used at Egyptalum during the period 1993 to 1997
- To reduce mass, decrease cost and further improve the process, two other sets of Ni- Cr shunts were designed, manufactured and used during the period 1997-1998 to preheat more than 70 cells in new pot-line.
- Further Improvements have been implemented in shunt design and construction materials to obtain better preheating efficiency.

Acknowledgments

The authors would like to express their appreciation to Mr. Sayed Abdel-Wahab, Chairman of Egyptalum, Mr. Zaki Bassyouni, current Chairman of the Holding Company for Metallurgical Industries and ex-chairman of Egyptalum for their support during the course of the present work. Special thanks to the name of Late Mr. Fawaz M. Ahmed, the ex-head of the Laboratories and R & D sector at Egyptalum. Sincere thanks to Prof. Mohammad Megahed of Cairo University and Egyptalum R & D consultant for his efforts during the course of the present work and for revising this manuscript

References

1. Sorlie M. and Øye H. A., "Cathodes in Aluminium Electrolysis", 2nd edition, Aluminium Verlag, 1994, pp77-96.
2. Clelland H., Keniry J. T. and Welch B. J., " A Study of Some Aspects of The Influence of Cell Operation on Cathode Life ", Light Metals 1982, pp. 299-310.
3. Rickards W. B., Young P. A., Keniry J.T., and Shaw P., "Thermal Bake-out of Reduction Cell Cathodes Advantages and Problem Areas", Light Metals, 1983, pp. 857-866.
4. Clelland H., et Al. "A Study of Some Aspects of the Influence of Cell Operation on Cathode Life", Light Metals, 1982, pp 299-309.
5. Marc Dupius et Al, "Thermal Study of the Coke Preheating for Hall-Heroult Cell", CIM, Qubec city, Canada, 1993.
6. Marc Dupius and Imad Tabsh, "Evaluation of Thermal Stresses due to Coke Preheat of Hall-Heroult Cell", ANSYS Conference 1994, pp 3.15-3.23.
7. Dunn M. R., and Galadari Q. M. I., "An Analysis of the Electrical Technique based on the Start-Up of the CD200 Prototypes at Dubai Aluminium Co. Ltd", Light Metals 1997, pp247-251.
8. Mohamed Sh. A, Wahhab M. A., Aref A. A. and Dahab O. M., "Mathematical Modeling for coke bed preheating of Aluminium Reduction Cell", Light Metals 1997, pp 457.
9. Mohamed Sh. A., "Coke Bed Preheating of Aluminium Reduction Cells", M Sc. Thesis, Minia University, Egypt, 1999, pp 20-53
10. Zangiacomi C, et al, "Preheating Study of Smelting Cells", Light Metals 2005, pp 333-336.
11. P.Nikiforov and A.Derkach, "Preliminary Test Results of Group of 6 Prototype prebaked Anode Pots at Egyptalum",Nag-Hammadi, 1993.
12. Kvande H.," Cell Start-up", The International Course on Process Metallurgy of Aluminium, Trondheim, June 3-7, 1996, pp 7: 1-34
13. Kazantsev I., "Industrial Furnaces", Mir Publishers, Moscow,1977.
14. An offer from "Osab Trade" Agent of "KRUPP VDM GMBH" German Company on 20/1/1993, pp.8 (According to DIN 17470).
15. El-Maghrabi M. G., Ali K. F. and Mohamed Sh. A. "Busbar system design for Aluminium Reduction Cell", 6th Arab International Aluminium Conference, Arabal 93, 11-14 dec. 1993-Cairo, pp IV-3
16. El-Maghraby M. G. et. al., "Development of Aluminium Smelter At Nage-Hammady", Report No. 8, Experimentation on Prototype Section, Jan.1993, pp31.

Light Metals 2006 *Edited by Travis J. Galloway* ***TMS (The Minerals, Metals & Materials Society), 2006***

SIMULATION STUDY ON THE HEATING-UP RATE FOR COKE BED PREHEATING OF ALUMINUM REDUCTION CELL

Li Jie, Liu Wei, Zhang Qinsong, Lai Yanqing, Liu Yexiang, Hu Guorong
School of Metallurgical Science and Engineering, Central South University, Changsha, 410083, China

Keywords: aluminum reduction cell, coke-bed preheating, transient analysis, heating-up rate, ANSYS

Abstract

With the purpose of simulation study on the coke-bed preheating process of prebaked-anode aluminum reduction cell, a nonlinear transient heat transfer model was developed on the platform of ANSYS. It can be used to study the influence of different shunting schedules (with different starting current, or different shunting intervals or different number of shunts) on the heating-up rate. Simulation results show that whatever shunting schedules are used, the curves of heating-up rate vs temperature will go together after the loaded current reaches the same final value. In other words, if conditions except for the shunting schedule are kept fixed, the heating-up rate at certain location and certain temperature is dependent on finally-loaded current, not on shunting schedule after shunting is removed. Based on this conclusion, a method for the design of shunting schedule was proposed which can use the series of curves of heating-up rate vs temperature in the design of temperature-rising curves to meet different demands for desired heating-up rate in different temperature intervals.

Introduction

The life of prebaked aluminum reduction cell is affected by many aspects, for example, the cell structure design, technics of building cell, lining materials, preheating and starting-up, cell management. Among these affecting factors, it is generally accepted that the affecting extent of preheating methods on cell life can reach to 25%[1].

There are three kinds of preheating methods usually used at present, namely, liquid metal preheating, coke-bed preheating and flame burning preheating. The coke-bed preheating is widely used in large-scale prebaked anode aluminum reduction cells.

The key technology of the coke-bed preheating is how to make the heating-up process of the cathode lining to meet the requirements for desired heating-up curves. Materials used in the lining are subject to definite requirements on the heating-up rate, for example, the ramming mix has strict requirements on temperature and if not well controlled, it will affect the quality of preheating. The ramming mix therefore is a key factor to the heating-up process.

Transient models can be built to study the thermo-structural problems of the cell preheating process. X.H. Mao[2], M. Dupuis[3], F. Hiltmann[4] and G.V. Arkhipov[5] did some research work by 2D or 3D transient models that were focused on the distribution of temperature, thermal stresses and concentration of stresses. In this paper, some simulation research were made on the heating-up rate and how to design optimal shunting schedules to reduce potential thermo-stresses generated in the process of the coke-bed preheating.

Model Development

Mathematic model

Nowadays, 3-D transient heat transfer models are widely used for the thermal analysis of aluminum reduction cell, the basic equation of which is described as follows,

$$\frac{\partial}{\partial x}\left(k_x\frac{\partial T}{\partial x}\right)+\frac{\partial}{\partial y}\left(k_y\frac{\partial T}{\partial y}\right)+\frac{\partial}{\partial z}\left(k_z\frac{\partial T}{\partial z}\right)+q_s=\rho c\frac{\partial T}{\partial t} \quad (1)$$

Enq. (1) satisfies the following boundary conditions:
the initial boundary condition,

$$T(x,y,z)\big|_{t=0}=T_0(x,y,z) \quad (2)$$

the first kind of boundary condition,

$$T(x,y,z)\big|_{s1}=T_b(x,y,z) \quad (3)$$

the second kind of boundary condition,

$$k_n\frac{\partial T}{\partial n}\bigg|_{s2}=-q(x,y,z) \quad (4)$$

and the third kind of boundary condition.

$$k_n\frac{\partial T}{\partial n}\bigg|_{s3}=h(T-T_g) \quad (5)$$

where, T, T_0 and T_b are the nodal temperature, the initial temperature and the applied temperature respectively, in unit of K; t is the time, in unit of second; k_x, k_y and k_z are the thermo-conductivities, in unit of W/(m·K); q_s is the heat generation rate per unit volume, in unit of W/m^3; ρ is the density, in unit of kg/m^3; c is the specific heat, in unit of W/(kg·K); S_i is the heat exchange area; n is the normal direction of S_i.

Finite element model

A 1/4 thermo-electrical model was built according to a design of 160kA prebaked aluminum reduction cell, including the anode assembly, coke bed, cathode, steel conductors and lining. The anodes are covered with a layer of alumina. The space between the anode and the side are filled with solidified electrolyte crust, cryolite and fluorides with asbestos sheet on top (see Figure 1). There is nothing in the gaps of anodes. The structural parameters of reduction cell are showed in Table 1.

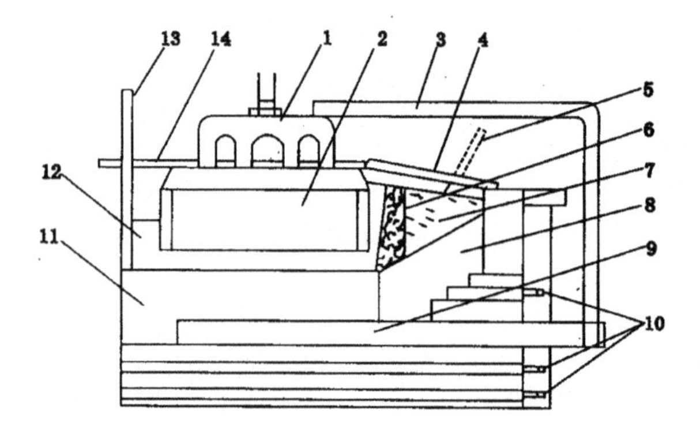

Figure 1. Schematic of the coke prebake cell

1—Anode rod & stubs; 2—Carbon anode; 3—Shunt; 4,14—Asbestos sheet; 5,10,13—Tubes with thermoelectric pair inserted; 6—Solidified electrolyte separation wall; 7—Fluoride, sodium carbonate and calcium fluoride; 8—Ramming paste; 9—Cathode conduction bar; 11—Carbon cathode; 12—Coke bed.

Table 1. Structural parameters of 160kA cell

Parameters	Value (mm)
Dimension of the anode	1400×660×540 (24)
Dimension of the stud	132×132×286 (4)
Depth of the steel stud	100
Height of the steel beam	150
Thickness of the alumina cover	16
Dimension of the cathode	3250×515×450(16)
Distance between the cathodes	35
Dimension of the steel bar	180×65×2150
Number of the steel bars	16×2
Thickness of the fire-resistant brick	65×2
Thickness of the bottom alumina layer	65
Thickness of the barrier brick	65×2
Thickness of the calcium silicate layer	65×2

The complete and reliable boundary conditions are the guarantee of solving the finite element model correctly. The boundary conditions are listed as follows:

(1) Voltage potential is set to zero at the end of the steel conductor bar.

(2) The ambient temperature is 25°C and the initial temperature of the reduction cell is also 25°C.

(3) Convection and radiation boundary conditions are applied to the outer surface of the wall, alumina cover, insulating parts (asbestos sheet) and anode stubs and rods. Heat transfer between anodes by radiation effect is also considered.

Results and analysis

In our simulation example, semi-graphite cathode carbon blocks and 20mm thick coke bed were chosen as the calculation base. The effects of the value of the starting current, shunting intervals and the number of shunts on the heating-up rate were studied respectively.

The Effect of Starting Current on the Heating-up Rate

The heating-up rate at the center of the cathode surface was calculated under the given conditions that the shunt-removing interval was 5 hours; the number of shunts was 5; the starting current was 10%, 30%, 50%, 70%, 90% and 100% of full current respectively; the current increased by the same step after removing every group of the shunting flakes and finally reached "full current" (100% loaded current) when the final group of the flakes was removed. But it should be pointed out that the case with the starting current being 100% loaded means no shunting is used and this is so called "directly current loading".

The heating-up rate curve is showed in Figure 2 and the maximal heating-up rate and the average rate are showed in Table 2.

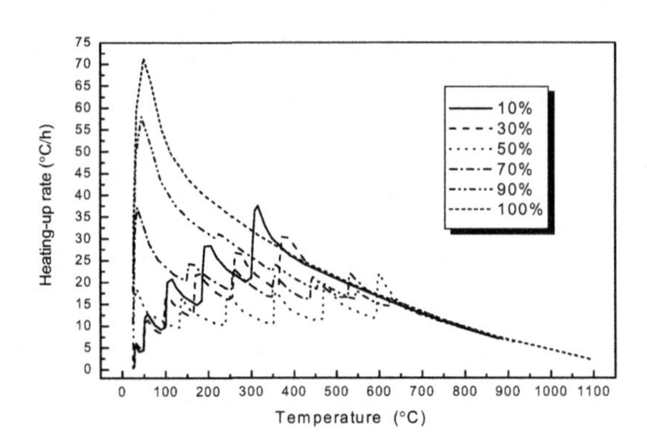

Figure 2. Heating-up rate curves with different starting current

Table 2. Heating-up rate with different starting current

Starting current %	Maximal heating-up rate °C /h	Average heating-up rate 25~200°C °C /h	Average heating-up rate 200~600°C °C /h
10	37.44	8.51	22.66
30	30.35	10.48	20.62
50	21.96	14.82	18.86
70	37.08	24.10	18.44
90	57.96	39.09	21.00
100	71.28	48.84	23.71

With the increase of starting current, the maximal heating-up rate and the average heating-up rate between 200°C and 600°C decreased at first and then increased, but the average heating-up between 25°C and 200°C increased all along. It was noticed that the heating-up rate varied quite small in the whole heating-up process when starting current was set to 50% of full current.

The Effects of Shunting Interval on the Heating-up Rate

In this case, the heating-up rate was calculated under the conditions that starting current was set on 50% of full current, the number of shunts was 5 and the shunting intervals (i.e. the shunting flake-removing intervals) were 0 hour (no shunting), 1 hour, 3 hours, 5 hours, 7 hours and 9 hours.

As shown in Figure 3 and Table 3, both the maximal heating-up rate and the average heating-up rate decreased with the increase of the intervals.

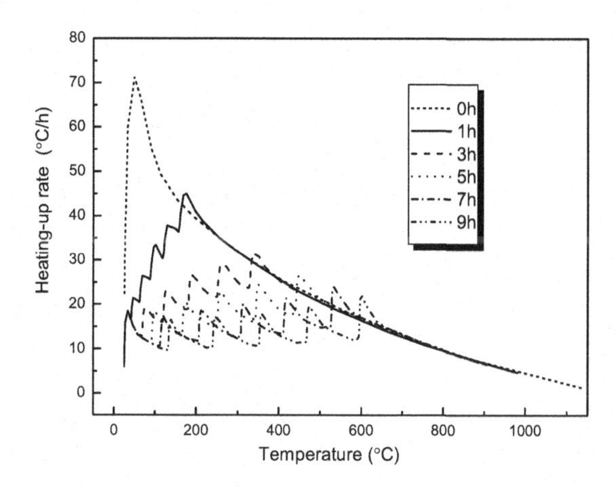

Figure 3. Heating-up rate curve with different shunting intervals

Table 3. Heating-up rate with different intervals

Interval h	Maximal heating-up rate °C /h	Average heating-up rate 25 ~ 200°C °C /h	Average heating-up rate 200 ~ 600°C °C /h
0	71.28	48.84	23.71
1	44.64	29.12	23.97
3	30.67	17.91	22.55
5	26.75	14.82	18.86
7	24.01	12.95	15.79
9	21.96	12.30	12.87

The Effect of the number of shunts on the Heating-up Rate

The heating-up rate was calculated under the conditions that shunting interval was 5 hours, the number of shunts was 0 (no shunting), 1, 3, 5, 7 and 9 and starting current was 50% of full current except for the case of no shunting and 100% for the case of no shunting.

As shown in Figure 4 and Table 4, both the maximal heating-up rate and the average heating-up rate decreased with the increase of the number of shunts.

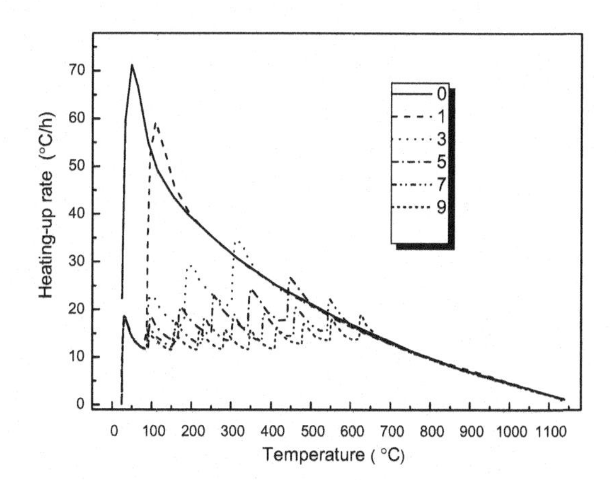

Figure 4. Heating-up rate curves with different number of shunts

Table 4. Heating-up rate with different number of shunts groups

Number of shunts	Maximal hearing-up rate °C /h	Average heating-up rate 25 ~ 200°C °C /h	Average heating-up rate 200 ~ 600°C °C /h
0	71.28	48.84	23.71
1	59.40	23.66	24.24
3	34.52	16.45	22.39
5	26.75	14.82	18.86
7	22.25	13.61	15.84
9	19.01	12.79	13.46

The Influence of Different Shunting Schedules on Heating-up Rate

From Figure 2 to Figure 4 it can be found that whatever shunting schedules are used (with different starting current, or different shunting intervals or different number of shunts), the curves of heating-up rate vs temperature will go together after the loaded current reaches the same final value (full current).

Furthermore, two sets of simulations were carried out: (1) directly loading current with 6 values: 50%, 60%, 70%, 80%, 90%, 100% of the full current (160kA) and no shunting is used, six curves of heating-up rate were got respectively; (2) loading current with starting current being 50% of the full current (160kA) and using 5 shunting schedules with the same number of shunts (the number is 5) and the same loading steps (60%, 70%, 80%, 90%,100%) but with different shunting intervals (1h, 3h, 5h, 7h, 9h), five curves of heating-up rate were got respectively. Then, by drawing the above curves acquired in the two sets of simulations together (see Figure 5), it can be found that, whatever shunting is used or not, or whatever shunting schedule is used, the curves of heating-up rate at a certain current will overlap. For example, when the current rise to 70% (of full current) for every shunting schedule, the corresponding curves of heating-up rate go down along that one of "directly-loading current with 70% of full current". The relative standard deviation between the two kinds of curves is less than 5%.

From the above discussion, it is concluded that if conditions except for the shunting schedule are kept fixed, the heating-up rate at certain location and certain temperature is dependent on finally-loaded current, not on shunting schedule after shunting is removed.

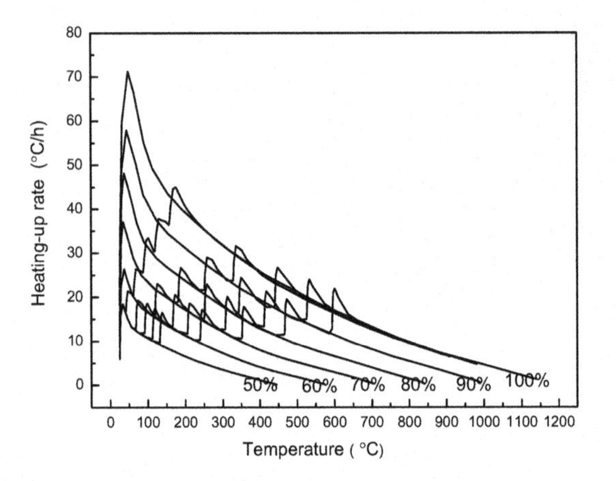

Figure 5. Comparison of directly current loading with shunt schedules

Design Method of Shunt Schedules

Based on the above conclusion, a method for the design of shunting schedule was proposed as follows.

Ten curves of "the reciprocal of heating-up rate -- temperature" corresponding to ten current values (10%, 20%, ..., 100% of full current) loaded directly (no shunting is used) can be acquired by simulation (see Figure 6). These curves can be called "reciprocal curves".

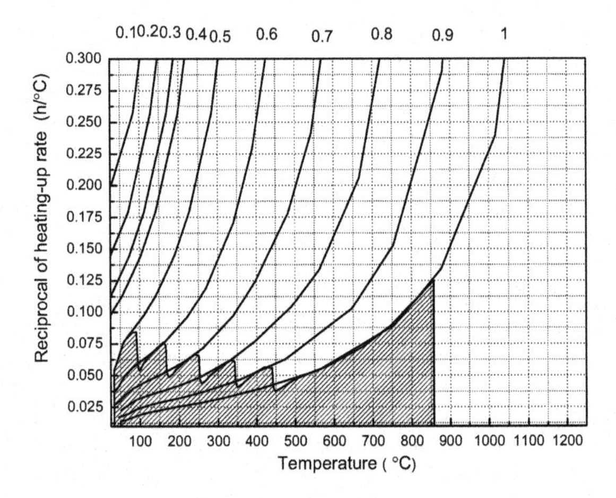

Figure 6. Curves of "the reciprocal of heating-up rate -- temperature" corresponding to ten current values

For every reciprocal curve, the integral from T_0 to T_1 is expressed as follows:

$$\int_{T0}^{T1} \frac{1}{\dot{T}} dT = \int_{T0}^{T1} \frac{dt}{dT} dT = t \Big|_{T0}^{T1} \qquad (6)$$

where $\dot{T}$ is the heating-up rate, T is the temperature and t is the time. So the shadow area under the curves from T0 and T1 (see the shadow area in Figure 6) is the required time for the heating-up process.

One procedure is suggested to choose an optimal heating-up curve: first select the starting current and the criteria of the heating-up rate for possible temperature intervals; then calculate the reciprocals of the max/min heating-up rate in every given temperature interval and mark them on the reciprocal curve corresponding to the given number of removed shunts (i.e. the percentage of full current); then calculate the shadow area under the curves and get the time of the heating-up process. Finally a shunt schedule can be made.

As shown in Figure 6, the shadow area denotes the shunt schedule where the starting current is 50%, shunts are removed every 5 hours till all the 5 shunts are removed.

Conclusion

With the purpose of simulation study on the coke-bed preheating process, a nonlinear transient heat transfer model of a 160kA prebaked cell was developed on the platform of ANSYS. It can be used to study the influence of different shunting schedules (with different starting current, or different shunting intervals or different number of shunts) on the heating-up rate.

Simulation results showed that with the prolonging of shunting interval or the increase of the number of shunts, the average heating-up rate and the maximal heating-up rate during the whole preheating process would go down; with the increase of starting current, the average heating-up rate between 25°C and 200°C would increase, but the average heating-up rate between 200 °C and 600°C and the maximum heating-up rate during the whole preheating process would decrease at first and then increase and their minimal points appeared at about 50% of full current

Whatever shunting schedules are used, the curves of heating-up rate vs temperature will go together after the loaded current reaches the same final value. In other words, if conditions except for the shunting schedule are kept fixed, the heating-up rate at certain location and certain temperature is dependent on finally-loaded current, not on shunting schedule after shunting is removed. Based on this conclusion, a method for the design of shunting schedule was proposed which can use the series of curves of the reciprocal of heating-up rate vs temperature in the design of temperature-rising curves to meet different demands for desired heating-up rate in different temperature intervals.

Acknowledgement: This work was supported by National Natural Science Foundation of China (No.50374081) and by National Basic Research Program of China (No. 2005CB623703).

References

[1] X.A. Liao, H.G. Wei, "Functional Requirements and Trend of Carbon Lining Materials for Aluminum Electrolysis cells ," *Light Metals*, (6)(2001), 51-53(in Chinese)

[2] X.H. Mao, "Mathematic Simulation of the Cell Bottom Prebake Process, " *Light Metals*, (11)(1991),60-63(in Chinese)

[3] M. Dupuis et al., "Thermal Study of the Coke Preheating for hall-héroult cell," *CIM Light Metals*, Québec, Canada: 1993:93-100

[4] F. Hiltmann, K.H. Meulemann, "Ramming paste properties and cell performance." *Light Metals*. Warrendale, Pennsylvania: TMS, 2000:405-411

[5] G.V. Arkhipov, V.V. Pingin, "Investigating thermoelectric fields and cathode bottom integrity during cell preheating, start-up and initial operating period." *Light Metals*. Warrendale, PA, USA: TMS, 2002:347-354

TITANIUM DIBORIDE AND WOLFRAM SILICIDE COMPOSITE USED AS ALUMINUM ELECTROLYSIS INERT CATHODE MATERIALS

Huimin Lu[1,2,3], Huanqing Han[2,4], Ruixin Ma[2], Yongheng Wang[2] and Dingfan Qiu[5]
[1] Beijing Univ. of Aeronautics & Astronautics, School of Material Sci. & Eng.; 37 Xueyuan Rd; Beijing 100083 China
[2]Univ. of Science and Technology Beijing, Light Metals Research Inst.; 30 Xueyuan Rd; Beijing 100083, China
[3]Univ. of Cambridge, Dept. of Mater. Sci. Metall.; Pembroke Street; Cambridge CB2, 3QZ, UK
[4]General Research Institute of Iron and steel; Beijing 100016, China
[5]General Research Institute of Mining and Metallurgy Beijing; Beijing 100041, China

Keywords: TiB_2- WSi_2 composite, aluminum electrolysis inert cathode, sintering performance

Abstract

It is well-known that TiB_2 is first-priority material for aluminum electrolysis inert cathode, but its limited application has been limited because of its difficult compact sintering. However, WSi_2 has good conductibility and easy compact sintering, TiB_2- WSi_2 composite could become a new metal-ceramic composite for aluminum electrolysis inert cathode.

In this paper, various performances of TiB_2-WSi_2 composite with WSi_2 content 10mass%, 30mass%, 50mass% and 70mass% were studied. The results show that WSi_2 can improve TiB_2 compact sintering performance apparently, when WSi_2 content of TiB_2-WSi_2 composite excesses 50mass%, the compactivity of the composite sintered 1hour at 1900°C is higher than 98%; its conductivity is prior to TiB_2; WSi_2 exhibits significantly better aluminum corrosion resistance but weaker cryolite corrosion resistance than TiB_2. Aluminum corrosion to TiB_2-WSi_2 composite is mainly that Al reacts with TiB_2 to produce TiAl and so on intermetallics, but cryolite corrosion to TiB_2-WSi_2 composite is mainly that WSi_2 dissolves and fractures in the cyolite solution. Because in aluminum electrolysis process, there is a thin aluminum liquid layer on the inert cathode surface, the inert cathode failure is mainly aluminum liquid corrosion, so TiB_2- WSi_2 composite is a better inert cathode material for aluminum electrolysis.

Introduction

In order to solve the serious pollution of industrial aluminum electrolysis cell, save the aluminum production cost and obtain the long life aluminum electrolysis cell, the research and development of the usage of inert cathode has an important significance. It is well-known that TiB_2 is first-priority material for aluminum electrolysis inert cathode because it has the high melting point, conductivity and hardness, significantly better molten aluminum and cryolite corrosion resistance. However its limited application has been limited because of its difficult compact sintering stemming from its high melting point and low diffusion coefficient and TiB_2 material poor performance originating from crystalline grains exceedingly crystal growth, which is owing to the growth tempo of crystalline grain along the C axis direction higher than other direction as the chemical bond difference of Ti and B [1 - 3]. Therefore, it is an urgent task to solve in the field of aluminum electrolysis cathode materials that developing new conducting electricity ceramic material for industrial application.

WSi_2 and common $MoSi_2$ both belong to cubic crystallization system, the chemical performance is extremely similar, but the conductibility of WSi_2 is better than $MoSi_2$, Table 1 shows the physical characteristic comparison of WSi_2 and $MoSi_2$ [4 - 7]. WSi_2 can solve the difficult task of TiB_2 compact sintering for WSi_2 easy compact sintering, and WSi_2 exhibits significantly good oxidation and corrosion resistance. Therefore, TiB_2-WSi_2 composite could become a new metal-ceramic composite for aluminum electrolysis inert cathode.

Table 1 The characteristic comparison of WSi_2 and $MoSi_2$

Mater.	Density (g/cm^3)	Melting point (°C)	Resistivity ($\mu\Omega \cdot cm$)	Tensile strength (MPa)		Micro hardness (MPa)
				20°C	1000°C	
WSi_2	9.25	2165	12.6	585	35	10531
$MoSi_2$	6.24	2030	21.6	1130	395	7210

Experimental

WSi_2 powder for experiments was prepared by self-propagating high-temperature synthesis (SHS) with reduction process. First adding 10mass%, 30mass%, 50mass% and 70mass% TiB_2 and some alcohols to a ball mill loaded with WSi_2 and fairly well-distributed by wet mixing method, then drying and loading the materials into a rubber mould for cold constant static pressure forming, the forming pressure is 200MPa. Under the atmosphere of Ar the forming TiB_2- WSi_2 composite is sintered for 1h at the temperature of 1750°C, 1800°C, 1850°C and 1900°C separately.

The molten electrolyte and aluminum corrosion resistance experiments of the TiB_2-WSi_2 composite are conducted in a box resistance furnace with the DWK – 702 model control temperature instrument, its precision is ± 1°C. The mole ratio of the electrolyte for experiments is 2.1. In the experiments, adding 5mass% CaF_2, the electrolysis temperature is 950°C, the electrolysis time is 48h. The crucible used is made of AlN material. The experiments are conducted under the atmosphere of Ar.

The densities and resistivity of the TiB_2- WSi_2 composite are measured by the Archimedes method and electric bridge method. The Regaku D/Max IIIB model X – ray spectrograph made in Japan is used for the powder phase analysis, the Hitachi S-450 model scanning electron microscope for SEM photomicrograph of the TiB_2-WSi_2 composite.

Results and discussion

The sintering and conductivity performance of the TiB_2- WSi_2 composite

WSi_2 has better compact sintering performance. When WSi_2 is sintered 1hour at 1750°C, its compactivity is 98%. But when TiB_2 is added to WSi_2, its compact sintering performance drops markedly and with the adding content increase, the compactivity of TiB_2-WSi_2 composite drops by a big margin. Raising the sintering temperature can strikingly enhance the compactivity of the composite. When TiB_2 excesses 50mass% in the composite, the compactivity of the composite sintered 1hour at 1900°C is higher than 98%. The composite meets the demands of aluminum electrolysis. Therefore, the results show that WSi_2 can well improve the compact sintering performance of TiB_2. In addition, the fact that the resistivity of the composite increases with TiB_2 adding content increase under the compactivity of TiB_2-WSi_2 composite is similar shows that the conductivity of WSi_2 is prior to TiB_2. Table 2 shows the performance of the composite. The interaction of WSi_2 and TiB_2 is marked. On the one hand, WSi_2 can drop the sintering temperature of TiB2, because WSi_2 phase mainly encircles around the TiB_2 particles or is in the place where the TiB_2 particles meet, which effectively stops the TiB_2 particles from crystal growth, see Figure 1; on the other hand, when the TiB_2 phase increases, the selecting crystal growth phenomenon of TiB_2 is remarkable, most particles appear long stick form, which enhances the mechanic performance of WSi_2. Figure 2 shows that the bridge joining action and drawing action of the long stick particles are remarkable. In the same time, the two phase materials close together, no crackle or no marked opening exists. The conductivity of WSi_2- TiB_2 composite is better listed in Table 2. The WSi_2- TiB_2 composites as cathode materials completely meet the demands of the aluminum electrolysis.

Table 2 The compactivity and conductivity of WSi_2-TiB_2 composite

Materials (mass%)	1750°C sintering 1h		1800°C sintering 1h	
	Compactiv -ity (%)	Resistivity (μΩ • Cm)	Compactivi -ty (%)	Resistivity (μΩ • Cm)
WSi_2	98	28	99	25
WSi_2+10%TiB_2	93	84	95	63
WSi_2+30%TiB_2	90	108	92	81
WSi_2+50%TiB_2	88	125	88	110
WSi_2+70%TiB_2	84	130	86	123
TiB_2	—	—	—	—

Materials (mass%)	1850°C sintering 1h		1900°C sintering 1h	
	Compactiv -ity (%)	Resistivity (μΩ • Cm)	Compactivi -ty (%)	Resistivity (μΩ • Cm)
WSi_2	—	—	—	—
WSi_2+10%TiB_2	98	46	99	37
WSi_2+30%TiB_2	95	73	98	68
WSi_2+50%TiB_2	92	103	96	92
WSi_2+70%TiB_2	89	110	94	100
TiB_2	81	165	84	142

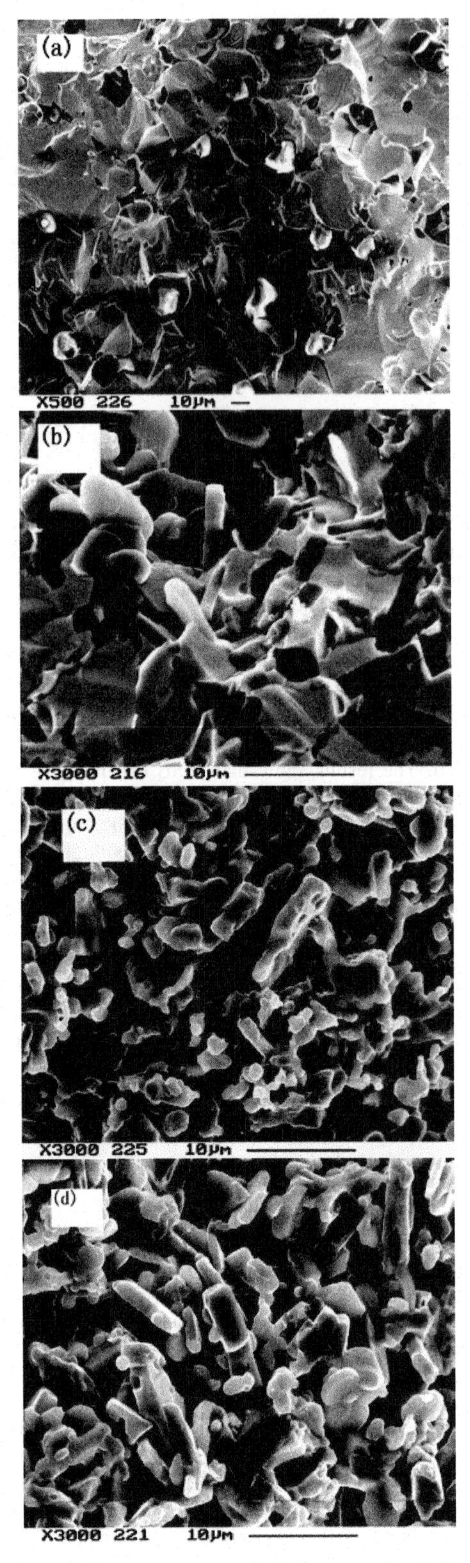

Fig. 1 SEM images of the fracture of the WSi_2- TiB_2 composites in 1850°C sintering 1h

(a)WSi_2+10mass%TiB_2; (b)WSi_2+30mass%TiB_2
(c)WSi_2+50mass%TiB_2; (d)WSi_2+70mass%TiB_2

The corrosion resistance performance of WSi_2- TiB_2 composite

A TiB_2-WSi_2 composite samples with dimensions of approximately 60×30× 20 mm prepared after sintering at 1850°C were conducted corrosion experiments in molten aluminum and electrolyte at 950°C for 48h and were taken separately, their fracture images were observed and the spread depth of Al^{+3} and Na^{+} were measured.

Through line scanning of aluminum element for the fractures of the molten aluminum corrosion samples, it was found that the total and depth of aluminum have much relation with the composite density, shown in Figure 2. It can be seen when the composite does not compact, the molten aluminum can penetrate into the composite; but with the compactivity gradual increase, the penetrating speed of molten aluminum can drop markedly. Raising the compactivity of the composite can effectively slow down the penetrating speed of molten aluminum into the composite. In addition, the fact of molten aluminum spread also shows that molten aluminum has a good wetting to the composite. Figure 3 is the SEM images of the fractures of the corrosion composite. It can be seen that under the molten aluminum corrosion, the stick form of TiB_2 particles become fine and the TiB_2 particles in the fracture easily come off but WSi_2 particles don't change markedly, which shows that molten aluminum mainly reacts with TiB_2 and hardly reacts with WSi_2.

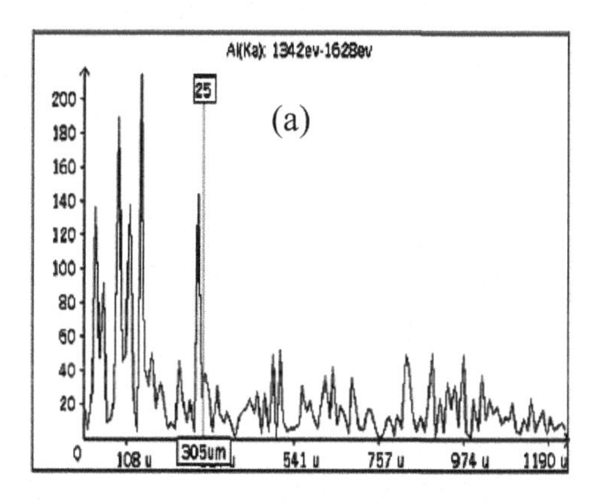

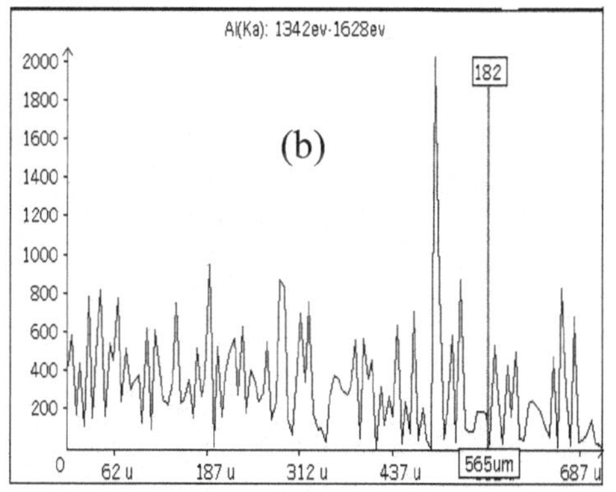

Fig. 2 The line scanning of aluminum element of the WSi_2-TiB_2 composites after molten aluminum corrosion
(a)WSi_2+30mass%TiB_2; (b)WSi_2+70mass%TiB_2

Through line scanning of sodium element for the fractures of the molten electrolyte corrosion samples, it was found that the total and depth of sodium have much relation with the composite density. It can be found when the composite does not compact, the molten sodium can spread along the opening to within the composite; but with the compactivity gradual increase, the spread speed of molten sodium can drop markedly. It can be seen that molten electrolyte also has a good soakage to the composite. Figure 4 is the SEM images of the fractures of the corrosion composite. It can be seen that under the molten electrolyte corrosion, the big WSi_2 particles become fine but the TiB_2 particles don't change markedly, which shows that molten electrolyte mainly reacts with WSi_2 and reacts with TiB_2 weaker than with WSi_2.

The corrosion experiments show that the conductivity and the resistance corrosion performance of the TiB_2 – WSi_2 composite can meet the demands of aluminum electrolysis, and the composite can be used as the inert cathode materials. However, it is necessary that the composite has to be tested by electrolysis process for using as the inert cathode materials.

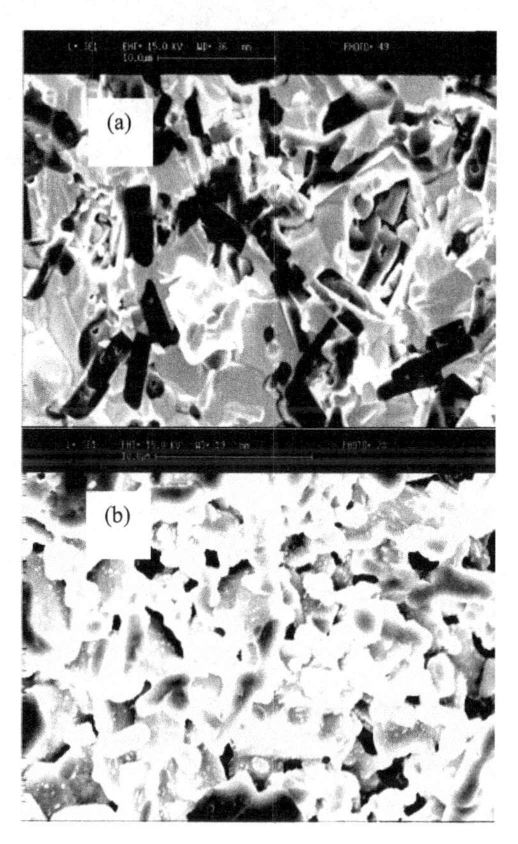

Fig. 3 The fracture scanning images of the WSi_2-TiB_2 composites after molten aluminum corrosion
(a)WSi_2+30mass%TiB_2; (b)WSi_2+70mass%TiB_2

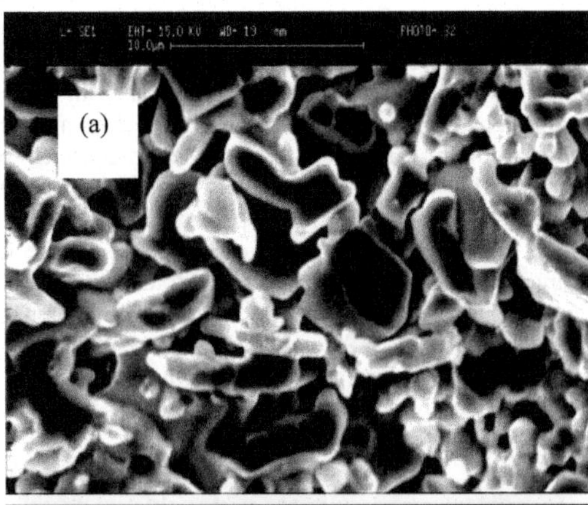

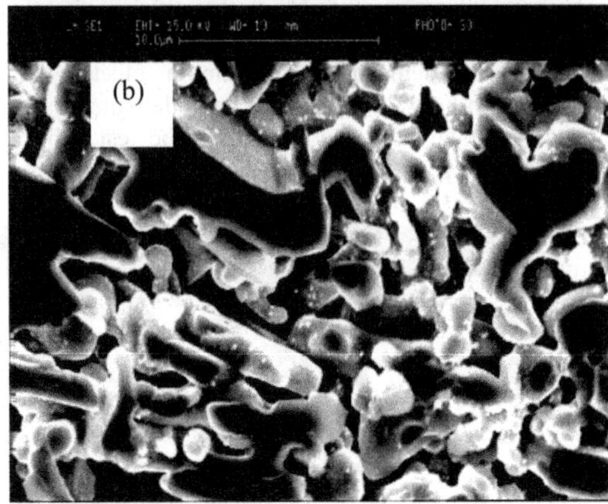

Fig.4 SEM images of the WSi_2-TiB_2 composites after molten electrolyte corrosion
(a) WSi_2+30mass%TiB_2; (b) WSi_2+70mass%TiB_2

Conclusions

WSi_2 can well improve the compact sintering performance of TiB_2. When the WSi_2 content of TiB_2-WSi_2 composite excesses 50mass%, the compactivity of the composite sintered 1hour at 1900°C is higher than 98%.

The resistivity of TiB_2-WSi_2 composite drops with the compactivity of the composite rising, and the composite can meet the demands of aluminum electrolysis cathode materials.

The molten aluminum corrosion resistance of WSi_2 is stronger than TiB_2, but the molten electrolyte corrosion resistance of WSi_2 is weaker than TiB_2.

Acknowledgement

The authors wish to thank the National Natural Science Foundation of China for the financial support to the project 50374013.

References

1 Zhuxian Qiu. The Study Development of the Inert Cathode Materials Used in Aluminum Electrolysis Cell. Light Metals (in Chinese), 1991, 11: 28 – 31.

2 Zhuxian Qiu. The Study Development of the Inert Cathode and Anode Materials Used in Aluminum Electrolysis Cell. Light Metals (in Chinese), 2001, 9: 30 – 34.

3 Yexiang Liu. The Progress of the Inert Anode and the Soakage Cathode in Aluminum Electrolysis. Light Metals (in Chinese) 2001, 5: 28 – 29.

4 Rongjiu Li. The Ceramic – Metals Composite. Beijing: Metallurgical Industry Press. 2004: 299 – 302.

5 Fei Zhou. The Electron Throety Study on the Phase Constructure and Performance of $MoSi_2$ and WSi_2, Journal of Silicate (in Chinese), 2000, 28(5): 462 – 464.

6 Zhaowen Wang, Shuping Sun, Bing Li and Zhuxian Qiu. The effect of $MoSi_2$ on Performance of TiB_2 Inert Cathode Materials. Journal of Northeastern University,1999, 20(6): 619 – 621.

7 Zhaowen Wang, Bingliang Gao and Zhuxian Qiu.Study on Resistance Corrosion of Improved TiB_2 Cathode. Light Metals (in Chinese), 2001, 12: 35 – 37.

STABILITY OF TiB_2–C COMPOSITE COATINGS

M. O. Ibrahiem, T. Foosnæs and H. A. Øye
Department of Materials Science and Engineering, Norwegian University of Science and Technology (NTNU)
7491 Trondheim, Norway

Keywords: Pitch-bonded TiB_2, Furan resin-based TiB_2, Wettability, Electrolysis and Aluminium carbide

Abstract

Several recipes of pitch-bonded TiB_2 were studied with respect to adherence, cracking and stability during electrolysis. A successful recipe for a crack-free coating was obtained. The coating showed good adherence and stability after 34 hours of electrolysis. The wetting of the coating by aluminium in the presence of cryolite melt was time dependent and almost complete wetting was observed after 120 minutes.

Furan resin-based TiB_2 coatings were not wetted by aluminium during electrolysis due to the presence of a carbon layer at the coating surface. Coating samples were polished by SiC paper after curing. The polished samples showed good stability and aluminium wetting.

Introduction

One way to save energy in aluminium reduction cells is to reduce the cell voltage drop by narrowing the anode-cathode distance (ACD). Excessive reduction in ACD in conventional cells does not result in improved energy efficiency due to increased back reaction when the anode is moved closer to the mobile, uneven surface of the molten aluminium cathode. Hence, the poor wettability of carbon cathodes limits the reduction of ACD, because it is needed to maintain a metal pad of a certain height. Modifying the cathode surface to make it more wettable would allow lower ACD with a decreased thickness of the aluminium pad. One centimetre lower ACD lowers the energy consumption by 1.8 kWh/kg Al **[1]**.

Titanium diboride and composites based on TiB_2 have been investigated as wettable cathode materials for several decades **[2-11]**. These materials have the unique combination of beneficial properties such as high electrical conductivity, good wettability by liquid aluminium, and acceptable stability in the corrosive cryolite-aluminium environment.

Experimental

Materials and the coating process

A TiB_2 powder, coal tar pitch, anthracite and carbon fibre were used to prepare the pitch-bonded TiB_2 coating on graphitized carbon. The TiB_2 particles had a size less than 45 μm (-325 mesh) with a bulk density of 4.38 g/cm^3. The softening point of the coal tar pitch used in the coating was around 120 °C. The carbon fibre Pyrograph® III (PR24-AG, Pyrograph®Products Inc, US) and electrocalcined anthracite (ECA) from Elkem AS were used. Table I lists mixing, forming and baking conditions which were used to prepare different coating compositions.

The starting materials were mixed by two different methods: a conventional mixing method and a ball-milling dispersion method. Conventional mixing meant mixing the raw materials for 15 min by a glass agitator (samples 1, 2, 7 and 8). Samples 3, 4, 5, 6 and 9 were prepared using the ball dispersion method. Raw materials and the dispersion medium, tetrahydrofuran (THF), were placed in the ball mill. After milling for 24 hrs, the mixture was vacuum distilled in order to remove the dispersion medium. Afterwards, the mixture was pulverized to a size between 250 and 600 μm.

The first coating method was to brush a 2.5 mm thick layer on the graphitized carbon after heating the mixed powders above the softening point of the pitch. The mix and the carbon cathode sample were heated to about 185 °C and 160 °C, respectively, to yield good adhesion between the coating and the carbon surface.

Table I. Pitch-bonded TiB_2 coating preparation.

Sample	Composition, %				Mixing method	Coating method	Coating thickness, mm	Electrolysis time, hr	Remarks
	TiB_2	pitch	ECA	Carbon fibre					
1	75	25	-	-	conventional	brushing	2.5	4	cracks
2	70	20	10	-	conventional	brushing	2.1	24	cracks
3	70	20	7.5	2.5	ball mill	brushing	2.2	30	cracks
4	70	20	7.5	2.5	ball mill	hot-pressing	2.0	8	crack-free
5	70	20	7.5	2.5	ball mill	multi-layered	2.0	34	crack-free
6	70	20	10	-	ball mill	hot-pressing	2.1	24	crack-free
7	70	20	10	-	conventional	hot-pressing	2.2	24	crack-free
8	70	20	7.5	2.5	conventional	multi-layered	2.1	24	crack-free
9	70	20	7.5	2.5	ball mill	multi-layered*	2.2	30	cracks

* : Each layer has been dried at room temperature

The second coating method was a multi-layered technique. Ten cm^3 of THF was added to 25 g of the powder mixture to form a slurry. The TiB_2 coating was applied on graphitized carbon in several layers to form a 2 mm thick coating. Each layer was dried at 100 °C for 1 hr. For sample no. 9 each layer was dried at room temperature for 2 hrs.

The third coating method was coating compaction by hot-pressing (23 g powder, 150 °C) at 10 MPa to form a 2 mm thick layer on the carbon substrate.

Finally, in the electrolysis experiments the samples were baked in-situ at 985 °C (heating rate 40 C/hr in argon atmosphere). For adhesion tests, the coating was applied on 15 mm Ø carbon samples and heat-treated in a coke bed up to 1000 °C with a heating rate of 40 C/hr followed by a 4 hours holding period.

Electrolysis experiments

Electrolysis experiments were carried out in an apparatus as shown in Figure 1. The samples were used as cathodes. The coated sample (Ø = 60 mm and L = 20 mm) was placed in an alsint tube. The cathode was protected by a layer of aluminium foil and a 20 mm thick aluminium disk. The alsint tube was mounted in a graphite crucible (inner diameter 71 mm) and filled with crushed pre-melted bath (500 g, cryolite ratio (CR) = 2.21). The assembly was placed inside the furnace and positioned on a steel support which was connected to the negative pole of a direct current source. A graphite anode (Ø = 30 mm and L = 120 mm) was fastened to the upper lid of the furnace. The furnace was flushed with argon (99.99 %) prior to heating, and a small flow of argon was maintained during the experiments.

After reaching the electrolysis temperature and ensuring that the electrolyte was completely melted, the anode was lowered until electrical contact was achieved, and then further 20 mm to obtain the desired depth. A constant current (21 A) was supplied during electrolysis by means of a HP 654321 DC power supply. The anode-cathode distance was adjusted to keep the cell voltage between 3 and 5 volts. The voltage increased gradually with time due to the anode consumption and the anode was lowered accordingly.

Alumina was fed to the bath using an automatic feeding system. The dosage was calculated and calibrated as 1.75 g Al_2O_3 per 15 min to maintain a constant level and to keep a reasonably constant concentration during the experiment. To make up for the evaporation losses of the electrolyte, 50 g of electrolyte and aluminium fluoride per day were also added using the same automatic feeding system. The duration of the experiments was between 4 and 34 hrs. After electrolysis the anode was raised clear of the melt and the furnace with its contents was left to cool to room temperature. The crucible was then cut open and the sample was removed and cleaned mechanically. The sample was cut perpendicular to the coating surface with a water-cooled diamond saw and prepared for microscopy examination by embedding in an epoxy resin. The exposed surfaces were polished using metal bonded diamond discs and diamond spray (15, 9, 6, 3 and 1 µm) using pure alcohol as lubricant and for 3 min at each stage.

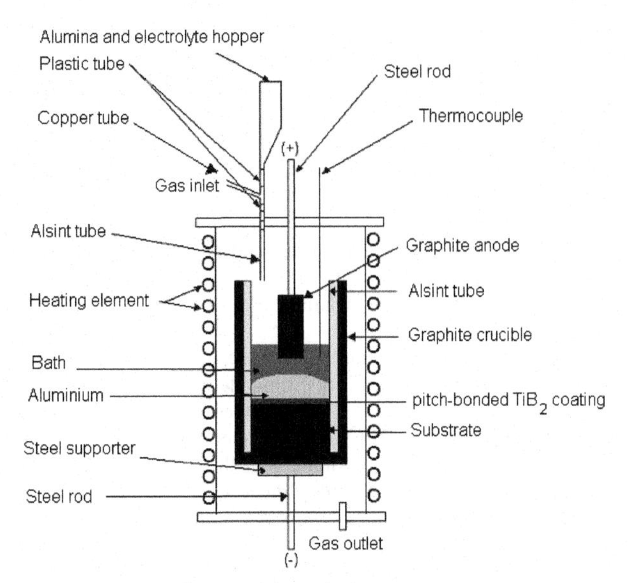

Figure 1. Electrolysis apparatus.

Wettability experiments

Wetting of baked coatings (15, 25, 50 and 70 % TiB_2) and graphitized carbon by liquid aluminium in cryolite melts was investigated using a molten drop cooling method. The experiments were conducted in a vertical tubular furnace in argon atmosphere. The test samples (17 mm x 17 mm x 2 mm) were placed in a quartz tube filled with super pure aluminium shot of nominal weight 0.5 ± 0.05 g and 20 g of crushed pre-melted bath (CR = 2.21). The surface of the samples was polished to a 1 µm finish with diamond paste. The assembly was placed inside the furnace which attained 1000 C within 40 min. This time was designated time zero for the test. Test times varied from 0 to 120 min. At the completion of the test, the quartz tube was cut open and the sample was removed and cleaned mechanically from the solid electrolyte on the surface of aluminium. Afterwards, wetting angles were measured by the tangent method.

Results and discussion

Coating adherence before electrolysis

The baked coating was fastened to a copper rod with rapidly polymerizing Araldit. A 3 mm horizontal hole was drilled in the copper and in the carbon rod. Rods were put through the holes and the copper rod was fastened with wires to a rack while a bucket was fastened to the carbon sample. Tension was measured until sample rupture.

Table II. Adherence test results.

Sample	Max. load, kg	Area, cm^2	Tension, MPa
Graphite	33	1.77	>1.82
TiB_2 coating on graphite	32		>1.78

Table II lists adhesion test results. After testing, it was noticed that in all cases the Araldit was the weakest material. This means that the adhesion strength between the coating and the carbon substrate is larger than the measured values.

Wettability results

Results from the contact angle measurements for the TiB_2 samples and graphitized carbon are plotted in Figure 2. The values presented are an average of five contact angles measured at the drop periphery.

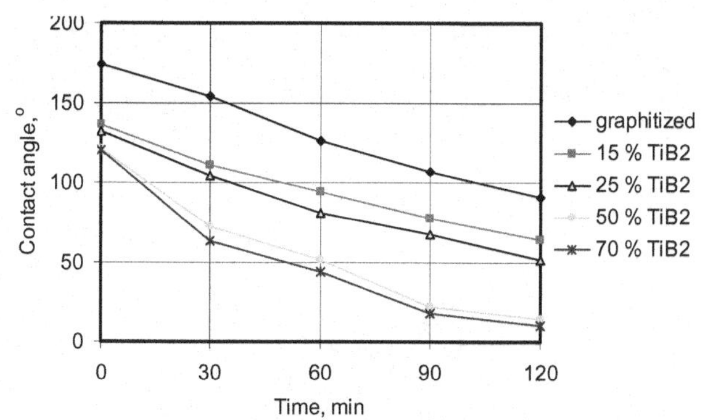

Figure 2. Contact angle vs. time for the tested samples.

The wetting of aluminium improved with time in all cases. The samples with 70 % and 50 % TiB_2 were similar giving nearly complete wetting after 120 minutes. The wetting angle decreased with 0, 15 and 25 % TiB_2. The improved wetting with time is probably due to dissolution of the oxide layer on the aluminium and the TiB_2. Error could be introduced as the wetting angle was measured for cooled-down samples. The consistency of the measurements, however, makes the presence of this error unlikely. Comparison of wettability observations from the drop cooling method and the electrolysis experiments revealed that the coating material is more easily wetted by nascent aluminium generated during electrolysis than by a passive metal drop.

Performance of the coating under electrolysis

Three concepts were tried to minimize coating cracks. The first concept is the use of a fibre-reinforced coating to possibly increase crack resistance. The fibres can act as crack arrestors while the coating material undergoes shrinkage during carbonization. The second idea of a multi-layered coating is based on the phenomenon that there is a critical coating thickness below which no coating cracks appear. The third concept is the coating compaction by hot-pressing to avoid gaps during the coating process. It was noticed that there is a relationship between the crack formation in the coating and the coating method. Multi-layered and compaction by hot pressing methods gave crack-free coatings as in samples no. 4 to 8 (Table I). On the other hand, neither the mixing methods nor addition of carbon fibres had any effect on avoiding cracks.

Figure 3 shows a polarized light micrograph of sample no. 1 after 4 hrs of electrolysis. Figure 3-a represents the microstructure of the sample at the coating-carbon interface. The coating showed good adherence to the carbon material. Aluminium carbide did not form at the coating-substrate interface. Figure 3-b represents the microstructure of the sample within the coating. It shows the presence of a large crack developed within the coating. The crack is filled with aluminium carbide and electrolyte. The width of the crack is 50 μm. Figure 3-c represents the microstructure of the sample at the aluminium-coating interface. Aluminium appeared to wet the coating surface well after 4 hrs. An aluminium carbide layer of 25 μm thickness was formed at the interface between aluminium and the surface of the coating. It should be noted that some of the TiB_2 particles were removed from the coating surface and was dispersed in the aluminium above the coating. Removal of the TiB_2 particles is probably due to the dissolution of the carbon matrix by the electrolytic bath.

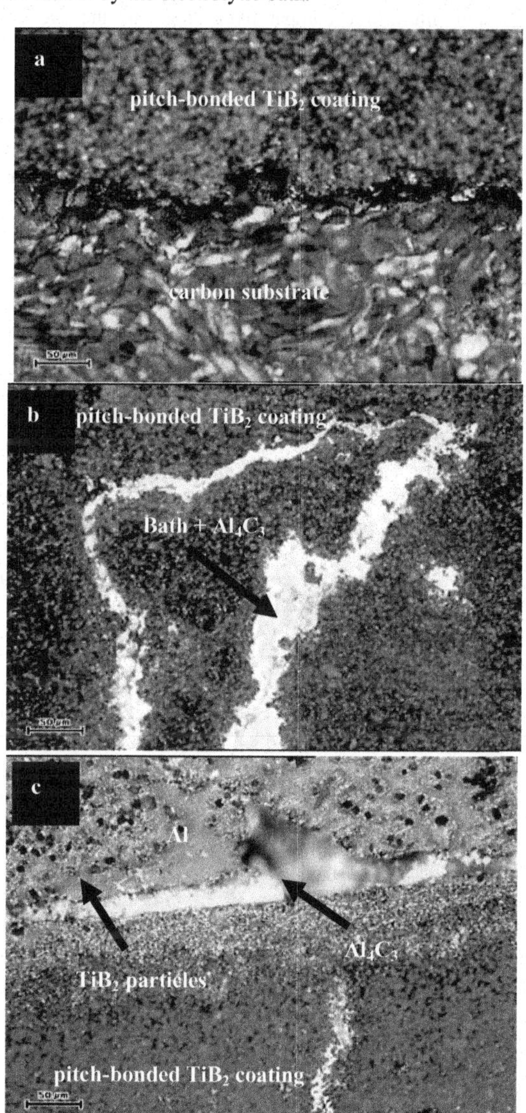

Figure 3. A polarized optical micrograph of sample no. 1 after 4 hrs of electrolysis; a, coating-substrate interface; b, aluminium-coating interface; c, within the coating

The penetration of the electrolyte is thought to contribute to the formation of aluminium carbide in one of two ways, either by direct reaction with carbon to form the carbide (reaction 1) or by aluminium metal transport into the interior to react with the carbon (reaction 2).

$$Na_3AlF_6(l) + 3Na + 0.75\ C\ (s) = 0.25Al_4C_3\ (s) + 6NaF(l) \quad (1)$$

$$4Al\ (l) + 3C\ (s) \rightarrow Al_4C_3\ (s) \quad (2)$$

The standard Gibbs' free energy change for reaction 1 is –79.4 kJ at 970 °C **[12]**. Aluminium carbide is not stable in the presence of N_2 or CO, and Al_4C_3 will only be formed in argon atmosphere **[13]**. The standard Gibbs' free energy change for reaction 2 is –147 kJ at 970 °C **[12]**. Reaction 2 is thermodynamically favoured at all temperatures of concern for electrolytic aluminium production. The solubility of aluminium carbide is very low in molten aluminium but comparatively high in the electrolyte, which is the main reason that more aluminium carbide is formed when electrolyte is present **[13]**. Aluminium carbide formation at the interface of Al-graphite and Al-TiB_2/C composite was studied by Xue and Øye **[14]**. They found that the kinetic stages for interfacial carbide formation are: onset by Al-diffusion through the oxide film to meet the carbon; continuous growth at the interface extending into the aluminium; and final volume expansion breaking the interfacial bond. The continued reaction is primarily controlled by C-diffusion through Al_4C_3. The aluminium carbide preferentially forms in the pores **[14]**.

Figures 4, 5 and 6 show electron probe micro analyzer (EPMA) micrographs and the corresponding x-ray maps of sample no. 5 at the coating-carbon interface, within the coating and at the aluminium-coating interface, respectively, after 34 hrs of electrolysis. The coating showed good adherence to the carbon material (Figure 4). A crack-free interface between the coating and the substrate was observed as well. The corresponding sodium x-ray map shows that sodium has completely penetrated the coating. Some aluminium and electrolyte have penetrated the coating and reached the surface of the carbon substrate. Figure 5 shows that the coating was crack-free. Aluminium, sodium and some electrolyte are present. Figure 6 shows that aluminium has wetted the surface of the coating and part of aluminium has penetrated inside. Aluminium did not penetrate the anthracite particles found in the coating. Some of the TiB_2 particles have separated from the coating into the aluminium above the coating.

Several trials were performed to apply the coating by the multi-layered method where each layer was dried at room temperature for 2 hr instead of drying in an oven at 100 °C (sample no. 9). Figure 7 shows a polarized micrograph of sample no. 9 at the coating-substrate interface after 30 hrs of electrolysis. Large cracks were found at the coating-substrate interface and within the coating. The width of the crack is 100 μm. A mixture of electrolyte and aluminium carbide filled the cracks. The cracks served as paths for aluminium and electrolyte into the coating resulting in the formation of aluminium carbide. Drying the layers at room temperature did not prevent cracks even when the carbon fibres were added and the ball-milling method was used.

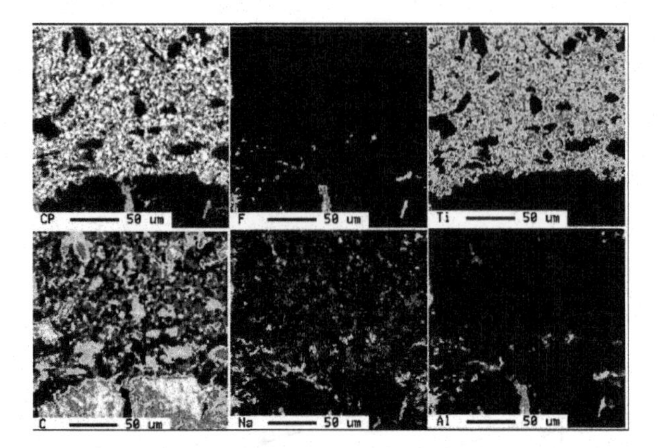

Figure 4. EPMA micrograph and the corresponding x-ray maps of sample 5 after 34 hrs of electrolysis at the coating-carbon interface.

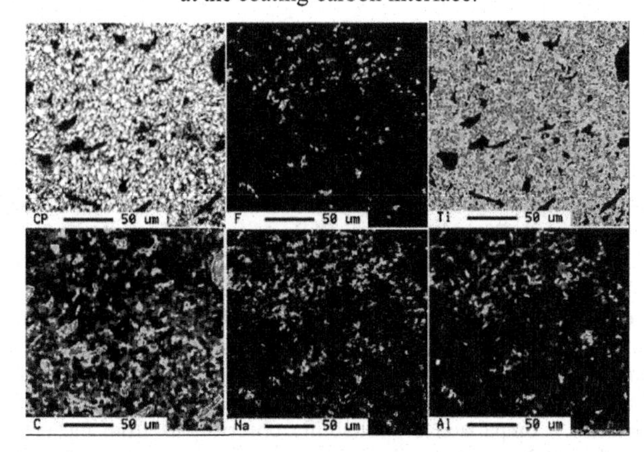

Figure 5. EPMA micrograph and the corresponding x-ray maps of sample 5 after 34 hrs of electrolysis within the coating.

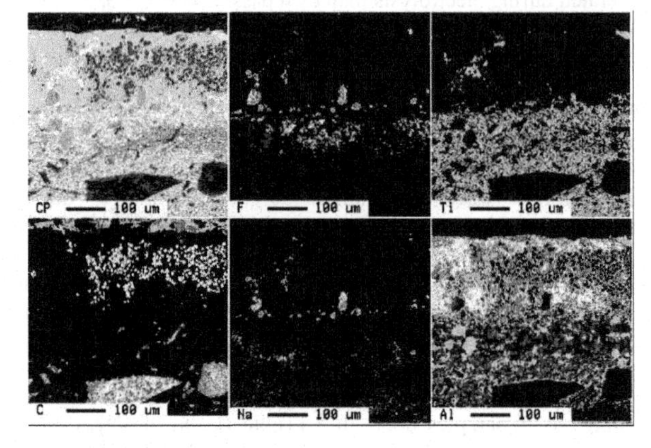

Figure 6. EPMA micrograph and the corresponding x-ray maps of sample 5 after 34 hrs of electrolysis at the Al-coating interface.

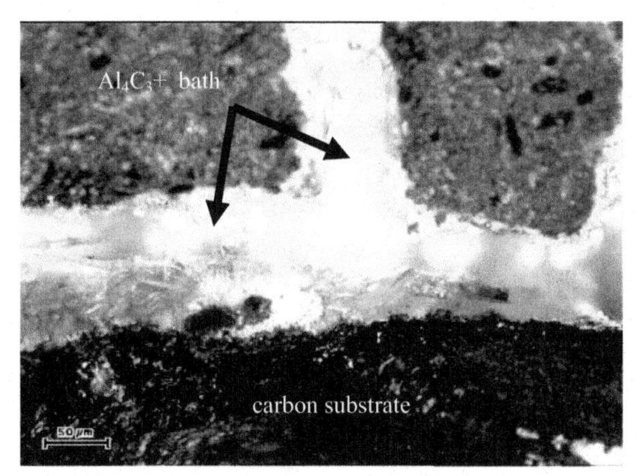

Figure 7. A polarized optical micrograph of sample no. 9 at the coating-substrate interface after electrolysis for 30 hrs.

The stability of the coating

The solubility of the TiB_2 coating was calculated by the titanium loss rate into the aluminium after electrolysis experiments. The effective TiB_2 content in aluminium is calculated from the total titanium analysis in the metal. The titanium concentration in aluminium after 24 hrs electrolysis equals 0.0018 % (difference after-before).. Using the analysed content of Ti and by assuming dissolution of TiB_2, the solubility product is:

$$(\text{ppm Ti})(\text{ppm B})^2 = 18(2\text{x}18\text{x}10.8/47.9)^2 = 0.12\text{x}10^4$$

This is lower than the solubility product given by Finch **[15]** equal to $1.21\text{x}10^4$ at 960 °C.

The measured TiB_2 loss from the coating appears to be almost exclusively due to dissolution and not erosion **[7]**. Some trials have been done earlier in this work to measure the stability of the coating using optical microscopy. The method depends on measuring the thickness of the coating after electrolysis. The coating thickness before electrolysis was measured by a calliper. The comparison between the two results has not shown any significant difference.

Furan resin-based TiB_2 coatings

Titanium diboride powder (70 %), anthracite (10 %), furfuryl alcohol (17 %) and zinc chloride solution as a curing agent (3 %) were used to prepare 1.2 mm thick crack-free coatings on graphitized carbon. The conventional mixing method at room temperature was used. The paste was applied to the substrate by painting and brushing as one layer. Curing the green coating was done at 80 °C for 24 hours in an air atmosphere furnace. For electrolysis experiments the samples were baked in-situ as previously described. The coating was protected by a layer of aluminium foil and a 20 mm thick aluminium disk.

Figure 8 shows polarized optical micrographs of the furan resin-based TiB_2 coating sample at the top coating surface after 4 and 24 hrs electrolysis. An aluminium carbide layer of 25 and 30 μm thickness was formed at the surface of the coating after 4 and 24 hrs, respectively. Some aluminium carbide formed inside the coating after 4 hrs of electrolysis. It formed around the anthracite particles mixed with the electrolyte after 24 hrs of electrolysis. The penetration of electrolyte through the coating increased with electrolysis time and the coating was completely penetrated by the electrolyte after 24 hrs. During preparation of the sample for the microscopy test, it was seen that electrolyte covered the top surface of the coating. Aluminium produced by electrolysis was easily removed from the surface and did not wet the coating surface well after electrolysis times. It is observed that a 100 μm layer of carbon covers the coating surface (Figure 9).

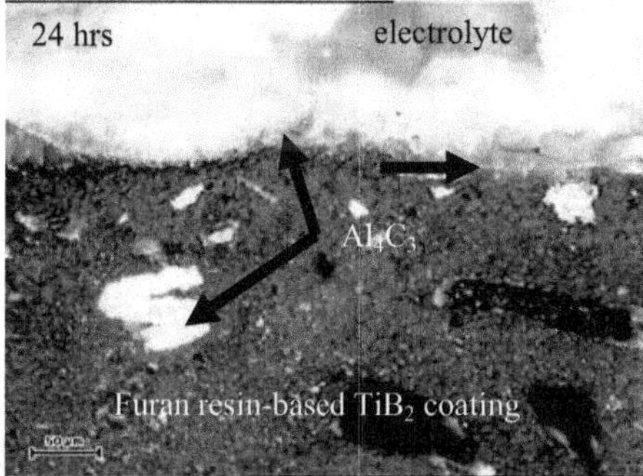

Figure 8. A polarized optical micrograph of the top surface of furan resin-based TiB_2 coating after 4 and 24 hrs electrolysis.

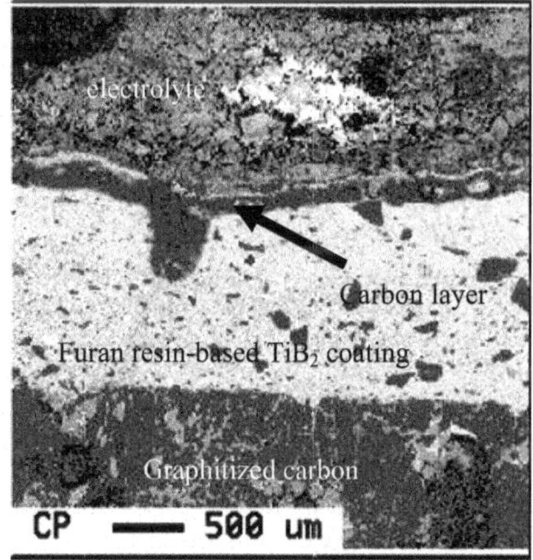

Figure 9. SEM micrograph showing a thin layer of carbon above the coating surface.

The presence of the carbon layer above the coating surface, due to the carbonization of the resin binder, prevented aluminium wetting. To overcome the poor wetting the coating was treated after the curing process. The coating surface was polished using SiC paper. The polished coating was washed by a solvent to remove dust. Figure 10 shows an optical micrograph of polished coating after 8 hrs electrolysis. The wettability of the polished by aluminium was achieved but, electrolyte drops still found in some areas above the coating surface and within aluminium (Figure 11). Aluminium carbide formed at the aluminium-coating interface and within the coating.

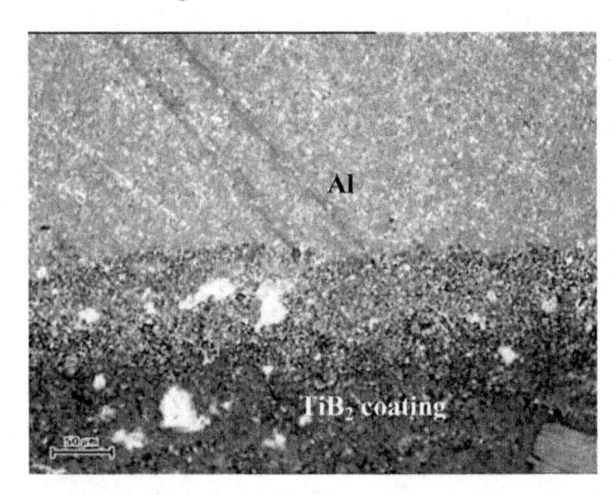

Figure 10. A polarized optical micrograph of the top surface of polished coating after 8 hrs of electrolysis; Al wetted the coating.

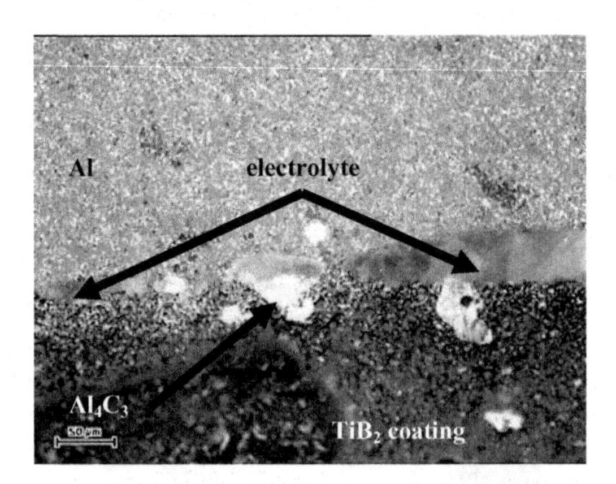

Figure 11. A polarized optical micrograph of the top surface of polished coating after 8 hrs of electrolysis; electrolyte drops and aluminium carbide found at the interface.

Conclusion

Prebaked pitch-bonded TiB_2 coating adhered strongly to the carbon substrate. Adhesion tests resulted in breakage in the araldit region rather than at the coating-substrate interface.

Wettability experiments showed that aluminium wets the coating well in the presence of molten electrolyte. The contact angle was time dependent. It decreases with the increase of TiB_2 content in the coating. From electrolysis experiments it was also observed that aluminium completely wetted the coating.

Obtaining crack-free coatings depends strongly on the coating method. Mixing methods and the addition of carbon fibres did not affect the suppression of cracks within the coating.

Electrolysis tests with a pitch-bonded TiB_2 coating resulted in the formation of aluminium carbide and electrolyte inside the cracks. Delamination between the coating and the substrate did not occur for all samples even with the presence of cracks which filled with aluminium carbide and electrolyte at the interface. The coating was penetrated by electrolyte, aluminium and sodium for all samples.

The life span of the pitch-bonded TiB_2 coating is related to its dissolution rate. Slow dissolution rate during electrolysis was noticed.

During electrolysis, aluminium did not wet the furan-resin TiB_2 coating in the presence of a carbon layer at the coating surface. The coated samples were grinded and polished to overcome the non-wetting. Treated samples showed good wetting by aluminium but some drops of electrolyte were found at the surface as well.

Acknowledgement

Financial support has been provided by the Norwegian Research Council and the Norwegian Aluminium Industry through the CarboMat program, for which the authors are grateful.

References

[1] H. Kvande " Energy Balance", Fundamentals of Aluminium Production 2004, Volume 1, Trondheim, Norway, May 10-21, 2004, p. 197-216.

[2] S. K. Das, P. A. Foster and G. J. Hildeman: U. S. Patent 4 308 114, 1981.

[3] A. Y. Sane: U. S. patent 4 595 545, 1986.

[4] B. Mazza, A. Bonfiglioli, F. Gregu and G. Serravalle, "Process Aspects in Aluminium Reduction Cells with Wettable Cathodes", Aluminium, 60 (1984), p. 760-763.

[5] C.J. McMinn, "A Review of RHM Cathode Development", Light Metals 1992, p. 419-425.

[6] R. C. Dorward, "Energy Consumption of Aluminium Smelting Cells Containing Solid Wetted Cathodes", J. Appl. Electrochem, 1(1983), p. 569-575.

[7] L. G. Boxall, A. V. Cooke, and H. W. Hayden, "TiB_2 Cathode Material: Application in Convential VSS Cells", J. Metals, 36 (1984), p. 35-40.

[8] A. J. Gesing and D. J. Wheeler, "Screening and Evaluation Methods of Cathode Materials for Use in Aluminium Reduction Cells in Presence of Molten Aluminium and Cryolite up to 1000 C", Light Metals 1987, p. 327-333.

[9] L. G. Boxall, W. M. Buchta, A. V. Cooke, D. C. Nagle and D. W. Townsend: U. S. Patent 4 466 996, 1984.

[10] A. Y. Sane, D. J. Wheeler and C. S. Kuivila: U. S. Patent 4 560 448, 1985.

[11] J. J. Duruz: U. S. Patent 5 004 524, 1991.

[12] D. R. Stull and H. Prophet, "JANAF Themochemical Tables", 2rd Edition, NSRDS-NBS 37, July 1970.

[13] M. Sørlie and H. A. Øye, "Cathodes in Aluminium Electrolysis", 2nd Edition, Aluminium-Verlag, 1994.

[14] J. Xue and H. A. Øye, "Al_4C_3 Formation at the Interface of Al-Graphite and Al-Carbon/TiB_2 Composite", Light Metals 1994, p. 211-217.

[15] N. S. Finch, "The Mutual Solubility of Titanium and Boron in Pure Aluminium", Metall Trans, 4, 1973, p. 1948-1952.

A NEW APPROACH TO ESTABLISHING THERMAL SHOCK RESISTANCE OF CATHODE BLOCKS

G.Vergazova[1], G.Apalkova[2]
[1] Engineering and Technology Center Ltd., Krasnoyarsk, Russia
[2] JSC "Ural Electrode Institute", Chelyabinsk, Russia

Keywords: carbon cathode block; laboratory test; thermal stress; thermal shock resistance

Abstract

Specifications for cathode materials do not contain a property reflecting cracking resistance of the material subjected to thermal shock. Thermal shock resistance (TSR) can be defined as ultimate strength of the material under mechanical stress created in a solid body by a temperature gradient or obstruction to its thermal expansion.
A laboratory test method and apparatus simulating extreme operating conditions were developed to evaluate TSR of electroconductive carbon materials, e.g. cathode bottom blocks used in aluminium reduction cells.
Variation in TSR and other properties of carbon materials is partly explained by imperfection of their crystalline structures. To evaluate this imperfection a new parameter was developed as a relative deviation of real density from theoretical density of the graphite which reflects the extent of crystalline structure defects in cathode blocks. The study showed correlation of crystalline structure defectiveness level with physical and mechanical properties of cathode blocks.

Introduction

In order to reduce the destruction of carbon lining in modern high amperage aluminium reduction cells the experts give preference to the following cathode block properties [1]:
- high thermal shock resistance,
- high durability,
- high thermal conductivity for uniform temperature distribution,
- low sodium expansion,
- low resistance and low cathode voltage drop.

One of the key properties of cathode materials is thermal shock resistance (TSR) which as a rule determines usability and service life of the material exposed to high temperatures.
Crack formation in the parts of aluminium reduction cell, including carbon materials, operating at high temperature still remains an issue for modern aluminium production industry.
Specifications for cathode materials and anodes do not contain a property reflecting cracking resistance of the material subjected to thermal shock.
TSR can be defined as ultimate strength of the material under thermal and mechanical stress created in a solid body by a temperature gradient across its parts or obstruction to its thermal expansion in general.
TSR has gained grounds as a physical property of materials serving as integral indicator interrelated with fundamental properties of the matter.

TSR researchers have noted the challenge of not having any scientific criterion for evaluation or any universal method for measuring this property. Depending on the objectives various formulas are used for calculating TSR criteria. The formulas stem from the theoretical premises developed by W.D.Kingery [2] and consider a homogenous isotropic structure with linear elasticity and brittleness typical of refractory materials.
Some researchers and manufacturers of carbon products use a well known formula for Index Thermal Shock Resistance valuation when it comes to selecting new types of carbon material aggregate [3]:

$$\text{Index TS} = \frac{\text{Thermal conductivity} \bullet \text{Bending strength}}{\text{Coefficient of thermal expansion} \bullet \text{Y-modulus}} \quad (1)$$

However formula (1) is practically useless for comparison of cathode blocks of the same grade from different manufacturers since specific values of the properties included in the formula are determined by specification for a particular grade. Moreover there is no uniform opinion on the mechanical strength property. It could be compressive, tensile, or bending strength that is used for calculations. Thus validity of the results will in each case be determined by the initial terms and conditions set forth.
There are some direct known methods for evaluating the thermal strength: J.A.Brown, P.J.Rhedy, D.Belitskus, S.Sato, T.Log et al., which test carbon samples with flame or electric arch. The thermal strength is characterized by the time it takes to form a crack in the sample subjected to oxygen-propane flame or arch or by the energy emitted by a disk-shaped sample, transient radial temperature up to the moment of crack formation [4-7].

Test method

Carbon cathode blocks belong to heterogeneous composite anisotropic materials containing components of various nature and structure: anthracite, artificial graphite, and binder coke formed from coal tar pitch.
The objective of this work was creation of a direct test method evaluating TSR of carbon materials for a comparative testing of cathode blocks from different manufacturers. In developing this method dimensions of a test sample were studied. Sample thickness (20 mm) and diameter (100 mm) were selected with account for coarseness of the cathode block material. Also sampling and preparation procedures were developed and refined.
A direct TSR evaluation method is based on a controlled simulation of extreme operating conditions through creation of internal transient thermal gradients in a disk sample using an electric arch [8].

The view of the apparatus during testing is given in Fig.1. The sample after testing is shown in Fig.2.
The essence of the method is to measure the time it takes to form a crack in the sample subjected to thermal stress created by a radial thermal gradient generated by a local heat up of the sample when electrical current of 1 kA is applied to two contact points on the sample sides.

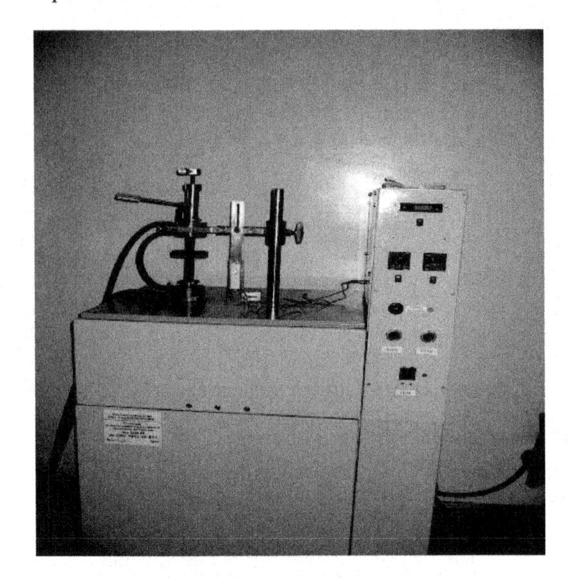

Fig.1. The AGN-01 apparatus for measuring TSR of carbon materials

Fig.2. Sample after thermal failure (with arrows pointing to the cracks).

The radial thermal stress distribution in the sample disk is shown in Fig.3.

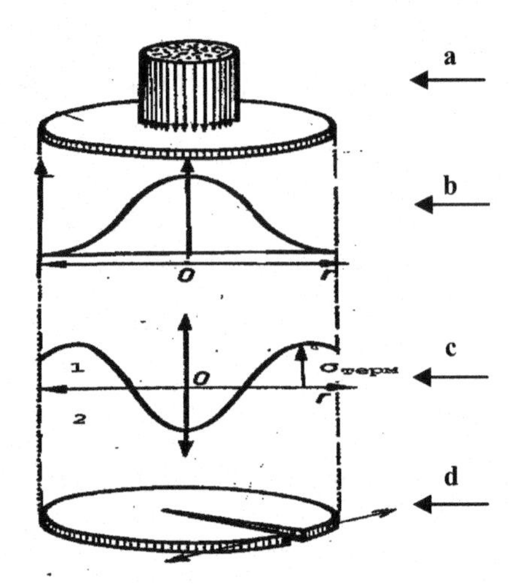

Fig.3. Radial distribution of thermal stress in a sample disk when electrical contact method is used for measuring TSR:
a – electrical current applied to the sample;
b – radial temperature distribution;
c – thermal stress in the sample:
1 – stretched area;
2 – compressed area;
d – crack formation (σ thermal > σ tensile)

The measured time to crack formation τ (seconds) is a numerical value representing TSR of the tested material.
Six different types of cathode bottom blocks from different producers numbered 1 through 6 were tested. The block grades ranged from semi-graphitic (30% graphite content) to graphitized.
A Kingery Criterion method was used together with the developed one to evaluate TSR.

Test results

The test results are given in Fig.4, 5 and Table 1. The presented data shows that despite similar nature the changes in TSR and Kingery Criterion significantly vary in magnitude for the tested cathode blocks: Kingery Criterion shows threefold inferiority of Block 3 to Block 1 whereas TSR values for these blocks are virtually the same.
Thus there is a weak correlation between experimental and calculation methods yet a visible correlation between experimental TSR and tensile strength, TSR and electrical resistivity of cathode blocks, with properties having similar trends. The same applies to TSR and heat conductivity. These results allow to forecast TSR based on the electrical resistivity change pattern in correlation with tensile strength of the material. Whereas correlation of TSR with tensile strength and heat conductivity is known and pretty well forecasted, that of TSR with electrical resistivity for carbon materials has not been discussed before. In part it can be explained by the fact that approaches to TSR problems based on W.D.Kingery's theoretical premises are founded on the research of non-electroconductive refractory materials which have heat conductivity as the most important property.

Table 1. Relationship equations and correlation with TSR for key properties of various cathode blocks

Property	Relationship equation	Correlation coefficient
Specific electrical resistance (ρ), μΩm	$\tau = 29{,}7 - 4{,}1\ \rho$	0.92
Tensile strength, (σ_{tens}), MPa	$\tau = -15{,}2 + 11{,}3\ \sigma_{tensile}$	0,98
Young's Modulus (E), GPa	$\tau = 24{,}7 - E$	0,38
Linear thermal expansion, (CTE), 10-6/K	$\tau = 45{,}8 - 9{,}6\ \alpha$	0,80
Thermal conductivity (λ), W/mK	$\tau = 5{,}3 + 0{,}59\ \lambda$	0,93
Flexural strength (σ_{Flex}), MPa	$\tau = 42{,}8 - 3{,}0\ \sigma_{flex}$	0,42

Electrical resistivity of electrode products commonly used as conductors in various electric furnaces not only determines the electrical loss but also creates local overheating in the higher resistance areas generating thermal gradients that lead to cracking.

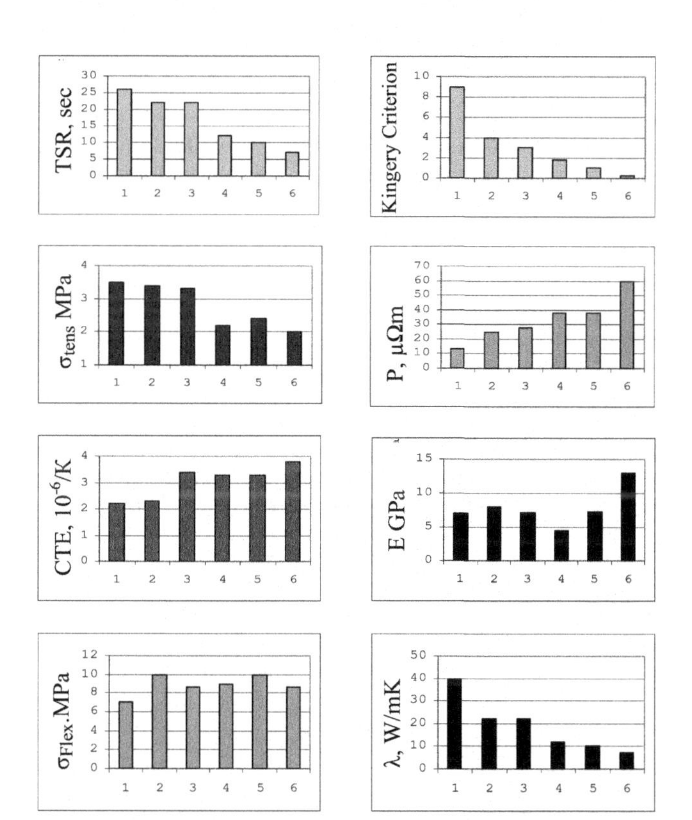

Fig.4. Measured and calculated TSR values related to key properties of various cathode bottom blocks

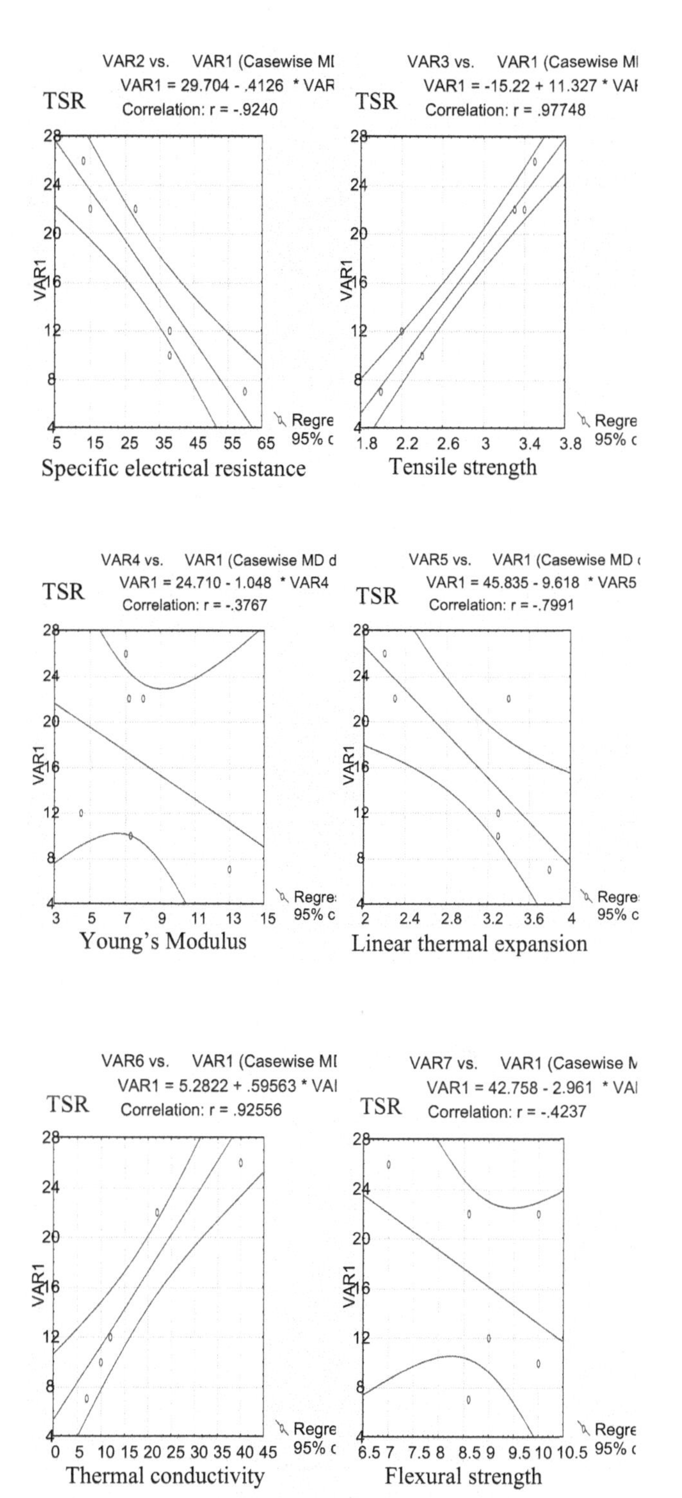

Fig.5. Relationship equations and correlation with TSR for key properties of various grades of cathode bottom block

Of major interest to us was testing of semi-graphitic cathode bottom blocks with 30% graphite content. Normally our customers receive such blocks from several suppliers. Table 2 contains properties of regular cathode bottom blocks from three suppliers.

Table 2. Key properties of semi-graphite cathode blocks from various manufacturers.

№	Property	Block 1 (Extrusion)	Block 2 (Vibroform)	Block 3 (Extrusion)
1	Real density, g/cm3	1.89	2.00	1.95
2	Apparent density, g/cm3	1.55	1.63	1.57
3	Open porosity, %	18	15	18
4	Compressive strength, (σcomp), MPa	34	30	32
5	Flexural strength, (σFlex), MPa	8.0	8.3	8.2
6	Specific electrical resistance (ρ), μΩm	34	28	30
7	Coefficient of linear thermal expansion (CTE), 10-6/K	3.4	3.8	3.6
8	Thermal conductivity (λ), W/mK	9	18	12
9	Young's modulus (E), GPa	9	8	10
10	Thermal Shock Resistance Index (ITS)	2.6	4.9	2.7
11	Experimental Thermal Shock Resistance (TSR), sec	9	22	15

Comparison of the properties for Block 1 and 3 shows their similarity in physical, mechanical properties and TSR Index. However the experimental TSR proves Block 1 and 3 to have different cracking resistance when exposed to high temperature.

Obviously cracking is affected by the material pore structure, aggregate grain orientation determined by the forming approach, and all that is not taken into consideration in TSR Index computation.

Service life analysis of reduction cells lined with various cathode bottom blocks with different TSR values has shown a correlation between electrical resistivity and TSR.

Variation in TSR of carbon materials is partly explained by imperfections or defects in their crystalline structures. The level of defects in carbon materials can be viewed in terms of a degree of crystal structure perfection. It is known, that real carbon bodies naturally contain various defects.

The divergence from a perfect model can be classified into following:

- defects of layer stacking;
- spiral dislocations;
- twinning of graphite crystals;
- edge dislocations;
- hole defects manifested as bond breaking or absence of atoms in hexagonal lattice;
- chemical defects - substitution of foreign atoms in the graphite lattice which affects physical properties and chemical reactivity of graphite.

To evaluate the defectiveness level of a carbon material structure a method was developed to measure difference between X-ray density based on crystalline lattice parameters and real density [9]. However one has to account for complex composition of the cathode block material and small sample size for X-ray diffraction analysis. It therefore seems reasonable to use relative divergence between the measured real density D_r and theoretical density of a perfect graphite crystalline structure D_{theor} as a measure of block crystal structure defectiveness level or Defectiveness Index (DI) following the formula:

$$DI = \frac{D_{theor} - D_r}{D_{theor}} \times 100\ \% \qquad (2)$$

where D_{theor} is theoretical density of graphite material with ideal crystalline structure and equals 2.267 g/cm^3 [9]; D_r is real measured density of the material.

Real density is a standard parameter for cathode blocks. Theoretical density however is virtually never discussed in research circles, manufacturing and use of cathode blocks.

A test was conducted on 10 cathode blocks of various grades and manufacturers numbered one through ten. Typical properties and parameters of carbon material structure DI for various grades of cathode blocks were calculated using formula (2) and are presented in Fig.6. These bar-graphs show a link (on a comparable scale) between the measured values of physical and mechanical properties and the DI of the cathode block crystalline structure.

The industry practice shows that increase in the number of cathode block grades demands a greater number of service durability parameters. The conducted study suggests that the group of informative parameters reflecting service durability of cathode blocks should be appended by a crystalline structure DI

Summary

The use of Defectiveness Index together with the developed experimental method for TSR evaluation allows for comparative quality analysis of the cathode blocks. Moreover they provide for better modeling of a cathode block behavior in the cell. That helps finding the maximum allowed current intensity that the cathode blocks can be used with which is useful for testing new raw materials used for production of these cathode blocks.

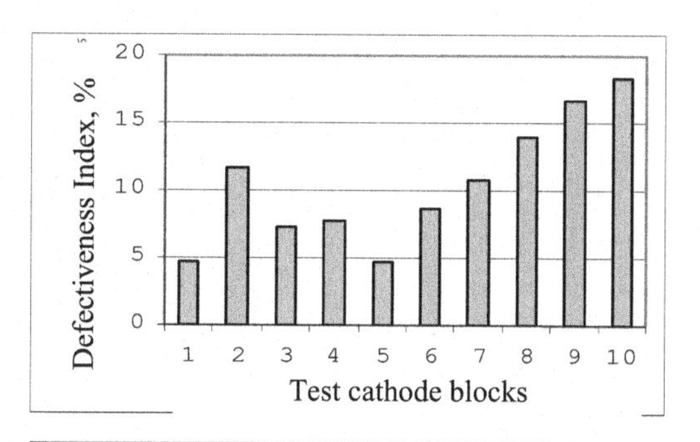

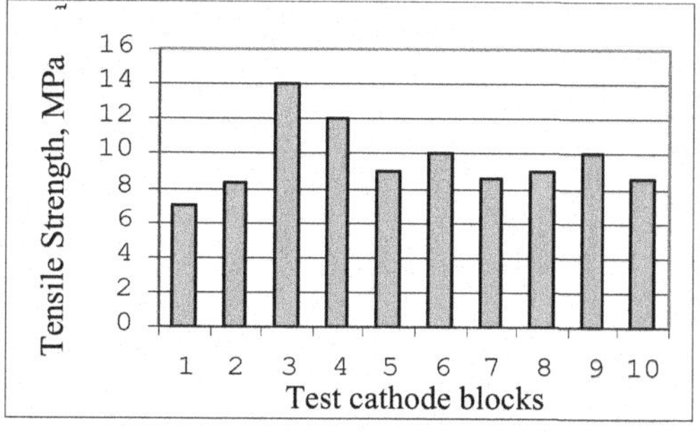

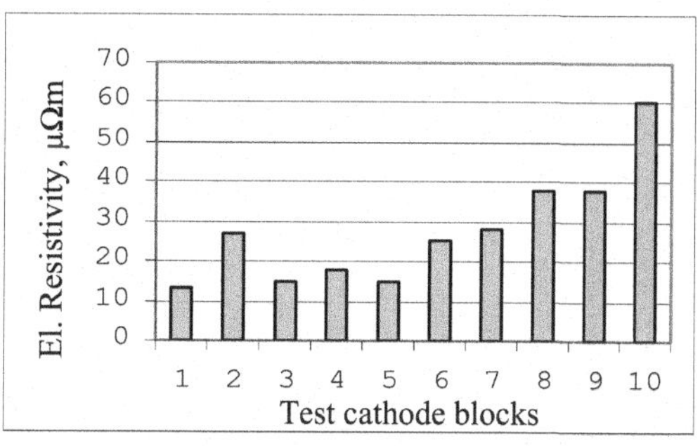

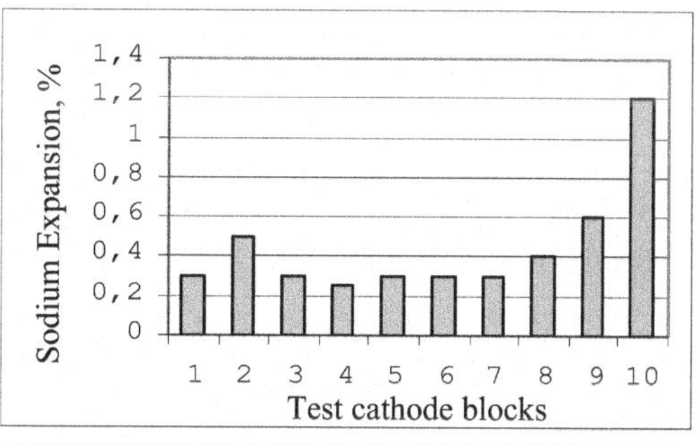

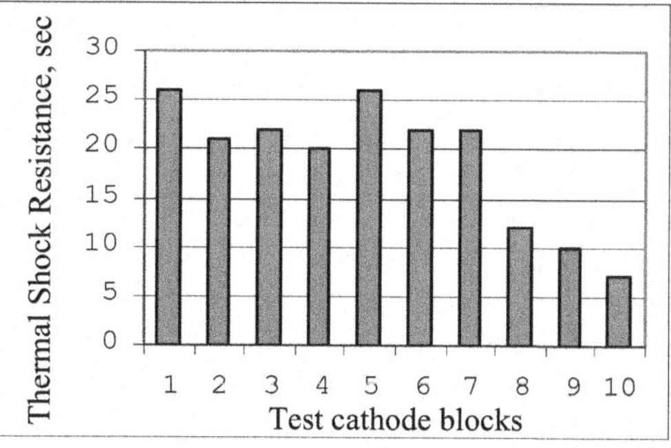

Fig.6. Structure DI and key cathode block properties for 10 tested cathode blocks

References

1. Øye, H., "Materials used in aluminium smelting", *TMS Light Metals,* (2000) 3-15.
2. Kingery, W.D., "Factors affecting thermal stress resistance of ceramic materials", *American Ceramic Soc.* **38** (1955) 3-5.
3. Piel, K., "Development in field of graphite cathodes", *TMS Light Metals* (1998) 697-702.
4. Brown, J.A., Rhedey, P.J., "Characterization of prebaked anode carbon by mechanical and thermal properties", *TMS Light Metals,* (1975) 253-269.
5. Sato, S. & al., "Determination of the thermal shock resistance of graphite by arc discharge heating". *Carbon,* **13** (1975) 309-316.
6. Belitskus, D., *TMS Light Metals,* (1978) 341-345.
7. Log, T., Melås, J., Larsen, B., "Improved technique for determinating thermal shock resistance of industrial carbon materials", *TMS Light Metals,* (1992) 717-723.
8. Vergazova, G., Apalkova, G., Glushkov, N., "Thermal deterioration of aluminum cell cathode blocks: an experimental study", *Aluminium of Siberia 2005, Scientific papers digest, Krasnoyarsk, Russia,* (2005) 228-231.
9. Vyatkin, S. & al., "Nuclear graphite", *Atomizdat, Moscow* (1967).

AUTHOR INDEX

Subject Index